AF389329

# Mineralogy of Noble Metals and "Invisible" Speciations of These Elements in Natural Systems

# Mineralogy of Noble Metals and "Invisible" Speciations of These Elements in Natural Systems

Special Issue Editor

**Galina Palyanova**

MDPI • Basel • Beijing • Wuhan • Barcelona • Belgrade

*Special Issue Editor*
Galina Palyanova
V.S. Sobolev Institute of Geology
and Mineralogy of the
Siberian Branch of the RAS
Russia
Novosibirsk State University
Russia

*Editorial Office*
MDPI
St. Alban-Anlage 66
4052 Basel, Switzerland

This is a reprint of articles from the Special Issue published online in the open access journal *Minerals* (ISSN 2075-163X) from 2018 to 2020 (available at: https://www.mdpi.com/journal/minerals/special_issues/noble_metals).

For citation purposes, cite each article independently as indicated on the article page online and as indicated below:

LastName, A.A.; LastName, B.B.; LastName, C.C. Article Title. *Journal Name* **Year**, *Article Number*, Page Range.

ISBN 978-3-03928-634-8 (Pbk)
ISBN 978-3-03928-635-5 (PDF)

Cover image courtesy of Galina Palyanova.

# Contents

# About the Special Issue Editor

**Galina Palyanova** Ph.D. is a leading scientist at V.S. Sobolev Institute of Geology and Mineralogy of the Siberian Branch of the Russian Academy of Sciences. She graduated from Novosibirsk State University with a degree in geochemistry. Her main area of research is ore mineralogy as well as thermodynamic and experimental modeling of ore-forming processes. She has published over 82 research articles in international journals. Her monograph "Physical and Chemical Features of the Behavior of Gold and Silver in the Processes of Hydrothermal Ore Formation" was awarded a medal and a diploma from the Russian Mineralogical Society. She has been awarded medals from the Russian Ministry of Higher Education and the Russian Academy of Sciences, and an honorary diploma from the Russian Ministry of Natural Resources and Energy Resources for her great contribution to the development of the mineral resource base of Russia.

# Preface to "Mineralogy of Noble Metals and "Invisible" Speciations of These Elements in Natural Systems"

This Special Issue book covers a broad range of topics related to the mineralogy of noble metals (Au, Ag, Pt, Pd, Rh, and Ru) and the occurrence, formation, and distribution of these elements in natural ore-forming systems. The eleven research articles contained within deal with various problems and topics, and can be divided into four themes: (1) three articles dedicated to the study of specific features and genesis of mustard gold; (2) four articles on thermodynamic and experimental modeling of systems containing noble metals; (3) two articles on research into Au-bearing arsenopyrite and pyrite; (4) two articles devoted to study of the ages of gold deposits in China in Nibao and Chaoyangzhai (Southeast Guizhou).

I hope this Special Issue will contribute to a better understanding of the genesis of gold, silver, and other noble metal deposits as well as the behavior of these elements in endogenic and supergene environments, and suggest ways forward to solving the problem of their full extraction from ores.

**Galina Palyanova**
*Special Issue Editor*

*Editorial*

# Editorial for Special Issue: "Mineralogy of Noble Metals and "Invisible" Speciations of These Elements in Natural Systems"

Galina Palyanova [1,2]

[1] V.S. Sobolev Institute of Geology and Mineralogy, Siberian Branch of the Russian Academy of Sciences, Akademika Koptyuga pr., 3, Novosibirsk 630090, Russia; palyan@igm.nsc.ru

[2] Novosibirsk State University, Pirogova str., 2, Novosibirsk 630090, Russia

Received: 11 February 2020; Accepted: 25 February 2020; Published: 26 February 2020

This Special Issue of *Minerals* covers a broad range of topics related to the mineralogy of noble metals (Au, Ag, Pt, Pd, Rh, Ru) and the forms of occurrence, formation and distribution of these elements in natural ore-forming systems. It contains eleven research articles on various problems and topics, which can be divided into four parts.

The first part of the issue includes three articles dedicated to the study of the specific features and genesis of mustard gold [1–3]. The typical features of mustard gold include low reflectivity, porous or colloform texture, and rusty, reddish, orange-red and brown-yellow colors in reflected light. The characteristics of mustard gold have been studied by many researchers [4–9], but its genesis is not fully understood and the mechanism of its formation is not completely clear. Tolstykh, Kalinin, Anisimova and coauthors [1–3] studied mustard gold from three deposits located in different regions of Russia: Kamchatka Peninsula (Maletoyvayam Ore Field) [1], Central Aldan district of Yakutia (Khokhoy ore field) [2], and Kola Peninsula (Oleninskoe Deposit) [3].

The main ore components of gold mineralization of the Gaching high-sulfidation (HS) epithermal Au–Ag deposit (part of the Maletoyvayam ore field) are native gold, tellurides, selenides, and sulphoselenotellurides of Au and oxidation products of Au-tellurides [1]. Two types of mustard gold were identified by Tolstykh et al. [1]: (a) mixtures of Fe-Sb(Te,Se,S) oxides and fine gold particles and (b) spotted and colloform gold consisting of aggregates of gold particles in a goethite/hydrogoethite matrix. This study examined different types of native gold in this ore deposit and the mechanisms and sequential transformation of calaverite ($AuTe_2$) into mustard gold.

Kalinin and coauthors [2] reported the results of the study of mustard gold from the Oleninskoe intrusion-related gold–silver Precambrian deposit of the Fennoscandian Shield. These authors showed that micropores in the mustard gold are filled with iron, antimony or thallium oxides, silver chlorides, bromides, and sulfides. They concluded that halogens (Cl, Br) played an important role in the remobilization of noble metals in the Oleninskoe deposit.

Anisimova and coauthors [3] investigated the features of native gold in karst cavities at the newly discovered Au-Te-Sb-Tl occurrence within the Khokhoy gold field of the Aldan-Stanovoy auriferous province (Aldan shield, East Russia). This was the first time the relationships between residual monolithic gold and unnamed tellurates, thallium carbonates and avicennite ($Te_2O_3$) had been described. Along with this native gold, secondary (sponge and "mustard") gold was investigated. The occurrence of monolithic, spongy and mustard gold was discussed.

The second part of the issue includes four articles on thermodynamic [10] and experimental modeling [11–13] of systems containing noble metals. Some of these articles were a continuation of the research [14,15] on the topics covered by the Special Issue "Experimental and Thermodynamic Modeling of Ore-Forming Processes in Magmatic and Hydrothermal Systems" [16].

Murzin and coauthors [10] constructed physicochemical models for the formation of magnetite–chlorite–carbonate rocks with copper–gold in the Karabash ultramafic massif (Southern Urals, Russia).

These authors showed that the metasomatic interaction of metamorphic fluid with serpentinites is responsible for the gold-poor mineralization (1st type), and that the gold-rich mineralization (2nd type) was formed during the mixing of metamorphic fluid and meteoric water in the open space of cracks in serpentinites.

Sinyakova and coauthors [11] carried out the crystallization of a Fe-Cu-S melt with the impurities of Ni, Sn, As, Pt, Pd, Rh, Ru, Ag, Au, Se, Te, Bi, and Sb. The cylindrical crystallized sample consisted of monosulfide solid solution (mss), nonstoichometric isocubanite (icb), and three modifications of intermediate solid solution (iss1, iss2, iss3). The simultaneous formation of two types of liquid separated during the cooling of the parent sulfide melt was observed. Noble metals associated with Bi, Sb, and Te were concentrated in inclusions in the form of $RuS_2$, $PdTe_2$, $(Pt,Pd)Te_2$, $PtRhAsS$, and $Ag_2Se$, doped with Ag, Cu, and Pd, in a monosulfide solid solution. Nobel metals form $PtAs_2$, gold alloys doped with Ag, Cu, and Pd, $Ag_2Te$ and $Pd(Bi,Sb)_xTe_{1-x}$ in nonstoichometric isocubanite and intermediate solid solutions. Rhodium is present in intermediate base metal solid solutions.

The surface species formed upon the contact of the pyrite, pyrrhotite, galena, chalcopyrite and valleriite with the solutions of $H_2PtCl_6$ (pH 1.5, 20 °C) were studied using X-ray photoelectron spectroscopy by Romanchenko et al. [12]. The highest rate of Pt deposition was observed on galena and valleriite and the lowest, on pyrite and pyrrhotite. Pt(IV) chloride complexes adsorb onto the mineral surface, and then the reduction of Pt(IV) to Pt(II) and the substitution of chloride ions with sulfide groups occur, forming sulfides of Pt(II) and then Pt(IV).

Vorobyev and coauthors [13] studied the reactions and species formed at different proportions of $HAuCl_4$, $H_2Se$ and $H_2S$ at room temperature. Metal gold colloids arose at the molar ratios $H_2Se(H_2S)/HAuCl_4$ less than 2. At higher ratios, pre-nucleation "dense liquid" species followed by fractional nucleation in the interior and coagulation of disordered gold chalcogenide occurred. The reactions proceed via the non-classical mechanism involving "dense droplets" of supersaturated solution which produce $AuSe1-xS_x/Au$ nanocomposites.

The third part of the issue contains two articles on Au-bearing arsenopyrite and pyrite [17,18]. A significant number of studies are devoted to Au and other noble metals in these minerals.

Sazonov and coauthors [17] used the Mössbauer spectroscopy method to study the ligand microstructure of natural arsenopyrites from the ores of major gold deposits of the Yenisei Ridge (Eastern Siberia, Russia). The elevated gold concentrations typical of arsenopyrites occur with an elevated content of sulfur or arsenic and correlate with the increase in the occupation degree of configurations {5S1As}, {4S2As}, {1S5As}, the reduction in the share of {3S3As}, and the amount of iron in tetrahedral cavities.

Tauson and coauthors [18] studied the forms of occurrence and distribution of "invisible" noble metals (Au, Ag, Pt, Pd, Ru) in the coexisting pyrite and arsenopyrite from the Natalkinskoe, Degdekan and Zolotaya Rechka deposits (Magadan region, Russia). They used a combination of methods of local analysis and statistics of the compositions of individual single crystals of different sizes to analyze the distribution coefficients of the structural (str) and surficial (sur) forms of noble metals. The data on Ag mostly indicate its fractionation into pyrite ($D_{str}$ Py/Asp = 17). Surface enrichment was considered a universal factor in the distribution of "invisible" noble metals. A number of elements (i.e., Pt, Ru, Ag) in pyrite and arsenopyrite tended to increase in abundance with a decrease in the crystallite size. This may be due to both the phase size effect and the intracrystalline adsorption of these elements at the interblock boundaries of a dislocation. Arsenopyrite with excess As over S has a tendency to have greater abundance of Pt, Ru and Pd. Sulfur deficiency was a favorable factor for the incorporation of Ag and platinoids into the structures of the studied minerals.

The last part of the issue includes two articles [19,20] devoted to the study of ages of gold deposits in China: Nibao and Chaoyangzhai (Southeast Guizhou). These deposits are an important part of the Yunnan–Guizhou–Guangxi "Golden Triangle" region. Zheng, Tsang and coauthors [19,20] also discussed possible sources of gold mineralization.

The Nibao gold deposit includes both fault-controlled and strata-bound gold orebodies. Zheng and coauthors [19] determined the mineralization age of these gold orebodies and provided additional evidence for constraining the formation ages of low-temperature orebodies and their metallogenic distribution in South China. The results confirm the Middle-Late Yanshanian mineralizing events of the Carlin-type gold deposits in Yunnan, Guizhou, and Guangxi Provinces of Southwest China.

Tsang and coauthors [20] determined the possible source of the newly discovered medium- to large-scale turbidite-hosted Chaoyangzhai gold deposit, Southeast Guizhou, South China, using LA-ICP-MS zircon U–Pb dating, whole-rock geochemistry and in situ sulfur isotopes. Together with the evidence of similar geochemical patterns between the tuffaceous- and sandy-slates and gold- bearing quartz, these authors proposed that gold might be sourced from the sandy slates.

Overall, I hope this Special Issue will contribute to a better understanding of the genesis of gold, silver and other noble metal deposits, the behavior of these elements in endogenic and supergene environments and suggest ways forward to solve the problem of their full extraction from ores.

**Author Contributions:** G.A.P. wrote this editorial. Author has read and agreed to the published version of the manuscript.

**Funding:** Work is done on state assignment of V.S. Sobolev Institute of Geology and Mineralogy of Siberian Branch of Russian Academy of Sciences financed by Ministry of Science and Higher Education of the Russian Federation.

**Acknowledgments:** I thank the authors of the articles included in this Special Issue and the organizations that have financially supported the research in the areas related to this topic. I would like to thank the Editors-in-Chief, Editors, Assistant Editors, and Reviewers for their important comments and constructive suggestions, which helped the contributing authors to improve the quality of the manuscripts.

**Conflicts of Interest:** The author declares no conflict of interest.

## References

1. Tolstykh, N.D.; Palyanova, G.A.; Bobrova, O.V.; Sidorov, E.G. Mustard Gold of the Gaching Ore Deposit (Maletoyvayam Ore Field, Kamchatka, Russia). *Minerals* **2019**, *9*, 489. [CrossRef]

2. Kalinin, A.A.; Savchenko, Y.E.; Selivanova, E.A. Mustard Gold in the Oleninskoe Gold Deposit, Kolmozero–Voronya Greenstone Belt, Kola Peninsula, Russia. *Minerals* **2019**, *9*, 786. [CrossRef]

3. Anisimova, G.S.; Kondratieva, L.A.; Kardashevskaia, V.N. Characteristics of Supergene Gold of Karst Cavities of the Khokhoy Gold Ore Field (Aldan Shield, East Russia). *Minerals* **2020**, *10*, 139. [CrossRef]

4. Petersen, S.O.; Makovicky, E.; Juiling, L.; Rose-Hansen, J. 'Mustard gold' from the Dongping Au-Te deposit, Hebei province, People's Republic of China. *N. Jb. Miner. Mh.* **1999**, 337–357.

5. Makovicky, E.; Chovan, M.; Bakos, F. The stibian mustard gold from the Kriváň Au deposit, Tatry Mts., Slovak Republic. *N. Jb. Miner. Abh.* **2007**, *184*, 207–215.

6. Okrugin, V.M.; Skilskaya, E.; Etschmann, B.; Pring, A.; Li, K.; Zhao, J.; Griffiths, G.; Lumpkin, G.R.; Triani, G.; Brugger, J. Microporous gold: comparison of textures from nature and experiments. *Am. Miner.* **2014**, *99*, 1171–1174. [CrossRef]

7. Zacharias, J.; Mate, J.N. Gold to aurostibite transformation and formation of Au-Ag-Sb phases: The Krásná Hora deposit, Czech Republic. *Miner. Mag.* **2017**, *81*, 987–999. [CrossRef]

8. Saunders, J.A.; Burke, M. Formation and aggregation of gold (electrum) nanoparticles in epithermal ores. *Minerals* **2017**, *7*, 163. [CrossRef]

9. Zhao, J.; Pring, A. Mineral Transformations in Gold–(Silver) Tellurides in the Presence of Fluids: Nature and Experiment. *Minerals* **2019**, *9*, 167. [CrossRef]

10. Murzin, V.; Chudnenko, K.; Palyanova, G.; Varlamov, D. Formation of Au-Bearing Antigorite Serpentinites and Magnetite Ores at the Massif of Ophiolite Ultramafic Rocks: Thermodynamic Modeling. *Minerals* **2019**, *9*, 758. [CrossRef]

11. Sinyakova, E.; Kosyakov, V.; Palyanova, G.; Karmanov, N. Experimental Modeling of Noble and Chalcophile Elements Fractionation during Solidification of Cu-Fe-Ni-S Melt. *Minerals* **2019**, *9*, 531. [CrossRef]

12. Romanchenko, A.; Likhatski, M.; Mikhlin, Y. X-ray Photoelectron Spectroscopy (XPS) Study of the Products Formed on Sulfide Minerals Upon the Interaction with Aqueous Platinum (IV) Chloride Complexes. *Minerals* **2018**, *8*, 578. [CrossRef]

13. Vorobyev, S.; Likhatski, M.; Romanchenko, A.; Maksimov, N.; Zharkov, S.; Krylov, A.; Mikhlin, Y. Colloidal and Deposited Products of the Interaction of Tetrachloroauric Acid with Hydrogen Selenide and Hydrogen Sulfide in Aqueous Solutions. *Minerals* **2018**, *8*, 492. [CrossRef]
14. Murzin, V.; Chudnenko, K.; Palyanova, G.; Kissin, A.; Varlamov, D. Physicochemical Model of Formation of Gold-Bearing Magnetite-Chlorite-Carbonate Rocks at the Karabash Ultramafic Massif (Southern Urals, Russia). *Minerals* **2018**, *8*, 306. [CrossRef]
15. Tauson, V.L.; Lipko, S.V.; Smagunov, N.V.; Kravtsova, R.G. Trace Element Partitioning Dualism under Mineral–Fluid Interaction: Origin and Geochemical Significance. *Minerals* **2018**, *8*, 282. [CrossRef]
16. Palyanova, G. Editorial for the Special Issue: Experimental and Thermodynamic Modeling of Ore-Forming Processes in Magmatic and Hydrothermal Systems. *Minerals* **2018**, *8*, 590. [CrossRef]
17. Sazonov, A.M.; Silyanov, S.A.; Bayukov, O.A.; Knyazev, Y.V.; Zvyagina, Y.A.; Tishin, P.A. Composition and Ligand Microstructure of Arsenopyrite from Gold Ore Deposits of the Yenisei Ridge (Eastern Siberia, Russia). *Minerals* **2019**, *9*, 737. [CrossRef]
18. Tauson, V.; Lipko, S.; Kravtsova, R.; Smagunov, N.; Belozerova, O.; Voronova, I. Distribution of "Invisible" Noble Metals between Pyrite and Arsenopyrite Exemplified by Minerals Coexisting in Orogenic Au Deposits of North-Eastern Russia. *Minerals* **2019**, *9*, 660. [CrossRef]
19. Zheng, L.; Yang, R.; Gao, J.; Chen, J.; Liu, J.; Li, D. Quartz Rb-Sr Isochron Ages of Two Type Orebodies from the Nibao Carlin-Type Gold Deposit, Guizhou, China. *Minerals* **2019**, *9*, 399. [CrossRef]
20. Tsang, H.; Cao, J.; Yang, X. Source of the Chaoyangzhai Gold Deposit, Southeast Guizhou: Constraints from LA-ICP-MS Zircon U–Pb Dating, Whole-rock Geochemistry and In Situ Sulfur Isotopes. *Minerals* **2019**, *9*, 235. [CrossRef]

Article

# Mustard Gold of the Gaching Ore Deposit (Maletoyvayam Ore Field, Kamchatka, Russia)

**Nadezhda D. Tolstykh** [1,2,*], **Galina A. Palyanova** [1,2], **Ol'ga V. Bobrova** [3] and **Evgeny G. Sidorov** [4]

[1]   VS Sobolev Institute of Geology and Mineralogy of Siberian Branch of Russian Academy of Sciences, Koptyuga Ave., 3, 630090 Novosibirsk, Russia

[2]   Department of Geology and Geophysics, Novosibirsk State University, Pirogova Ave., 2, 630090 Novosibirsk, Russia

[3]   Coralina Engineering, Poslannikov per., 5, 105005 Moscow, Russia

[4]   Institute of Volcanology and Seismology, Far East Branch of Russian Academy of Sciences, Piipa Blvd., 9, 683006 Petropavlovsk-Kamchatsky, Russia

*   Correspondence: tolst@igm.nsc.ru

Received: 1 July 2019; Accepted: 8 August 2019; Published: 15 August 2019

**Abstract:** The Gaching high-sulfidation (HS) epithermal Au–Ag deposits, part of the Maletoyvayam ore field, which is located in the volcanic belts of the Kamchatka Peninsula (Russia). The main ore components are native gold, tellurides, selenides, and sulphoselenotellurides of Au and oxidation products of Au-tellurides. This study examines the different types of native gold in this ore deposit and the mechanisms and sequential transformation of calaverite ($AuTe_2$) into mustard gold. The primary high fineness gold (964‰–978‰) intergrown with maletoyvayamite $Au_3Te_6Se_4$ and other unnamed phases (AuSe, Au(Te,Se)) differ from the secondary (mustard) gold in terms of fineness (1000‰) and texture. Primary gold is homogeneous, whereas mustard is spongy. Two types of mustard gold were identified: (a) Mixtures of Fe-Sb(Te,Se,S) oxides and fine gold particles, which formed during the hypogenic transformation stage of calaverite due to the impact of hydrothermal fluids, and (b) spotted and colloform gold consisting of aggregates of gold particles in a goethite/hydrogoethite matrix. This formed during the hypergenic transformation stage. Selenides and sulphoselenotellurides of gold did not undergo oxidation. Pseudomorphic replacement of calaverite by Au-Sb(Te,Se,S,As) oxides was also observed.

**Keywords:** Gaching ore deposit; mustard gold; calaverite; maletoyvayamite; Fe-Sb(Te,As,Se,S)-oxides; Au-Sb(Te,Se,S,As)-oxides

---

## 1. Introduction

One of the fundamental tasks in the study of the epithermal deposits is identifying the mechanisms of gold formation. The traditional gravity method can sometimes give negative results in the identifying of the cause of the high Au concentrations. Occasionally sample analyses show high concentrations of Au, while the presence of visible gold is absent or is only represented by single grains where the amount of Au is not capable the providing such high concentrations. It appears that the finest fraction (<0.06 mm) is the most enriched in Au, which is almost beyond visible detection. The high concentrations of Au in the hypergenesis zones of epithermal deposits is due to the presence of so-called mustard gold, which is difficult to detect.

The term "mustard gold" was introduced by W. Lindgren [1]. Typical features of mustard gold are low reflectivity, porous or colloform texture, and rusty, reddish, orange-red, and brown-yellow colors in reflected light. Mustard gold is easily scattered during crushing and the resulting fine powdered fraction may become progressively enriched in gold. Mustard gold was previously studied by numerous researchers [2–8] and has been shown to be a characteristic feature in gold-telluride deposits [9,10] and antimony-gold deposits [11,12]. Mustard gold was also documented in the Ozernovskoye and Aginskoye deposits (Kamchatka) as rims surrounding primary Au tellurides or as micro-veinlets in the altered matrix [4,6]. These deposits belong to the alunite-quartz group of the gold-silver deposits (Figure 1) and are associated with metasomatic rocks (secondary quartzites) that are confined to the volcanic centers of volcanic-tectonic structures within the Vetrovayam volcanic zone. Mustard gold is formed due to the oxidation of primary Au minerals and is the result of exposure to acidic solutions leaching tellurium or their transformation into tellurates [2,3]. This replacement may take place under hydrothermal conditions, such as during the late-stages of ore deposit formation [5]. The development of microporous gold indicates that the deposits may have experienced overprint after mineralization [13]. Mustard gold is heterogeneous and represented by multiphase aggregates consisting of two, three, or more phases, which differ in their chemical composition. Typomorphic elements in these compounds within the pores of mustard gold include Sb, Te, Pb, Fe, Cu, Ag, Hg, and others, which are dependent on the composition of the primary mineral assemblage of the deposit. Mustard gold intergrowths with Au-sulfoselenotellurides, sulfosalts (tetrahedrite, goldfieldite), and the products of oxidation of Au-tellurides from the Gaching deposit of the Maletoyvayam ore field in the volcanic belts of the Kamchatka Peninsula (Figure 1) are presented in this study [14].

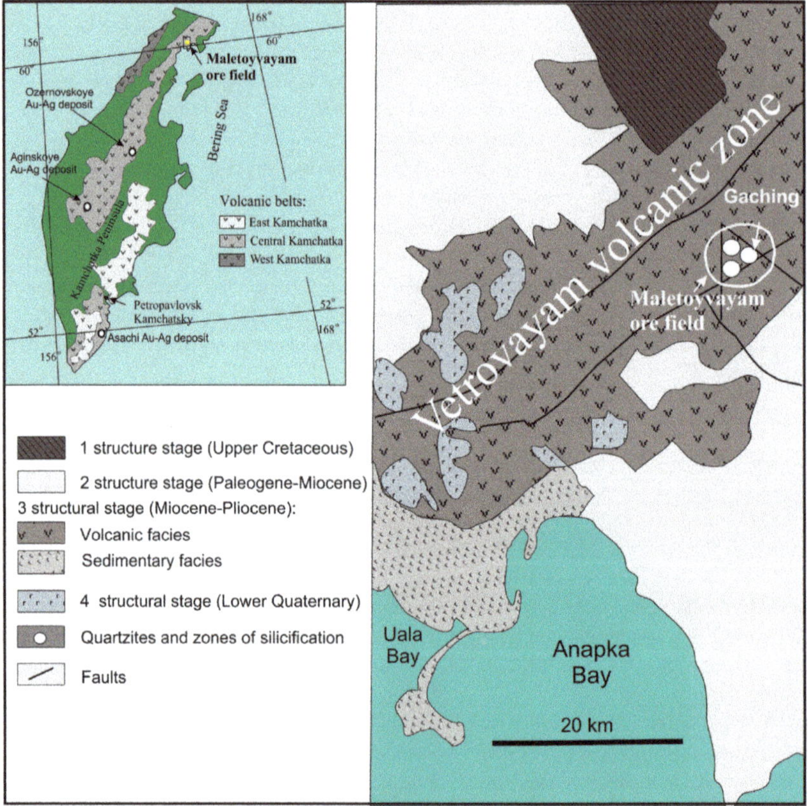


**Figure 1.** Volcanic belts of the Kamchatka Peninsula [15], the tectonic scheme of the Vetrovayam volcanic zone [16] and location of Maletovvayam ore field on the base [17] (reproduced with permission from reference [14]).

This study presents detailed textural descriptions, chemical compositions and reconstructs the sequence of replacement of primary minerals in order to understand its origin. Our research shows a sequential transformation of the primary mineralization, which is represented by calaverite to mustard gold.

## 2. Materials and Methods

The studied grains of Au-minerals were obtained by crushing mineralized rocks and panning the resulting material into a heavy fraction using hydroseparation, followed by concentration in the heavy liquid. Mustard gold was found in polished sections made from the heavy mineral concentrate, which was obtained from a 20-kilogram sample of alunite-quartz rock. The chemical compositions of the minerals, mineral aggregate textures, and separated grains were examined at the Analytical Center for Multi-Elemental and Isotope Research at the VS Sobolev Institute of Geology and Mineralogy SB RAS in Novosibirsk (Russia) using a LEO-413VP scanning electron microscope (SEM) with INCA Energy 350 microanalysis system (Oxford Instruments Ltd., Abingdon, UK) equipped with EDS (analysts Dr. N. Karmanov, M. Khlestov, and V. Danilovskaya), operating at an accelerating voltage of 20 kV, beam current of 0.4 nA, 50 s measuring time, and beam diameter of ~1 μm. The following standards were used: Pure substances (Ag, Au, Bi, Se, Sb, Fe, and Cu), pyrite (S), synthetic HgTe (Te), and sperrylite (As). The detection limit was 0.02%. The following X-ray lines were selected: L$\alpha$ for Ag, Te, As, Sb, and Se; K$\alpha$ for S, Fe, Cu, and O; and M$\alpha$ for Au and Bi. All the compositions of minerals in this study were performed using an EDS spectrometer. A compositional comparison Au–Te–Se–S minerals determined by the EDS and WDS methods is presented in a previous study [14], which shows their complete convergence.

## 3. Results

### 3.1. Types of Mustard Gold

Previous studies of the Gaching ore deposit that refer to the epithermal Au–Ag high-sulfidation (HS) type have identified native gold, barite, anglesite, quartz, pyrite, Au-telluride (Se-bearing calaverite), Au-sulphoselenotellurides, sulphosalts (tetrahedrite, goldfieldite, tennantite), and other rare minerals (famatinite, enargite, watanabeite, senarmontite, tripuhyite, rooseveltite, tiemanite, antimonsilite, and guanajuatite) in heavy fractions [14]. Over 200 grains of Au–Te–Se–S compounds were additionally extracted from the ore to obtain new data.

The studied grains of primary Au–Ag alloys (10–50 μm in size) are always found as intergrowths with sulfosalts or unique unnamed phases of the Au–Te–Se–S system that are potentially new minerals with unique compositions: $Au_3Te_6(Se,S)_4$, $Au_2Te_4(Se,S)_3$, AuSe, $Au_2TeSe$, and other (Figure 2a,b) [14,18]. One of these minerals, $Au_3Te_6Se_4$, was recently approved by a commission on new minerals and named maletoyvayamite [19]. The possibility of natural occurrences of compounds of Te, Se, and S and Ag was previously shown in the Prasolovskoye deposit on Kunashir Island [3,20].

Primary gold in the Gaching deposit occurs in high grades and Ag does not exceed 2.5 wt. % [14]. In the studied samples (Figure 2), Ag-content varied in range 1.47–1.98 wt. % (Tables 1 and 2). Neither primary gold, nor these compounds undergo replacement or oxidation during hypogene processes. It should be noted that direct intergrowths of primary gold with calaverite ($AuTe_2$) or mustard gold was not found. Primary gold (Au–Ag alloys) in the Gaching deposit accounts for no more than 5% when the porous gold reaches up to 60% of the amount of the Au-bearing minerals in the ore assemblage. Mustard gold exhibited more yellowish, reddish, and brown colors in reflected light (Figure 3) when compared to primary gold (Figure 3c) and porous texture gold, as observed in the other deposits worldwide. Irregular grains of mustard gold are between 10 and 60 μm in size and exhibited spongy textures, which were either empty or imbedded with microscopic inclusions in the pore spacings (Figure 4), similar to mustard gold from the Dongping Au-Te deposit [9,10,13]. Mustard gold varied from microporous (i.e., spongy) to colloform and zoned. However, microporous aggregates of gold filled by compounds of Fe, Sb, Te, As, Bi, and S with oxygen (antimonate/tellurate of iron; Figure 4) were prevalent. The contents of these compounds in microporous gold were due to the weathering of minerals (mainly sulfosalts). The presence of Ag in these compounds was likely due to inheritance from primary calaverite. At this stage of mustard gold formation, Ag is a minor element that enters oxides, since analyses of larger fragments of secondary mustard gold showed the complete absence of Ag in its composition (Table 3), sample d_9-6). Analyses of these compounds mixed with gold particles are presented in Table 3. The types of mustard gold were identified by contents of various antimonate/tellurate/hydroxides in the submicroscopic pores of the mustard gold aggregate. As shown in Table 1, all elements showed great compositional variation. The oxygen concentration also changed due to (1) different degrees of oxidation of the primary products, and (2) different ratios of secondary gold and Fe-Sb(Te,As,Bi,S) oxide. Due to the microscopic size of the particles, it is difficult to determine whether Au belongs to the native reduced form of mustard gold, or is still is part of the complex oxides as it suggested by [14]. This type of mustard gold was often associated with iron antimonates and antimonites, such as tripuhyite $Fe^{2+}Sb^{5+}_2O_6$ or $Fe^{3+}Sb^{5+}O_4$ [21] (Figure 4d). If Au and Ag were assumed to belong to the native phase, then the total of the remaining elements (Table 4) had variable ratios forming a trend towards iron oxides/hydroxides, which was most likely limonite (Figure 5). These compounds were the products of the successive oxidation of tripuhyite.

**Table 1.** The composition of the primary gold and associated minerals (in wt. %) shown in Figure 2.

| No. | Sample | Sp. | Cu | Au | Ag | Bi | Sb | Te | As | Se | S | Total |
|-----|--------|-----|------|--------|------|------|-------|-------|------|-------|-------|--------|
| 1 | 3_3 | 1 | - | 98.45 | 1.86 | - | - | - | - | - | - | 100.31 |
| 2 | 3_3 | 2 | - | 96.23 | 1.98 | - | - | - | - | - | - | 98.21 |
| 3 | 3_3 | 3 | 41.81 | - | 0.37 | - | 19.4 | - | 6.51 | 4.64 | 27.76 | 100.49 |
| 4 | 3_3 | 4 | - | 36.25 | - | - | - | 49.35 | - | 5.05 | 8.6 | 99.25 |
| 5 | 4_1 | 1 | - | 98.54 | 1.23 | - | - | - | - | - | - | 99.77 |
| 6 | 4_1 | 2 | - | 100.33 | 1.61 | - | - | - | - | - | - | 101.94 |
| 7 | 4_1 | 3 | - | 35.92 | - | - | - | 47.67 | - | 15.35 | 2.05 | 100.99 |
| 8 | 4_1 | 4 | - | 36.61 | - | - | - | 48.48 | - | 11.16 | 4.29 | 100.54 |
| 9 | 13_1 | 1 | - | 98.1 | 1.88 | - | - | - | - | - | - | 99.98 |
| 10 | 13_1 | 2 | 1.06 | 34.8 | - | 1.28 | - | 47.35 | - | 12.24 | 3.38 | 100.11 |
| 11 | 13_1 | 3 | - | 35.41 | - | - | - | 45.48 | - | 15.77 | 1.03 | 97.69 |
| 12 | 13_1 | 4 | - | 96.05 | 1.55 | - | - | - | - | - | - | 97.6 |
| 13 | 13_1 | 5 | - | 35.45 | - | - | - | 45.64 | - | 14.61 | 1.71 | 97.41 |
| 14 | 13_1 | 6 | - | 37.41 | - | 0.92 | - | 46.47 | - | 9.91 | 3.94 | 98.65 |
| 15 | 13_1 | 7 | 0.19 | 35.58 | - | 0.75 | - | 46.39 | - | 13.28 | 2.65 | 98.84 |
| 16 | 13_1 | 8 | 0.25 | 35.26 | - | - | - | 45.67 | - | 14.66 | 1.77 | 97.61 |
| 17 | 13_1 | 9 | 41.89 | - | - | - | 19.29 | - | 4.72 | 4.36 | 26.97 | 98.38 |
| 18 | 11_3 | 1 | - | 98.52 | 1.52 | - | - | - | - | - | - | 100.04 |
| 19 | 11_3 | 2 | - | 97.6 | 1.47 | - | - | - | - | - | - | 99.07 |
| 20 | 11_3 | 4 | - | 35.09 | - | 0.74 | - | 45.96 | - | 14.56 | 1.32 | 97.67 |
| 21 | 11_3 | 5 | - | 69.91 | - | - | - | 1.09 | - | 26.04 | 2.37 | 99.41 |
| 22 | 11_3 | 6 | - | 35.93 | - | - | - | 47.78 | - | 13.6 | 3.32 | 100.63 |
| 23 | 11_3 | 7 | - | 35.53 | - | 0.81 | - | 47.12 | - | 13.23 | 2.76 | 99.45 |
| 24 | 11_3 | 9 | - | 37.31 | - | 0.74 | - | 46.6 | - | 10.99 | 3.84 | 99.48 |
| 25 | 11_3 | 10 | - | 66.71 | - | - | - | 18.2 | - | 14.78 | - | 99.69 |

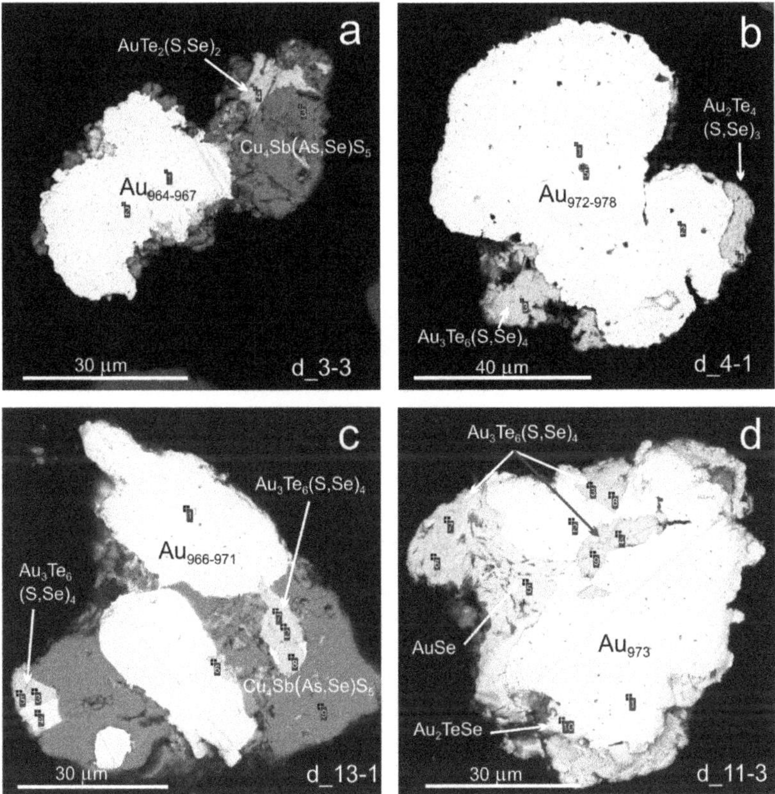



**Figure 2.** Secondary electron images of scanning electron microscopy (SEM). Primary gold (Au$_N$, where N is gold fineness) intergrown with the unnamed minerals of Au–Te–Se–S system in an intergrowth with (**a**)—unnamed phases Cu$_4$Sb(As,Se)S$_5$ and AuTe$_2$(S,Se)$_2$; (**b**)—maletoyvayamite Au$_3$Te$_6$(Se,S)$_4$ and unnamed Au$_2$Te$_4$(Se,S)$_3$; (**c**)—unnamed Cu$_4$Sb(As,Se)S$_5$ and maletoyvayamite; and (**d**)—maletoyvayamite and unnamed AuSe and Au$_2$TeSe. Compositions of minerals and unnamed phases shown in Table 1. The numbers in microphotographs are analitical spots, here and in other figures.

**Table 2.** Formulas to Table 1.

| No. | Formula | Abbr. | No. | Formula | Abbr. |
|---|---|---|---|---|---|
| 1 | Au$_{0.97}$Ag$_{0.03}$ | gd | 14 | Au$_{3.06}$Te$_{5.87}$(Se$_{2.02}$S$_{1.98}$ Bi$_{0.07}$)$_{4.07}$ | Mt |
| 2 | Au$_{0.97}$Ag$_{0.03}$ | gd | 15 | (Au$_{2.93}$Cu$_{0.05}$)$_{2.98}$Te$_{5.90}$(Se$_{2.73}$S$_{1.34}$ Bi$_{0.06}$)$_{4.13}$ | Mt |
| 3 | Cu$_{3.95}$Sb$_{0.96}$(As$_{0.52}$Se$_{0.35}$)$_{0.87}$S$_{5.20}$ | wt | 16 | (Au$_{2.98}$Cu$_{0.07}$)$_{3.05}$Te$_{5.95}$(Se$_{3.09}$S$_{0.92}$)$_{4.01}$ | Mt |
| 4 | Au$_{1.02}$Te$_{2.14}$(S1.48Se$_{0.35}$)$_{1.87}$ | unn | 17 | Cu$_{4.08}$Sb$_{0.98}$(As$_{0.39}$Se$_{0.34}$)$_{0.73}$S$_{5.21}$ | Wt |
| 5 | Au$_{0.98}$Ag$_{0.02}$ | gd | 18 | Au$_{0.97}$Ag$_{0.03}$ | Gd |
| 6 | Au$_{0.97}$Ag$_{0.03}$ | gd | 19 | Au$_{0.97}$Ag$_{0.03}$ | Gd |
| 7 | Au$_{2.91}$Te$_{5.95}$(Se$_{3.10}$S$_{1.02}$)$_{4.12}$ | mt | 20 | Au$_{3.02}$Te$_{6.10}$(Se$_{3.12}$S$_{0.70}$ Bi$_{0.06}$)$_{3.88}$ | Mt |
| 8 | Au$_{1.99}$Te$_{4.07}$(Se$_{1.51}$S$_{1.43}$)$_{2.94}$ | unn | 21 | Au$_{0.93}$(Se$_{0.86}$S$_{0.19}$Te$_{0.02}$)$_{1.07}$ | Unn |
| 9 | Au$_{0.97}$Ag$_{0.03}$ | gd | 22 | Au$_{1.97}$Te$_{4.05}$(Se$_{1.86}$S$_{1.12}$)$_{2.98}$ | Unn |
| 10 | (Au$_{2.76}$Cu$_{0.26}$)$_{3.02}$Te$_{5.81}$(Se$_{2.43}$S$_{1.65}$ Bi$_{0.10}$)$_{4.18}$ | mt | 23 | Au$_{2.91}$Te$_{5.95}$(Se$_{2.70}$S$_{1.39}$ Bi$_{0.06}$)$_{4.15}$ | Mt |
| 11 | Au$_{3.04}$Te$_{6.03}$(Se$_{3.38}$S$_{0.54}$)$_{3.92}$ | mt | 24 | Au$_{3.01}$Te$_{5.81}$(Se$_{2.21}$S$_{1.91}$ Bi$_{0.06}$)$_{4.18}$ | Mt |
| 12 | Au$_{0.97}$Ag$_{0.03}$ | gd | 25 | Au$_{2.03}$Te$_{0.85}$Se$_{1.12}$ | Unn |
| 13 | Au$_{3.02}$Te$_{5.99}$(Se$_{3.10}$S$_{0.89}$)$_{3.99}$ | - | - | - | - |

Note. Sp—analysis spot in figures, hereinafter. Signs: gd—gold; wt—watanabeite; mt—maletoyvayamite; and unn—unnamed phase. Abbr.—minerals abbreviation.

**Figure 3.** Reflected light photomicrograph of mustard gold (**a–c**) compared with primary gold (yellow) and pyrite (white) (**c**) and with Fe–Sb oxide (gray; **b,d**).

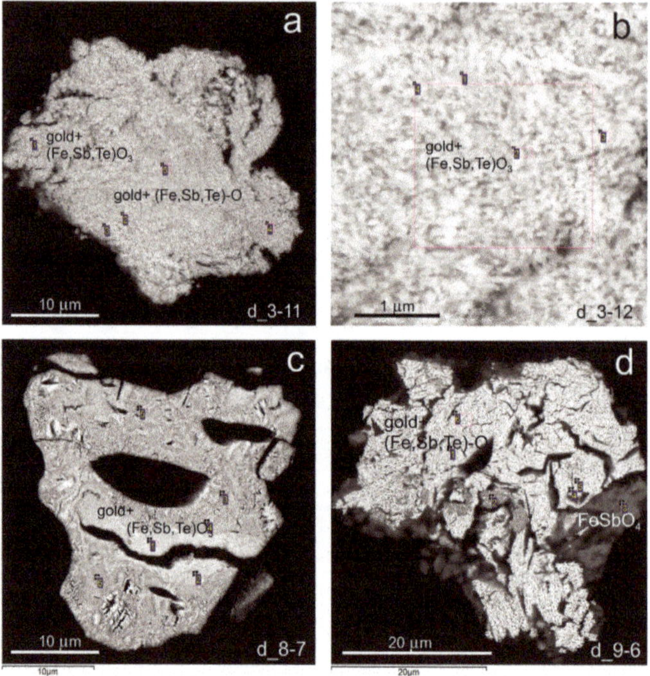

**Figure 4.** SEM image of mustard (secondary) gold of porous (**a,b,d**) and colloform (concentrically banded), (**c**) textural intergrowth with Fe-(Sb,Te,As,Bi,S)-O compounds. The compositions of compounds or mixtures in the analyzed sites are presented in Table 3. d_9-6 spot 3 (FeSbO$_4$)—tripuhyite. The purple square is scan of analytical area.

**Table 3.** The mixed compositions of mustard gold and oxides/hydroxides of Fe, Sb, Te, As, Bi, and S (in wt. and at. %) localized in the pores of mustard gold shown in Figure 4.

| Sample | Sp. | Fe | Au | Ag | Bi | Sb | Te | As | S | O | Total |
|---|---|---|---|---|---|---|---|---|---|---|---|
| 3_11 | 2 | 5.93 | 76.70 | 2.33 | - | 2.23 | 2.32 | 1.40 | 0.59 | 7.13 | 98.63 |
| 3_11 | 3 | 2.18 | 88.66 | 0.70 | - | 1.00 | 1.28 | 0.80 | - | 5.61 | 100.23 |
| 3_11 | 4 | 6.81 | 78.07 | 0.84 | - | 1.13 | 1.72 | 0.79 | - | 9.03 | 98.39 |
| 3_11 | 5 | 4.76 | 79.85 | 0.78 | - | 1.69 | 2.06 | 1.46 | - | 9.62 | 100.22 |
| 3_12 | 2 | 4.45 | 82.36 | 0.72 | - | 2.12 | 1.97 | 1.05 | - | 6.74 | 99.41 |
| 3_12 | 4 | 4.62 | 81.42 | 1.00 | - | 2.16 | 2.12 | 0.80 | - | 6.05 | 98.17 |
| 8_7 | 1 | 4.58 | 70.20 | 1.40 | 2.60 | 9.02 | 1.00 | 1.43 | 0.30 | 9.42 | 99.95 |
| 8_7 | 2 | 5.17 | 69.63 | 1.47 | 2.53 | 9.58 | 0.70 | 1.72 | 0.39 | 9.53 | 100.72 |
| 8_7 | 3 | 6.63 | 62.91 | 1.46 | 2.02 | 11.42 | 1.00 | 2.28 | 0.41 | 12.70 | 100.83 |
| 8_7 | 4 | 5.73 | 67.88 | 1.32 | 2.58 | 10.26 | 1.21 | 2.01 | 0.34 | 9.57 | 100.90 |
| 8_7 | 5 | 6.62 | 63.82 | 1.32 | 2.07 | 11.11 | 1.10 | 2.15 | 0.35 | 10.58 | 99.12 |
| 8_7 | 6 | 5.44 | 63.65 | 1.58 | 2.72 | 11.41 | 1.22 | 2.06 | - | 11.85 | 99.93 |
| 9_6 | 1 | 5.32 | 86.15 | - | - | - | - | 0.63 | - | 6.66 | 98.76 |
| 9_6 | 2 | 15.83 | 41.44 | - | 1.53 | 14.45 | 1.44 | 2.06 | 0.33 | 23.23 | 100.31 |
| 9_6 | 4 | 4.34 | 85.16 | - | - | - | - | - | - | 8.78 | 98.28 |
| 9_6 | 5 | 1.58 | 96.52 | - | - | - | - | - | - | 2.88 | 100.98 |
| 9_6 | 6 | 4.39 | 85.92 | - | - | 0.70 | - | 0.61 | - | 8.80 | 100.42 |
| Sample | Sp. | Fe | Au | Ag | Bi | Sb | Te | As | S | O | Total |
| at. % | | | | | | | | | | | |
| 3_11 | 2 | 9.87 | 38.18 | 2.12 | - | 1.80 | 1.78 | 1.83 | 0.73 | 43.69 | 100 |
| 3_11 | 3 | 4.24 | 51.55 | 0.74 | - | 0.94 | 1.15 | 1.22 | - | 40.16 | 100 |
| 3_11 | 4 | 10.34 | 35.47 | 0.70 | - | 0.83 | 1.21 | 0.94 | - | 50.51 | 100 |
| 3_11 | 5 | 7.06 | 35.43 | 0.63 | - | 1.21 | 1.41 | 1.70 | - | 52.55 | 100 |
| 3_12 | 2 | 7.80 | 43.18 | 0.69 | - | 1.80 | 1.59 | 1.45 | - | 43.50 | 100 |
| 3_12 | 4 | 8.48 | 44.73 | 1.00 | - | 1.92 | 1.80 | 1.16 | - | 40.91 | 100 |
| 8_7 | 1 | 6.71 | 30.76 | 1.12 | 1.07 | 6.39 | 0.68 | 1.65 | 0.81 | 50.81 | 100 |
| 8_7 | 2 | 7.42 | 29.91 | 1.15 | 1.02 | 6.66 | 0.46 | 1.94 | 1.03 | 50.40 | 100 |
| 8_7 | 3 | 8.07 | 22.92 | 0.97 | 0.69 | 6.73 | 0.56 | 2.18 | 0.92 | 56.95 | 100 |
| 8_7 | 4 | 8.13 | 28.82 | 1.02 | 1.03 | 7.05 | 0.79 | 2.24 | 0.89 | 50.02 | 100 |
| 8_7 | 5 | 8.92 | 25.73 | 0.97 | 0.79 | 7.25 | 0.68 | 2.28 | 0.87 | 52.51 | 100 |
| 8_7 | 6 | 7.02 | 24.58 | 1.11 | 0.99 | 7.13 | 0.73 | 2.09 | - | 56.34 | 100 |
| 9_6 | 1 | 9.48 | 45.93 | - | - | 0.00 | 0.00 | 0.88 | - | 43.71 | 100 |
| 9_6 | 2 | 12.76 | 9.99 | - | 0.35 | 5.64 | 0.54 | 1.31 | 0.49 | 68.94 | 100 |
| 9_6 | 4 | 6.98 | 40.99 | - | - | - | - | - | - | 52.03 | 100 |
| 9_6 | 5 | 3.85 | 70.32 | - | - | - | - | - | - | 25.83 | 100 |
| 9_6 | 6 | 6.93 | 40.59 | - | - | 0.54 | - | 0.76 | - | 51.18 | 100 |

**Table 4.** Compositions (at. %) of oxides/hydroxides of Fe, Sb, Te, As, Bi, and S (after removal of Au and Ag) localized in the pores of mustard gold (data of Table 3 recalculated to 100%).

| Sample | Sp. | Fe | Bi | Sb | Te | As | S | O | Total |
|---|---|---|---|---|---|---|---|---|---|---|
| 3_11 | 2 | 16.53 | - | 3.02 | 2.98 | 3.07 | 1.22 | 73.18 | 100 |
| 3_11 | 3 | 8.89 | - | 1.97 | 2.41 | 2.56 | - | 84.18 | 100 |
| 3_11 | 4 | 16.20 | - | 1.30 | 1.90 | 1.47 | - | 79.13 | 100 |
| 3_11 | 5 | 11.04 | - | 1.89 | 2.21 | 2.66 | - | 82.20 | 100 |
| 3_12 | 2 | 13.89 | - | 3.21 | 2.83 | 2.58 | - | 77.48 | 100 |
| 3_12 | 4 | 15.63 | - | 3.54 | 3.32 | 2.14 | - | 75.38 | 100 |
| 8_7 | 1 | 9.85 | 1.57 | 9.38 | 1.00 | 2.42 | 1.19 | 74.59 | 100 |
| 8_7 | 2 | 10.76 | 1.48 | 9.66 | 0.67 | 2.81 | 1.49 | 73.12 | 100 |
| 8_7 | 3 | 10.60 | 0.91 | 8.84 | 0.74 | 2.86 | 1.21 | 74.84 | 100 |
| 8_7 | 4 | 11.59 | 1.47 | 10.05 | 1.13 | 3.19 | 1.27 | 71.30 | 100 |
| 8_7 | 5 | 12.17 | 1.08 | 9.89 | 0.93 | 3.11 | 1.19 | 71.64 | 100 |
| 8_7 | 6 | 9.45 | 1.33 | 9.60 | 0.98 | 2.81 | - | 75.83 | 100 |
| 9_6 | 1 | 17.53 | - | - | - | 1.63 | - | 80.84 | 100 |
| 9_6 | 2 | 14.17 | 0.39 | 6.26 | 0.60 | 1.46 | 0.54 | 76.57 | 100 |
| 9_6 | 4 | 11.83 | - | - | - | - | - | 88.17 | 100 |
| 9_6 | 5 | 12.97 | - | - | - | - | - | 87.03 | 100 |
| 9_6 | 6 | 11.66 | - | 0.91 | - | 1.28 | - | 86.15 | 100 |

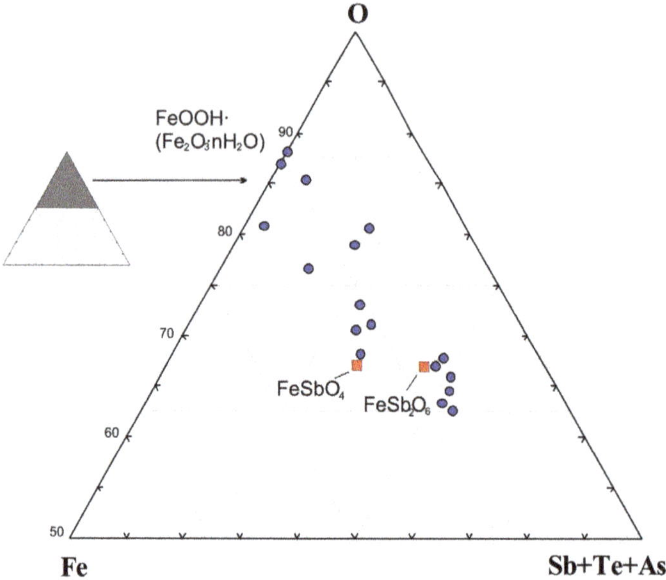



**Figure 5.** Ternary plot showing the composition of Fe–Sb–Te–As–Se–S-oxides localized in the pores of mustard gold, shown in the Figure 4 and Table 4. $FeSbO_4$ and $FeSb_2O_6$ are tripuhyite [21]. The blue circles are analytical data, the red square is stoichiometric composition of minerals.

Less commonly, mustard gold of the Gaching deposit occurred as close intergrowths with goethite/hydrogoethite/limonite, forming different colloform and spotted textures (Figure 6a,b). Goethite also sometimes formed a rims on mustard gold microaggregates (Figure 6c,d). Mustard gold was high fineness in this case and did not contain Ag. The presence of iron in the gold was the result of the imposition of Fe hydroxides during the analytical procedure (Table 5). Association of secondary gold and Fe hydroxides was the final oxidation product of the studied parageneses. Secondary high fineness gold in association with supergene minerals, including goethite, has been described in Au-bearing regolith in deposits in Kazakhstan [22], where the secondary gold is not the mustard species, but has a crystal shape and different genesis.

**Table 5.** The mixed compositions of mustard gold and goethite/hydrogoethite (wt. %) as shown in Figure 6.

| No. | Sample | Sp. | Fe | Au | O | Total |
|---|---|---|---|---|---|---|
| 1 | 3_1 | 1 | - | 97.18 | - | 97.18 |
| 2 | 3_1 | 2 | - | 97.62 | - | 97.62 |
| 3 | 3_1 | 3 | 52.57 | - | 45.39 | 97.96 |
| 4 | 4_1 | 1 | - | 98.62 | - | 98.62 |
| 5 | 4_1 | 2 | 13.26 | 77.8 | 9.65 | 100.71 |
| 6 | 4_1 | 3 | 51.19 | 0.92 | 45.18 | 97.29 |
| 7 | 4_2 | 1 | 4.95 | 90.18 | 3.99 | 99.12 |
| 8 | 4_5 | 11 | - | 97.57 | - | 97.57 |
| 9 | 4_5 | 2 | 14.11 | 70.06 | 17.48 | 101.65 |
| 10 | 4-5a | 1 | 12.87 | 72.42 | 15.28 | 100.57 |
| 11 | 4-5a | 2 | 12.18 | 72.23 | 16.12 | 100.53 |
| 12 | 4-5a | 3 | 13.14 | 71.36 | 16.06 | 100.56 |

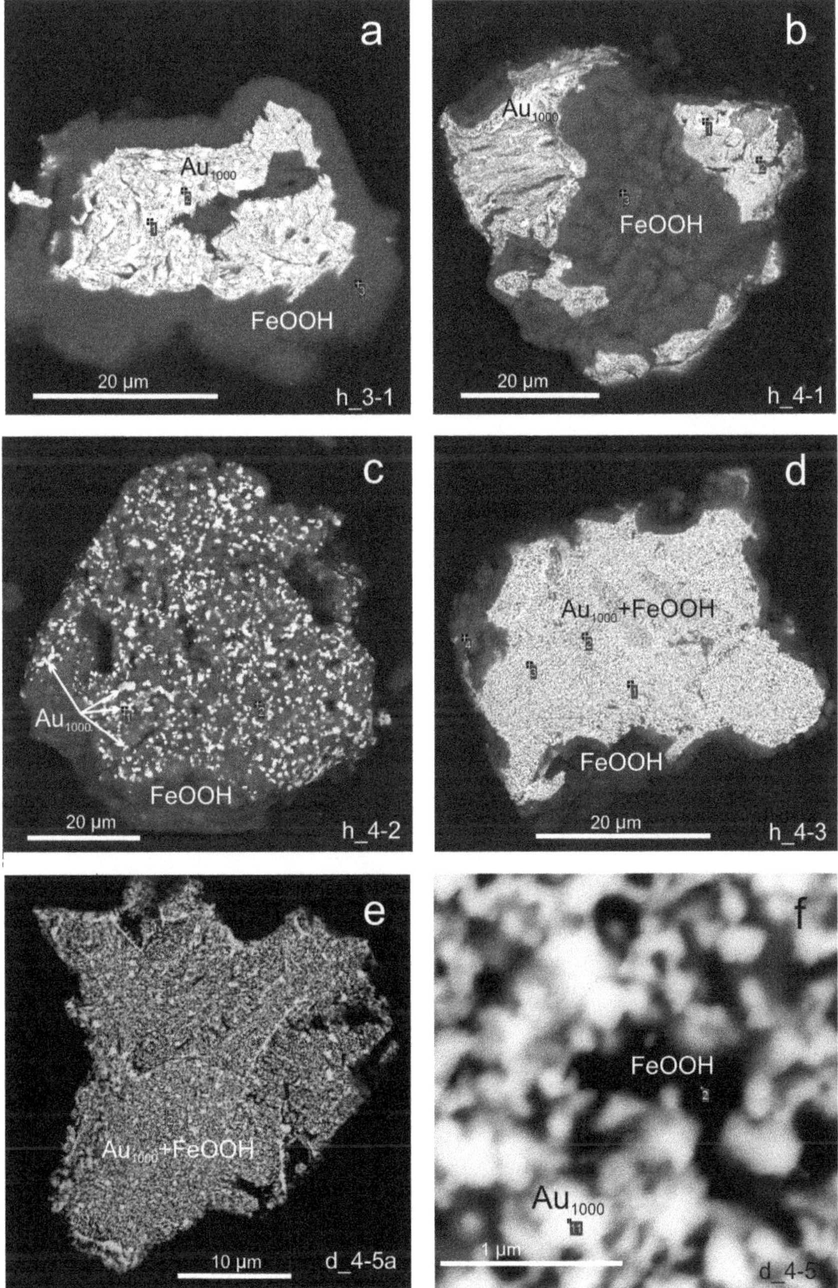

**Figure 6.** SEM image of mustard gold of colloform (**a**,**b**) and spotted (**c–f**) textures intergrown with goethite/hydrogoethite/limonite (FeOOH). The compositions of compounds and mixtures of the minerals are presented in Table 5. The purple square is scan analytical area.

### 3.2. Transformation Sequence from Calaverite to Gold

In addition Au–Te–Se–S system minerals and gold (primary and mustard), multicomponent grains were found that show the individual stages of the formation of mustard gold from calaverite (Figure 7). These grains had a vermicular texture, where secondary Fe–Sb–Te oxides with worm-like shapes developed in a matrix of calaverite. The smallest particles of gold were deposited in the marginal zones of these oxides (Figure 7, Table 6). All compositions of $AuTe_2$ relate to calaverite, since the concentration of Ag in these minerals did not exceed the 3.4 wt. % [23]. Krennerite $(Au,Ag)Te_2$ was not detected in the ore assemblage. At the same time, maletoyvayamite intergrown with calaverite did not undergo dissolution (Figure 7c,d). Calaverite was the Au-bearing mineral that oxidized due to reaction with infiltrating fluids containing Fe, Sb ± As, Se, S, Bi, along with the formation of secondary minerals (complex Au oxides, Fe antimonates, and stibiotellurates in the first stage and goethite/hydrogoetite in the final supergene stage).

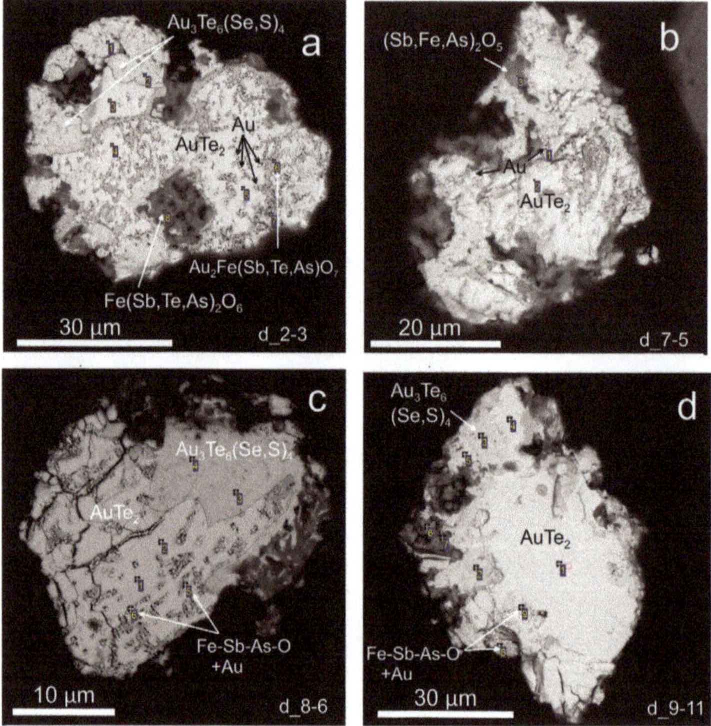


**Figure 7.** SEM images showing the reacted calaverite grains, the surface of microporous gold (**a–d**), and boundary between microporous gold and calaverite (**a–d**). $AuTe_2$—calaverite, $Au_3Te_6(Se,S)_4$—maletoyvayamite, and $Fe(Sb,Te,As)_2O_6$—tripuhyite. 1–9—the compositions of compounds or mixtures in the analyzed spots are presented in Table 6.

All analyses of mustard gold from the various samples (Figures 4 and 6–8) on the diagram O–Fe + Sb + As + Te + Se + S–Au (+Ag) were located along the line connecting the $(Fe,Sb,As,Te,Se,S)O_3$ and native gold, possibly showing the ratio of particles of fine gold and the matrix of Fe–Sb oxides (Trend I; Figure 9). The concentration of Fe varied greatly in these compounds and was not correlated with the sum of Sb, As, Te, Se, and S. This indicates that Fe,Sb ± As,Te,Se,S oxides (tripuhyite) were in turn also successively oxidized and contain hydroxides in the compound (Figure 5).

**Table 6.** Composition of calaverite grains, products of its replacement (mustard gold and Fe antimonate/tellurate) and associated minerals (maletoyvayamite) (wt. %) shown in Figure 7.

| Sample | Sp. | Fe | Au | Ag | Sb | Te | As | Se | S | O | Total | Abbr. | Formula |
|---|---|---|---|---|---|---|---|---|---|---|---|---|---|
| 2_3 | 1 | - | 35.25 | - | - | 45.61 | - | 16.63 | 0.34 | - | 97.83 | mt | $Au_{3.07}Te_{6.13}(Se_{3.61}S_{0.18})_{3.79}$ |
| 2_3 | 2 | - | 35.32 | - | - | 46.06 | - | 16.32 | 1.13 | - | 98.83 | mt | $Au_{2.98}Te_{6.00}(Se_{3.44}S_{0.59})_{4.03}$ |
| 2_3 | 3 | - | 43.52 | 0.41 | - | 52.91 | - | 1.52 | - | - | 98.36 | calv | $(Au_{1.01}Ag_{0.02})_{1.03}(Te_{1.89}Se_{0.09})_{1.98}$ |
| 2_3 | 4 | - | 44.79 | 0.37 | - | 53.38 | - | 1.67 | - | - | 100.21 | calv | $(Au_{1.02}Ag_{0.02})_{1.04}(Te_{1.87}Se_{0.09})_{1.96}$ |
| 2_3 | 5 | - | 43.88 | - | - | 51.97 | - | 1.4 | 0.27 | - | 97.52 | calv | $Au_{1.02}(Te_{1.86}Se_{0.08}S_{0.04})_{1.98}$ |
| 2_3 | 6 | 6.62 | 59.19 | 0.96 | 10.1 | 3.53 | 2.43 | - | - | 15.71 | 98.54 | Auox | $(Au_{0.97}Ag_{0.03})_{1.00}(Fe_{0.36}Sb_{0.27}Te_{0.10}As_{0.10})_{0.83}O_{3.17}$ |
| 7_5 | 2 | - | 44.7 | - | - | 53.45 | - | 1.77 | 0.68 | - | 100.6 | calv | $Au_{0.99}(Te_{1.82}Se_{0.10}S_{0.09})_{2.01}$ |
| 8_6 | 1 | - | 42.55 | 0.97 | - | 54.36 | - | 1.87 | 0.22 | - | 99.97 | calv | $(Au_{0.95}Ag_{0.04})_{0.99}(Te_{1.88}Se_{0.10}S_{0.03})_{2.01}$ |
| 8_6 | 2 | - | 41.26 | 0.82 | - | 54.5 | - | 1.93 | 0.27 | - | 98.78 | calv | $(Au_{0.93}Ag_{0.03})_{0.96}(Te_{1.89}Se_{0.11}S_{0.04})_{2.04}$ |
| 8_6 | 3 | - | 36.46 | - | - | 46.71 | - | 14.44 | 1.99 | - | 99.6 | mt | $Au_{3.02}Te_{5.98}(Se_{2.99}S_{1.01})_{4.00}$ |
| 8_6 | 4 | - | 37.39 | - | - | 46.89 | - | 10.37 | 3.83 | - | 98.48 | mt | $Au_{3.05}Te_{5.91}(Se_{2.11}S_{1.92})_{4.03}$ |
| 8_6 | 5 | 2.64 | 53.84 | - | 2.36 | 29.31 | 0.73 | 0.46 | - | 8.96 | 98.30 | Auox | $Au_{0.96}(Te_{0.80}Fe_{0.16}Sb_{0.07}As_{0.03}Se_{0.02})_{1.08}O_{1.96}$ |
| 8_6 | 6 | 1.98 | 58.09 | 0.85 | 1.41 | 30.09 | 0.43 | 0.64 | - | 6.51 | 100.00 | Auox | $(Au_{2.94}Ag_{0.08})_{3.02}(Te_{2.35}Fe_{0.33}Sb_{0.12}Se_{0.08}As_{0.06})_{2.94}O_{4.05}$ |
| 9_11 | 1 | - | 42.58 | - | - | 54.27 | - | 1.61 | - | - | 98.46 | calv | $Au_{0.98}(Te_{1.93}Se_{0.09})_{2.02}$ |
| 9_11 | 2 | - | 35.92 | - | - | 46.06 | - | 18.08 | - | - | 100.06 | mt | $Au_{3.07}Te_{6.08}Se_{3.85}$ |
| 9_11 | 3 | - | 35.24 | - | - | 45.54 | - | 17.6 | - | - | 98.38 | mt | $Au_{3.07}Te_{6.12}Se_{3.82}$ |
| 9_11 | 4 | - | 42.98 | - | - | 54.27 | - | 1.68 | - | - | 98.93 | calv | $Au_{0.98}(Te_{1.92}Se_{0.10})_{2.02}$ |
| 9_11 | 5 | 0.42 | 42.15 | - | - | 53.94 | - | 1.64 | - | - | 98.15 | calv | $Au_{0.97}(Te_{1.91}Se_{0.09}Fe_{0.03})_{2.03}$ |

Note. Auox—complex Au-antimonates and Au-tellurates, mt—maletoyvayamite, and calv—calaverite. Abbr.—minerals abbreviation.

**Table 7.** Compositions of Au-(Te,Fe,Se,Sb,Bi,S,As) and Sb-(Fe,Te,As,S,Au,Ag) oxides, maletoyvayamite and calaverite (in wt %) shown in Figure 8.

| Sample | Sp. | Fe | Au | Ag | Bi | Sb | Te | As | Se | S | O | Total | Abbr. | Formula |
|---|---|---|---|---|---|---|---|---|---|---|---|---|---|---|
| 8_10 | 1 | 4.36 | 56.34 | 2.3 | - | 5.72 | 11.92 | 1.58 | 5.06 | 1.74 | 10.53 | 99.55 | Auox | $(Au_{0.87}Ag_{0.06})_{0.93}(Te_{0.28}Fe_{0.22}Se_{0.19}S_{0.16}Sb_{0.14}As_{0.06})_{0.99}O_{2.00}$ |
| 8_10 | 2 | 2.91 | 55.94 | 2.62 | 1.58 | 6.11 | 13.67 | - | 5.36 | 2.03 | 9.68 | 99.9 | Auox | $(Au_{0.90}Ag_{0.08})_{0.98}(Te_{0.34}Se_{0.22}S_{0.20}Fe_{0.16}Sb_{0.16}Bi_{0.02})_{1.10}O_{1.92}$ |
| 8_10 | 3 | 3.92 | 59.55 | 2.24 | 0.92 | 4.61 | 8.19 | 0.94 | 3.03 | 1.3 | 9.23 | 93.93 | Auox | $(Au_{1.04}Ag_{0.07})_{1.11}(Fe_{0.23}Te_{0.22}S_{0.14}Sb_{0.13}As_{0.04}Bi_{0.02})_{0.91}O_{1.98}$ |
| 8_10 | 4 | 2.15 | 74.83 | 1.67 | 1.77 | 5.34 | 1.31 | - | - | 1.78 | 10.11 | 98.96 | Auox | $(Au_{1.29}Ag_{0.05})_{1.34}(S_{0.19}Sb_{0.15}Fe_{0.12}Te_{0.03}Bi_{0.03})_{0.52}O_{2.14}$ |
| 8_10 | 5 | 2.71 | 54.62 | 2.97 | 1.5 | 3.59 | 17.25 | - | 6.96 | 1.91 | 7.51 | 99.02 | Auox | $(Au_{0.97}Ag_{0.10})_{1.07}(Te_{0.47}Se_{0.31}S_{0.21}Fe_{0.16}Sb_{0.10}Bi_{0.03})_{1.28}O_{1.65}$ |
| 8_10 | 6 | 2.17 | 78.16 | 1.02 | 1.68 | 4.56 | 1.10 | - | - | 1.47 | 7.6 | 97.76 | Auox | $(Au_{1.56}Ag_{0.04})_{1.60}(S_{0.18}Sb_{0.15}Fe_{0.14}Te_{0.03}Bi_{0.03})_{0.53}O_{1.87}$ |
| 8_10 | 9 | 6.26 | 2.37 | 0.48 | - | 45.52 | 5.30 | 0.91 | - | | 24.37 | 85.21 | SbFeox | $(Sb_{0.72}Fe_{0.20}Te_{0.08}As_{0.02}Au_{0.02}Ag_{0.01})_{1.05}O_{2.94}$ |
| 8_11 | 1 | 4.11 | 52.68 | 1.99 | 1.54 | 5.64 | 12.35 | 1.19 | 4.88 | 1.12 | 9.3 | 94.8 | Auox | $(Au_{0.89}Ag_{0.06})_{0.95}(Te_{0.32}Fe_{0.23}Se_{0.21}Sb_{0.15}S_{0.12}As_{0.05}Bi_{0.02})_{1.10}O_{1.94}$ |
| 8_11 | 2 | 4.12 | 54.54 | 2.64 | 1.94 | 5.81 | 13.8 | 1.28 | 6.25 | 1.61 | 8.89 | 100.88 | Auox | $(Au_{0.89}Ag_{0.08})_{0.97}(Te_{0.35}Se_{0.26}Fe_{0.23}S_{0.16}Sb_{0.15}As_{0.06}Bi_{0.03})_{1.24}O_{1.79}$ |
| 8_11 | 3 | 4.23 | 53.55 | 2.50 | 1.98 | 5.8 | 11.84 | 1.58 | 5.39 | 1.31 | 9.25 | 97.43 | Auox | $(Au_{0.89}Ag_{0.08})_{0.97}(Te_{0.35}Se_{0.26}Fe_{0.23}S_{0.16}Sb_{0.15}As_{0.06}Bi_{0.03})_{1.24}O_{1.79}$ |
| 8_11 | 4 | 4.25 | 58.63 | 2.38 | 2.08 | 4.61 | 9.70 | 1.48 | 4.42 | 1.4 | 8.28 | 97.23 | Auox | $(Au_{1.03}Ag_{0.08})_{1.11}(Te_{0.26}Fe_{0.25}Se_{0.19}S_{0.15}Sb_{0.13}As_{0.07}Bi_{0.03})_{1.08}O_{1.80}$ |
| 8_11 | 5 | 2.71 | 50.98 | 2.39 | 1.51 | 2.9 | 20.88 | 0.98 | 8.59 | 1.91 | 5.89 | 98.74 | Auox | $(Au_{0.97}Ag_{0.08})_{1.05}(Te_{0.61}Se_{0.41}S_{0.22}Fe_{0.17}Sb_{0.09}As_{0.05}Bi_{0.03})_{1.58}O_{1.37}$ |
| 8_11 | 6 | 4.03 | 53.89 | 1.99 | 1.85 | 4.24 | 15.25 | 1.19 | 6.73 | 1.34 | 7.05 | 97.56 | Auox | $(Au_{0.99}Ag_{0.07})_{1.06}(Te_{0.43}Se_{0.31}Fe_{0.25}S_{0.15}Sb_{0.13}As_{0.06}Bi_{0.03})_{1.36}O_{1.59}$ |
| 8_11 | 7 | 4.47 | 53.75 | 1.42 | 2.07 | 7.21 | 12.97 | 1.37 | 5.63 | 1.45 | 9.03 | 99.37 | Auox | $(Au_{0.89}Ag_{0.04})_{0.93}(Te_{0.33}Fe_{0.25}Se_{0.23}Sb_{0.19}S_{0.15}As_{0.06}Bi_{0.03})_{1.24}O_{1.83}$ |
| 8_11 | 8 | 6.22 | 5.33 | 1.49 | - | 51.14 | 6.38 | 0.91 | - | 0.18 | 26.66 | 98.31 | SbFeox | $(Sb_{0.73}Fe_{0.18}Te_{0.09}As_{0.02}S_{0.01}Au_{0.05}Ag_{0.02})_{1.10}O_{2.90}$ |
| 11_7 | 1 | 6.93 | 58.62 | 1.37 | - | 12.92 | 1.74 | 2.13 | - | - | 14.17 | 97.88 | Auox | $(Au_{1.02}Ag_{0.04})_{1.06}(Fe_{0.40}Sb_{0.36}As_{0.10}Te_{0.05})_{0.91}O_{3.03}$ |
| 11_7 | 2 | 6.30 | 58.55 | 1.38 | - | 13.36 | 1.57 | 2.14 | - | - | 13.26 | 96.56 | Auox | $(Au_{1.06}Ag_{0.05})_{1.11}(Sb_{0.39}Fe_{0.38}As_{0.10}Te_{0.04})_{0.92}O_{2.97}$ |
| 11_7 | 3 | - | 35.55 | - | 1.19 | - | 47.63 | - | 10.4 | 5.41 | - | 100.18 | mt | $Au_{2.73}Te_{5.64}(S_{2.55}Se_{1.99}Bi_{0.09})_{4.63}$ |
| 11_7 | 4 | - | 36.77 | - | 0.74 | - | 47.39 | - | 7.92 | 4.89 | - | 97.71 | mt | $Au_{2.98}Te_{5.93}(S_{2.43}Se_{1.60}Bi_{0.06})_{4.09}$ |
| 11_7 | 5 | - | 41.95 | - | - | - | 53.77 | - | 1.21 | - | - | 96.93 | calv | $Au_{0.98}(Te_{1.95}Se_{0.07})_{2.02}$ |
| 11_7 | 6 | - | 44.06 | - | - | - | 54.28 | - | 1.16 | - | - | 99.50 | calv | $Au_{1.01}(Te_{1.92}Se_{0.07})_{1.99}$ |

Note. Auox—complex Au-antimonates/tellurates, SbFeox—Fe-antimonite, mt—maletoyvayamite, and calv—calaverite. Abbr.—minerals abbreviation.

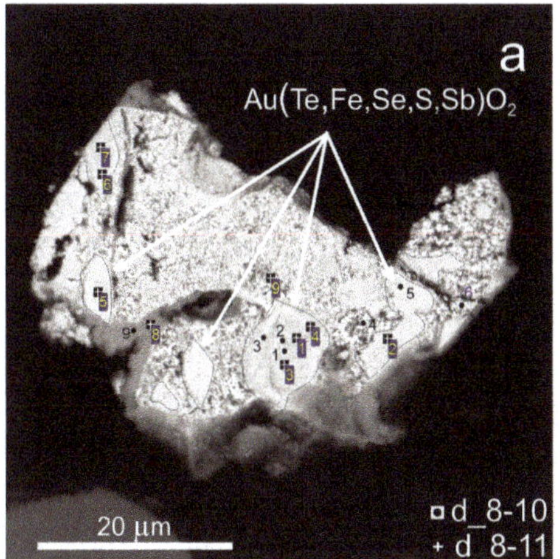
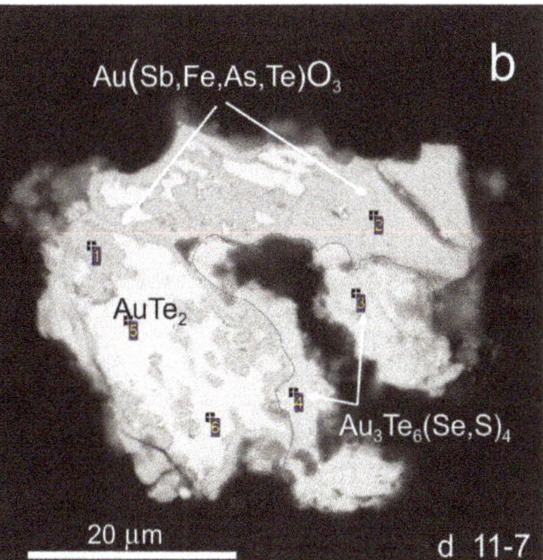



**Figure 8.** SEM images of complex oxides: zonal Au tellurates in association with mustard gold (**a**) and homogeneous Fe-rich auroantimonate Au(Sb,Fe)O₃ in association with calaverite (AuTe₂) and maletoyvayamite (Au₃Te₆(Se,S)₄ (**b**). 1–11—the compositions of compounds or mixtures in the analyzed sites are presented in Table 7.

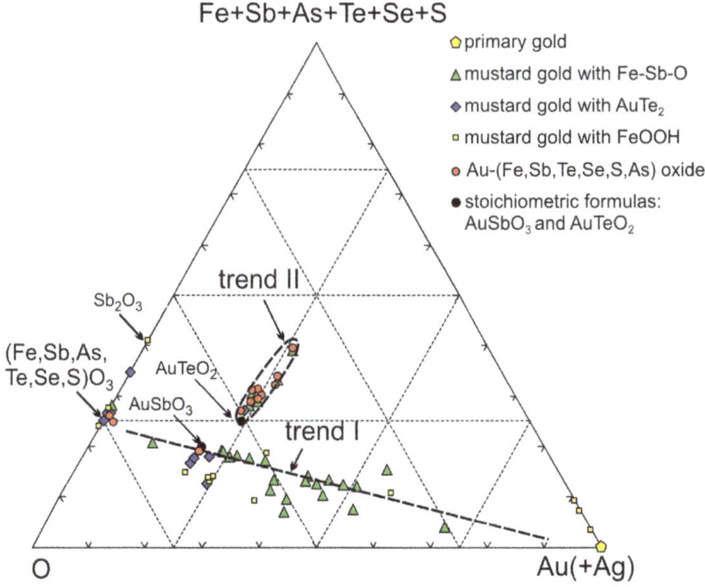



**Figure 9.** The composition of mustard gold mixed with Fe–Sb–Te–As–Se–S-oxide matrix (trend I) and Au-complex oxides (trend II) in O–Fe + Sb + As + Te + Se + S–Au (+Ag) diagram from the various associations shown in the Figures 2, 4 and 6–8.

The part of the analyses that correspond to homogeneous areas of mineral microaggregates (Figure 8) formed another compositional trend (Trend II), where the concentration of Au remained approximately constant, while the ratio of oxygen to the sum of Sb, As, Te, Se, and S (chalcogenides) varied within certain limits (Figure 9). The composition with the maximum amount of oxygen in this trend corresponded to the formulae Au(Te,Fe,Se,Sb,S,As)O₃ and Au(Fe,Sb)O₃ (Figure 9) with variable ratios of chalcogenides and Fe. These probably exist as chemical compounds rather than as mixtures of gold and Fe tellurate/antimonate. The existence of auroantimonate or Au–Sb-oxides in natural systems is highly questionable according to [24]. However, it is difficult to agree that the numerous compositions located along trend II were mixtures of gold and oxides, since the concentration of Au remained almost constant in different spots of the grains (Figure 9). The gray-zoned areas of these Au oxides (Figure 8a) were due to the different degrees of oxidation of tellurium, antimony, and other elements structurally related to Au. The Au, Sb-oxides, and auroantimonate were described by [25–27] also.

## 4. Discussion

Mustard gold is characteristic in gold-telluride deposits [9,10], antimony-gold deposits [11,12,24,28], and laterites [7,8]. The stibian mustard gold from the Kriván Au deposit, which formed from $Au_2Sb$ is a composite material consisting of submicroporous sponge of gold with pores infilled by oxidation products of Sb and Fe [24]. In contrast, the Krásná Hora deposit (Czech Republic) a reverse reaction is observed in the formation of aurostibite to gold via dissolution-precipitation and solid-state diffusion processes at temperatures <200 °C [29]. In all cases the formation of mustard gold occurred due to the decomposition of tellurides, antimonides, sulphides, bismuthides of Au(Ag), and low-grade gold under oxidation conditions. Since the Gaching deposit is gold-telluride, the transformation mechanism of Au telluride (calaverite) is important to reconstruct. It is likely that there were two mechanisms for the replacement of primary Au tellurides with different oxidation intensities and different degrees of removal of Te:

(1) $AuTe_2$ + Fe, Sb, Bi, As, Se, and S-containing solutions + $O_2 \rightarrow$ Au + Te,Se solid solution + $TeO_2$ + $Fe(Sb,As)O_3$ as a composite of a gold sponge and Sb–Fe oxide ± admixtures.

(2) $AuTe_2$ + Fe, Sb, Bi, As, Se, and S-containing solutions + $O_2 \rightarrow Au(Te,Fe,Se,Sb,S)O_2$.

The appearance of secondary high-grade mustard gold and tellurites/tellurates in the Aginskoye deposit (Kamchatka) [4,5] and placer region of Northeast Russia [8] are considered to be products of hypergene processes. Alternatively, mustard gold, Au antimonates/tellurates, and complex oxides of Au are products of hypogene processes [3,24] that formed from hydrothermal low-temperature solutions at a high oxidative potential.

Moreover, the auroantimonate ($AuSbO_3$) that was found by Z. Johan and co-authors [27] in Au–Sb ores of the Krâsnâ Hora gold deposit and by G.N. Gamyanin and co-authors [25] in Eastern Yakutia was characterized in detail by I.Ya Nekrasov [3], who obtained the compositions and X-ray images of this oxide. Our data confirmed the presence of Au oxides in the epithermal Gaching deposit, however, there are doubts regarding its existence [24]. It remains stable in hypogene conditions, but it is not stable in the hypergenesis zone, where it decomposes into sponge gold and antimony oxides, therefore, it is rarely found in nature [3].

Experimental studies by [29,30] have identified a diversity of reactionary textures and the transition mechanism of calaverite ($AuTe_2$) into metallic gold. During the replacement of calaverite by gold, there is the coupled calaverite dissolution-gold (re)precipitation (ICDR) mechanism [29,30]. This is a redox reaction controlled by the solution chemistry. The rate of replacement would be controlled by such major factors: The pH value, redox, and temperature. Natural mustard gold, which results from the weathering of Au-tellurides, may form via a similar dissolution-reprecipitation mechanism. It has been shown that the replacement of krennerite/calaverite by gold occurs only in hydrothermal solutions, whereas such reactions do not occur in anhydrous conditions [31]. These minerals follow a simple ICDR reaction path leading to pseudomorphic replacement by gold; and both minerals transform at similar rates. The presence of Ag in tellurides has an effect on the ICDR reaction path, since the solid-state reactions in the transformation of sylvanite $(Au,Ag)_2Te_4$ is also involved. The porosity is textural evidence for a CDR reaction, which leads to the negative volume changes [32]. The reaction is sustained by continuous mass transport through open pathways allowing the influx of

Fe,Sb,As,Se,Bi-solutes and the removal of oxidized Te from the reaction interface. Au(I) is controlled by the redox potential of the fluid at the reaction front. The decrease in oxygen activity favors the precipitation of gold since the oxygen is continuously removed as Te(IV) complexes by the oxidation of tellurium, and the soluble oxidation product leaves the reaction front by mass transport in the fluid, and precipitates far from the site of dissolution [31]. The dominant Te aqueous species is $H_2TeO_3(aq)$, which occurs under acidic to slightly basic (pH 2–7) conditions [33]. In low-sulfidation environments, telluride and native tellurium deposition may result from condensation of $H_2Te(g)$ and $Te_2(g)$ into saline waters. A minor amount of tellurium will be deposited by cooling or fluid mixing [34]. Native tellurium and tellurium-selenium solid solutions precipitate out of the mustard gold grains as separate grains or in intergrowths with the maletoyvayamite [14,19].

Microaggregates of mustard gold can be large in size, but they are unstable in the hypergenic process due to the loose texture and therefore have weak placer-forming potential. The concentration of Au increased from large size fractions to small ones, and its greatest value was reached in the finest grain concentrates (Figure 10). This circumstance requires attention when developing technological schemes for exploiting such deposits and explains the discrepancy between high concentrations of Au in samples and the absence (or undetected) of its mineral forms.

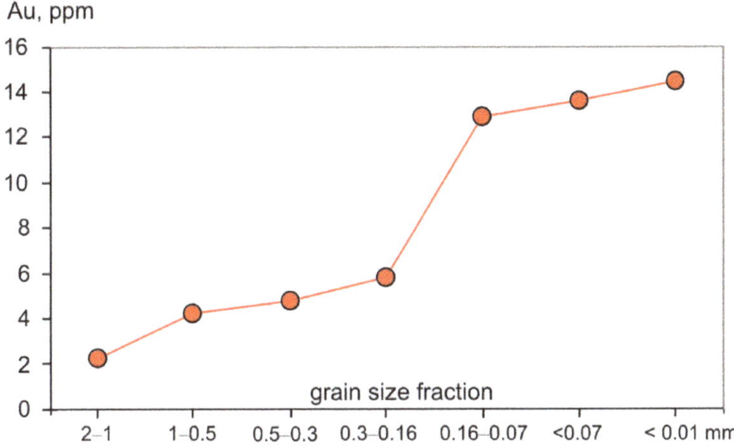


**Figure 10.** The graph of increasing Au concentration with decreasing size fraction of rock crushing.

Au(Ag) tellurides in gold deposits are considered refractory ores from a mineral processing perspective, as they are not efficiently leachable in cyanide solutions. Typically, tellurides are heated at temperatures ≥ 800 °C [35,36]. Au tellurides under hydrothermal conditions can be transformed into gold relatively fast (within hours) under all conditions at ~200 °C. This process can be used in preliminary ore processing before the addition cyanide instead of the toxic process of ore heating. This method has an advantage, as the dissolution of a gold telluride occurs over a wider range of solution conditions than Au–Ag alloys [31]. Porous gold obtained by the replacement of Au tellurides can present significant technological potential due to its low density, high strength and large surface area [33].

## 5. Conclusions

Gold in the epithermal Au–Ag Gaching deposit is present as primary gold (fineness 964‰–978‰) occurring as intergrowths with a numerous minerals (maletoyvayamite $Au_3Te_6Se_4$ and other rare unnamed phases of Au–Te–Se–S system), and as secondary porous mustard gold of fineness 1000‰.

There are two types of mustard gold:

(a) Mustard gold with inclusions of the oxides of Sb, Te(Se,S), and Fe (Fe-antimonate/tellurate) infilling the pores of spongy gold—the early (hypogene) transformation stage of calaverite due to the impact of Fe, Sb, Te, As, Se, and S-containing hydrothermal solutions and high oxidation potential.

(b) Spotted and colloform gold consisting of aggregates of small particles of gold in a goethite/hydrogoethite matrix—the late (possibly hypergene) transformation stage associated with the maximum degree of ore oxidation.

Among the numerous compounds of Au, only calaverite was oxidized and transformed into mustard gold. The other minerals (Au sulfoselenotelluride–maletiyvayamite) remained unchanged. This process is associated with hypogene conditions and proceeds in two directions: (a) Formation of a mixture of Fe–Sb oxides (tripuhyite) and gold particles, and (b) formation of Au–Sb(Te,Se,S,As) oxides by calaverite.

Mustard gold has a weak placer-forming potential in hypergene conditions because it disintegrates easily into dust particles that enrich the fine fractions of oxidized ores.

**Author Contributions:** N.D.T. conceived and designed the study, interpreted the results, and wrote this article. G.A.P. provided valuable ideas for the discussion and edited the manuscript. O.V.B. provided samples for research and determined the Au concentrations in rock fractions. E.G.S. provided the field work on the gold ore occurrences.

**Funding:** The studies were carried out within the framework of the state assignment of the VS Sobolev Institute of Geology and Mineralogy of Siberian Branch of Russian Academy of Sciences financed by Ministry of Science and Higher Education of the Russian Federation. Financial assistance from the Russian Foundation of Basic Research projects No 19-05-00316.

**Acknowledgments:** The authors thank N. Karmanov, M. Khlestov and V. Danilovskaya for help in carrying out a large amount of work on a scanning electron microscope and an X-ray spectral microanalyzer (VS Sobolev Institute of Geology and Mineralogy, Siberian Branch, Russian Academy of Sciences).

**Conflicts of Interest:** The authors declare no conflict of interest.

## References

1. Lindgren, W. *Mineral Deposits*; McGraw-Hill Book Company, Inc.: New York, NY, USA; London, UK, 1933; p. 930.

2. Sotnikov, V.I. Gold in the bedrock source—Placer system. *Soros Educ. J.* **1998**, *5*, 66–71. (In Russian)

3. Nekrasov, I.Y. *Geochemistry, Mineralogy and Genesis of Gold Deposits*; Nauka: Moscow, Russia, 1991; p. 302. (In Russian)

4. Okrugin, V.M.; Andreeva, E.D.; Yablokova, D.A.; Okrugina, A.M.; Chubarov, V.M.; Ananiev, V.V. The new data on the ores of the Aginskoye gold-telluride deposit (Central Kamchatka). In *"Volcanism and Its Associated Processes" Conference*; Petropavlovsk-Kamchatsky: Kamchatka Krai, Russia, 2014; pp. 335–341. (In Russian)

5. Okrugin, V.M.; Andreeva, E.; Etschmann, B.; Pring, A.; Li, K.; Zhao, J.; Griffiths, G.; Lumpkin, G.R.; Triani, G.; Brugger, J. Microporous gold: Comparison of textures from Nature and experiments. *Am. Mineral.* **2014**, *99*, 1171–1174. [CrossRef]

6. Kudaeva, A.L.; Andreeva, E.D. Mustard gold: Characteristics, types and chemical composition. In *The Natural Environment of Kamchatka*; Petropavlovsk-Kamchatsky: Kamchatka Krai, Russia, 2014; pp. 17–29. (In Russian)

7. Wilson, A.F. Origin of quartz-free gold nuggets and supergene gold found in laterites and soils—A review and some new observations. *Aust. J. Earth Sci.* **1984**, *31*, 303–316.

8. Litvinenko, I.S.; Shilina, L.A. Hypergene gold neomineralization in placer deposits of Nizhne-Myakitsky ore-placer field, North-East Russia. *J. Ores Met.* **2017**, *1*, 75–90. (In Russian)

9. Petersen, S.O.; Makovicky, E.; Juiling, L.; Rose-Hansen, J. Mustard gold from the Dongping Au-Te deposit, Hebei province, People's Republic of China. *N. Jb. Mineral. Mh.* **1999**, *8*, 337–357.

10. Li, J.; Makovicky, E. New studies on mustard gold from the Dongping Mines, Hebei Province, China: The tellurian, plumbian, manganoan and mixed varieties. *N. Jb. Mineral. Abh.* **2001**, *176*, 269–297.

11. Gamyanin, G.N.; Zhdanov, Y.Y.; Nekrasov, T.Y.; Leskova, N.V. "Mustard" gold from gold-antimony ores of Eastern Yakutia. *New Data Miner.* **1987**, *34*, 13–20. (In Russian)

12. Dill, H.G.; Weiser, T.; Benhardt, I.R.; Kilibarda, R. The composite gold—Antimony vein deposit at Kharma (Bolivia). *Econ. Geol.* **1995**, *90*, 51–66. [CrossRef]

13. Wang, D.; Liu, J.; Zhai, D.; Carranza, E.J.M.; Wang, Y.; Zhen, S.; Wang, J.; Wang, J.; Liu, Z.; Zhang, F. Mineral Paragenesis and Ore-Forming Processes of the Dongping Gold Deposit, Hebei Province, China. *Resour. Geol.* **2019**, *29*, 287–313. [CrossRef]

14. Tolstykh, N.; Vymazalova, A.; Tuhy, M.; Shapovalova, M. Conditions of formation of Au–Se–Te mineralization in the Gaching ore occurrence (Maletoyvayam ore field), Kamchatka, Russia. *Mineral. Mag.* **2018**, *82*, 649–674. [CrossRef]

15. Tsukanov, N.V. Tectono-stratigraphic terranes of Kamchatka active margins: Structure, composition and geodynamics. In *Materials of the Annual Conference "Volcanism and Related Processes"*; Petropavlovsk-Kamchatsky: Kamchatka Krai, Russia, 2015; pp. 97–103. (In Russian)

16. Golyakov, V.I. *Geological Map of the USSR Scale 1: 200,000*; Pogozhev, A.G., Ed.; Series Koryak; Sheets P-5 8-XXXIII, O-58-III; VSEGEI Cartographic Factory: St. Petersburg, Russia, 1980. (In Russian)

17. Melkomukov, B.H.; Razumny, A.V.; Litvinov, A.P.; Lopatin, W.B. New highly promising gold objects of Koryakiya. *Min. Bull. Kamchatka* **2010**, *14*, 70–74. (In Russian)

18. Palyanova, G.A.; Tolstykh, N.D.; Zinina, V.Y.; Kokh, K.A.; Seryotkin, Y.V.; Bortnikov, N.S. Synthetic chalcogenides of gold in the Au-Te-Se-S system and their natural analogs. *Dokl. Earth Sci.* **2019**, *487*, 929–934.

19. Tolstykh, N.D.; Tuhý, M.; Vymazalová, A.; Plášil, J.; Laufek, F.; Kasatkin, A.V.; Nestola, F. Maletoyvayamite, IMA 2019-021. CNMNC Newsletter No. 50. *Mineral. Mag.* **2019**, *31*. [CrossRef]

20. Kovalenker, V.A.; Nekrasov, I.Y.; Sandomirskaya, S.M.; Nekrasova, A.N.; Malov, V.S.; Danchenko, V.Y.; Dmitrieva, M.T. Sulfide-selenide-telluride mineralization of epithermal ores in the Kuril-Kamchatka volcanic belt. *Mineral. J.* **1989**, *11*, 3–18. (In Russian)

21. Basso, R.; Cabella, R.; Lucchetti, G.; Marescotti, P.; Martinelli, A. Structural studies on synthetic and natural Fe–Sb–oxides of $MO_2$ type. *N. Jb. Mineral. Mh.* **2003**, 407–420. [CrossRef]

22. Kalinin, Y.A.; Palyanova, G.A.; Naumov, E.A.; Kovalev, K.R.; Pirajno, F. Supergene remobilization of Au in Au-bearing regolith related to orogenic deposits: A case study from Kazakhstan. *Ore Geol. Rev.* **2019**, *109*, 358–369. [CrossRef]

23. Cabri, L.J. Phase relations in the Au-Ag-Te system and their mineralogical significance. *Econ. Geol.* **1965**, *60*, 1569–1605. [CrossRef]

24. Makovicky, E.; Chovan, M.; Bakos, F. The stibian mustard gold from the Krivánˇ Au deposit, Tatry Mts., Slovak Republic. *N. Jb. Mineral. Abh.* **2007**, *184*, 207–215.

25. Gamyanin, G.N.; Nekrasov, I.; Zhdanov, J.J.; Leskova, I.V. Auroantimonate—A new natural compound of gold. *Dokl. Akad Nauk SSSR Mineral.* **1988**, *301*, 947–950. (In Russian)

26. Tian, S.; Li, X. A preliminary study of some new minerals in Dongping gold deposit, Hebei province. *Gold Geol.* **1995**, *1*, 61–67.

27. Johan, Z.; Šrein, V. Un nouvel oxide naturel de Au et Sb. *Sci. Terre Planetes* **1998**, *326*, 533–538.

28. Zacharias, J.; Mate, J.N. Gold to aurostibite transformation and formation of Au-Ag-Sb phases: The Krásná Hora deposit, Czech Republic. *Mineral. Mag.* **2017**, *81*, 987–999. [CrossRef]

29. Zhao, J.; Brugger, J.; Pascal, V.; Gundler, P.; Xia, F.; Chen, G.; Pring, A. Mechanism and kinetics of a mineral transformation under hydrothermal conditions: Calaverite to metallic gold. *Am. Mineral.* **2009**, *94*, 1541–1555. [CrossRef]

30. Zhao, J.; Xia, F.; Pring, A.; Brugger, J.; Grundler, P.V.; Chen, G. A novel pretreatment of calaverite by hydrothermal mineral replacement reactions. *Miner. Eng.* **2010**, *23*, 451–453. [CrossRef]

31. Xu, W.; Zhao, J.; Brugger, J.; Chen, G.; Pring, A. Mechanism of mineral transformations in krennerite, Au3AgTe8, under hydrothermal conditions. *Am. Mineral.* **2013**, *98*, 2086–2095. [CrossRef]

32. Xia, F.; Brugger, J.; Chen, G.; Ngothai, Y.; O'Neill, B.; Putnis, A.; Pring, A. Mechanism and kinetics of pseudomorphic mineral replacement reactions: A case study of the replacement of pentlandite by violarite. *Geochim. Cosmochim. Acta* **2009**, *73*, 1945–1969. [CrossRef]

33. Zhao, J.; Pring, A. Mineral Transformations in Gold–(Silver) Tellurides in the Presence of Fluids: Nature and Experiment. *Minerals* **2019**, *9*, 167. [CrossRef]

34. Cooke, D.R.; McPhail, D.C. Epithermal Au–Ag–Te mineralization, Acupan, Baguio District, Philippines; numerical simulations of mineral deposition. *Econ. Geol.* **2001**, *96*, 109–131.

35. Kongolo, K.; Mwema, M.D. The extractive metallurgy of gold. *Hyperfine Interact.* **1998**, *111*, 281–289. [CrossRef]

36. Grosse, A.C.; Dicinoski, G.W.; Shaw, M.J.; Haddad, P.R. Leaching and recovery of gold using ammoniacal thiosulfate leach liquors (a review). *Hydrometallurgy* **2003**, *69*, 1–21. [CrossRef]

*Article*

# Mustard Gold in the Oleninskoe Gold Deposit, Kolmozero–Voronya Greenstone Belt, Kola Peninsula, Russia

**Arkadii A. Kalinin *, Yevgeny E. Savchenko and Ekaterina A. Selivanova**

Geological Institute, Kola Science Center, Russian Academy of Science, 184200 Apatity, Russia; evsav@geoksc.apatity.ru (Y.E.S.); selivanova@geoksc.apatity.ru (E.A.S.)
* Correspondence: kalinin@geoksc.apatity.ru; Tel.: +7-921-663-68-36

Received: 29 October 2019; Accepted: 12 December 2019; Published: 13 December 2019

**Abstract:** The Oleninskoe intrusion-related gold–silver deposit is the first deposit in the Precambrian of the Fennoscandian Shield, where mustard gold has been identified. The mustard gold replaces küstelite with impurities of Sb and, probably, gold-bearing dyscrasite and aurostibite. The mosaic structure of the mustard gold grains is due to different orientations and sizes of pores in the matrix of noble metals. Zonation in the mustard gold grains is connected with mobilization and partial removal of silver from küstelite, corresponding enrichment of the residual matter in gold, and also with the change in the composition of the substance filling the pores. Micropores in the mustard gold are filled with iron, antimony or thallium oxides, silver chlorides, bromides, and sulfides. The formation of mustard gold with chlorides and bromides shows that halogens played an important role in the remobilization of noble metals at the stage of hypergene transformation of the Oleninskoe deposit.

**Keywords:** Fennoscandian Shield; Kolmozero–Voronya belt; Oleninskoe deposit; mustard gold; dyscrasite; chlorargyrite

---

## 1. Introduction

Mustard gold (the term is used below regardless of the Au/Ag ratio in it, and terms "electrum" for Au/Ag = 1/1 and "küstelite" for Au/Ag = 1/3 are used when the composition of Au–Ag alloy is important) is a relatively rare mineral aggregate, formed in the zone of hypergenesis as a result of oxidation and decomposition of gold tellurides and antimonides. It represents a lacy network, or sponge of native gold with very fine (often less than a micrometer) pores either filled with Fe, Te, Pb, Cu, Au, Ag, Sb, Hg oxides or unfilled. These aggregates were named mustard gold because of their distinctive appearance—loose texture and yellow–brown to brown color caused by Fe-oxides filling the pores. Mustard gold has low reflectivity due to high porosity.

The history of the study of mustard gold is not long. The term was first used in geological literature, following a traditional miners' lexicon, by Waldemar Lindgren in 1933 [1] in the paragraph about the oxidation zone in Au–Te deposits. A more extensive study of mustard gold refers to the end of the XX and the beginning of the XXI century—to the time of fast development of local methods of mineral analysis.

Mustard gold is often found in oxidized ores of Au–Ag–Sb and Au–Ag–Te deposits [2], mainly in weathered quartz granular rocks in the zone of secondary sulfide enrichment. Individual mustard gold grains are rare. More often, mustard gold forms aggregates with native gold: It can grow on the surface or inside gold grains, or be overgrown by late high-grade gold [3].

The composition of mustard gold can vary within an individual deposit. For example, eight varieties were determined in the Dongping gold deposit, where 30–50% of Au is concentrated in the mustard gold [4]. The mustard gold grains in the Dongping deposit differ in the composition of gold, which makes the sponge—it can be high-grade gold and/or gold alloys with lead and with silver. The pores are filled either by goethite, or various tellurites and tellurates or remain unfilled [4]. Six varieties of mustard gold are described in Sb–Au deposits in Sakha–Yakutia [5].

The structure of individual mustard gold grains is often heterogeneous. Many grains are zonal or concentrically banded, other display spotted, or mosaic textures [6,7]. The heterogeneity appears due to gradual decomposition and re-deposition of the material.

As the size of the pores is usually less than 1 micrometer, and thickness of gold filaments ranges from 200 to 500 nm, the microprobe analysis of the mustard gold gives some summary results, averaging data on gold filaments and the substance filling the pore, and shows a presence of Au, Ag, Fe, Te, Pb, Cu, Sb, Hg, etc. It can be difficult to define mineral phases correctly, and microprobe analysis appears rather qualitative than quantitative. Total estimates from the microprobe analyses are often significantly less than 100%. The deficit appears due to unfilled pore space and the presence of oxides.

Mustard gold was found and studied in the deposits in the Far East in Russia (in the Aginskoe, Ozernovskoe, Asachinskoe deposits and Gaching ore occurrence in the Kamchatka Peninsula [6–9], in the Nizhne–Myakitskoe ore field in the Magadan Region [3], in the Tumannoe deposit in Chukotka [10], in the Kuranah ore field, and in the Sarylahskoe and Sentachanskoe gold–antimony deposits in Sakha–Yakutia [9,11]), in China (the Dongping [4,12] and Sandaowanzi [12], gold deposits), in Slovakia (the Kriváň deposit in the High Tatra Mts.) [13], and in Bolivia (Au–Sb deposit Kharma) [14]. All these deposits are located in the areas of Mesozoic–Cenozoic volcanism with numerous epithermal Au–Te and Au–Sb gold deposits. Mustard gold was not described earlier in the Precambrian gold deposits, and, probably, our finding in the Oleninskoe deposit is the first such occurrence described.

## 2. Geological Setting of the Oleninskoe Deposit

The Oleninskoe is a small gold deposit (~10 t or 0.32 Moz @ 7.6 g/t Au [15,16]), located in the Neoarchean Kolmozero–Voronya greenstone belt, which separates two major tectonic units of the Fennoscandian Shield—the Murmansk craton and the Kola Province (Figure 1A). The geological position and structure of the Oleninskoe deposit are described in [15–18], and here is a summary of the published data.

The location of the deposit is controlled by a shear zone of the NW strike in the amphibolite of the Oleny Ridge sequence (Figure 1) [15,16]. The amphibolite and high-alumina metasedimentary schist in the area of the deposit contain numerous granodiorite quartz porphyry sills 0.1–6.0 m thick (Figure 1). The rocks were low amphibolite (T = 600 °C, P = 3–4 kbar) metamorphosed in Neoarchean. The youngest rocks (2.45 Ga [19]) in the deposit area are granite–pegmatite veins, which cut all metamorphic rocks, including those hosting gold mineralization (Figure 1).

Amphibolite and granodiorite porphyry are intensely altered, the alteration zone is ~50 m thick and traced along the strike for 200–250 m. The whole zone of alteration is covered with early biotitization (potassium metasomatism) and formation of diopside–zoisite–carbonate mineral assemblage (calcium metasomatism) in the amphibolite [15,16]. Quartz-rich metasomatic rocks (quartz–muscovite–albite, quartz–tourmaline, and quartz rocks) formed later, replacing amphibolite and granodiorite porphyry as an echelon-like series of three lens bodies, cutting general schistosity in the host rocks at an acute angle of 10–15° (Figure 1): We interpret the structure of the deposit [15] as a strike-slip fault bridge structure (this kind of structures is described in [20]). The lenses are up to 3.5 m thick (1.5 m on the average) with the length up to 50 m. These quartz-rich rocks control the distribution of the gold–silver mineralization.

The thickness of the Quaternary glacial sediments is 0–2 m. The mineralized lenses are exposed with a number of trenches and drill holes (Figure 1). The zone of ore oxidation is less than 1 m thick.

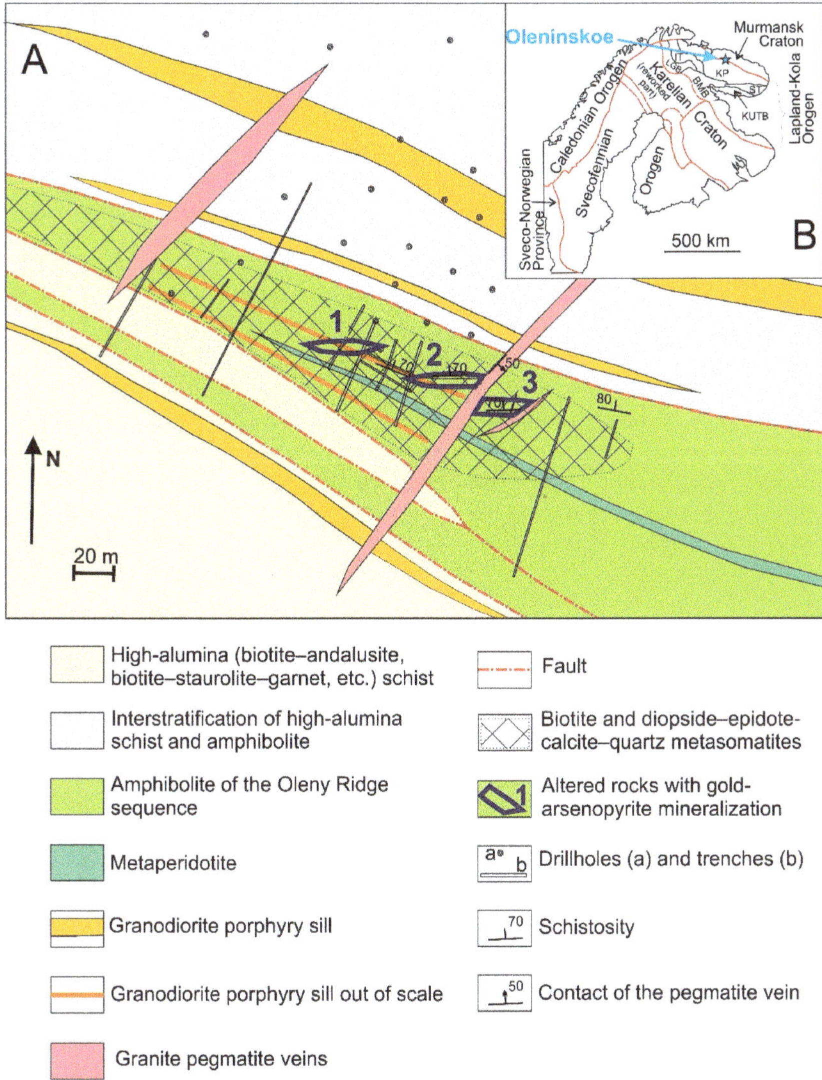



High-alumina (biotite–andalusite, biotite–staurolite–garnet, etc.) schist

Interstratification of high-alumina schist and amphibolite

Amphibolite of the Oleny Ridge sequence

Metaperidotite

Granodiorite porphyry sill

Granodiorite porphyry sill out of scale

Granite pegmatite veins

Fault

Biotite and diopside–epidote–calcite–quartz metasomatites

Altered rocks with gold–arsenopyrite mineralization

Drillholes (a) and trenches (b)

Schistosity

Contact of the pegmatite vein

**Figure 1.** Schematic geological map of the Oleninskoe gold deposit (**A**) and the position of the Oleninskoe deposit in the tectonic map of the Fennoscandian Shield (**B**). KP—Kola province; BMB—Belomorian mobile belt; LGB—Lapland granulite belt; KUTB—Kolvitsa–Umba–Tersk belt; IT—Inari terrane; ST—Strel'na terrane.

Arsenopyrite, pyrrhotite, and ilmenite are the most abundant ore minerals, which are present in all altered rocks. Quartz–tourmaline and quartz metasomatic rocks in lens #2 (Figure 1) contain rich Pb–Ag–Sb mineralization (galena, freibergite, dyscrasite, boulangerite, semseyite, diaphorite, pyrargyrite, and other Sb-sulfosalts of Pb, Ag, and Cu—totaling more than 40 mineral species [21]).

The age of mineralization is probably Neoarchean because mineralized rocks are cut by 2.45 Ga pegmatite veins, although Volkov and Novikov [17] reported some signatures of overprinting hydrothermal events of Paleoproterozoic age (1.9 Ga) in the deposit. The temperature of the formation of mineralization was estimated at 550–300 °C [18,22] with arsenopyrite geothermometer.

Four types (generations) of minerals from the Au–Ag series were recognized in the deposit [21]: (1) küstelite 25–32 wt% Au in association with arsenopyrite, löllingite, and pyrrhotite; (2) electrum 33–47 wt% Au in intergrowths with galena, dyscrasite, and sulfosalts; (3) gold (78–95 wt% Au) in quartz; (4) native silver (<7 wt% Au). Types 1 and 3 are widespread in the deposit, 2 and 4 relate only to the quartz and quartz–tourmaline rocks rich in Pb–Ag–Sb. Other gold minerals in the Oleninskoe deposit are gold-bearing dyscrasite (grains up to 1 mm), aurostibite, petzite, calaverite (inclusions <10 μm in electrum of the 2nd type), and yutenbogaardtite. The latter forms rims around the grains of küstelite and electrum. The gold content in dyscrasite varies from 0 to 17 wt.%, some grains are zonal with the outer parts enriched in S and poor in Au.

The genetic type of the Oleninskoe deposit is debatable. It was classified earlier as a greenstone orogenic deposit [23], but its geochemical characteristics, mineral composition of the ore [15,21], and the composition of fluid inclusions [24] correspond better to the class of intrusion-related deposits.

## 3. Materials and Methods

Gold and silver mineralization was studied in the specimens, collected by the authors in 1981–1983 and again in 2017. The collection included specimens of metasomatic rocks, formed after amphibolite and quartz porphyry, and heavy mineral concentrates from crushed samples of the mineralized rocks and from the glacial deposits, overlaying the ore bodies. Polished sections of mineralized specimens and of heavy mineral concentrates were studied with reflected light microscopy (optical microscope Axioplane) in the Geological Institute of the Kola Science Center.

Preliminary estimation of composition of mineral species, the study of element distribution and intraphase heterogeneity were conducted with LEO-1450 scanning electron microscope (SEM) (Carl Zeiss, Oberkochen, Germany) equipped with a Bruker XFlash-5010 Nano GmbH (Bruker, Bremen, Germany) energy-dispersive spectrometer (EDS). Microprobe analyses were performed for grains larger than 20 μm with MS-46 CAMECA, operating at an accelerating voltage 22 kV, beam current 30–40 nA. The following standards and analytical lines were used: $Fe_{10}S_{11}$ (FeK$\alpha$, SK$\alpha$), $Bi_2Se_3$ (BiM$\alpha$, SeK$\alpha$), $LiNd(MoO_4)_2$ (MoL$\alpha$), Co (CoK$\alpha$), Ni (NiK$\alpha$), Pd (PdL$\alpha$), Ag (AgL$\alpha$), Te (TeL$\alpha$), Au (AuL$\alpha$).

The specific features of analysis of mustard gold, described in the Introduction, made it possible to characterize the chemical composition of the mustard gold not only with the microprobe analysis but with qualitative data obtained using an energy-dispersive system Bruker XFlash-5010 (Tables 1 and 2).

Visually homogenous material from the areas close to the points of microprobe analysis, $50 \times 10$ μm or more in size, was extracted from the polished sections and examined with the X-ray powder diffraction (Debye–Scherer) on URS-1 operated at 40 kV and 16 mA with RKU-114.7 mm camera and FeK$\alpha$-radiation to identify mineral phases making the mustard gold grains.

**Table 1.** Estimation of composition of mustard gold in weight percent with an energy-dispersive spectrometer.

| Point # | 1 | 2 | 3 | 4 | 5 | 6 | 7 | 8 | 9 | Point # | 10 | 11 | 12 | 13 | 14 | 15 | 16 | 17 | 18 | 19 |
|---|---|---|---|---|---|---|---|---|---|---|---|---|---|---|---|---|---|---|---|---|
| Mg | n.a. | n.a. | 0.18 | n.a. | n.a. | n.a. | n.a. | n.a. | n.a. | S | 0.42 | 7.67 | 1.33 | 1.07 | 1.28 | 0.71 | 8.84 | 4.52 | 0.48 | 0.53 |
| Si | n.a. | n.a. | 1.18 | n.a. | n.a. | n.a. | n.a. | n.a. | n.a. | Cl | 9.23 | 0.09 | 0.36 | 0.13 | 0.32 | 0.13 | 0.15 | 0.11 | 11.33 | 0.09 |
| S | n.a. | n.a. | 0.99 | n.a. | 0.44 | 0.76 | n.a. | 0.61 | 0.75 | Fe | 0.14 | 0.62 | n.a. | n.a. | n.a. | 0.47 | 1.33 | 0.68 | n.a. | n.a. |
| Cl | n.a. | n.a. | n.a. | n.a. | 0.05 | 0.08 | n.a. | 0.18 | 0.13 | As | n.a. | 1.22 | n.a. | n.a. | n.a. | 0.26 | 0.75 | 0.31 | n.a. | n.a. |
| Fe | n.a. | n.a. | 18.76 | n.a. | 0.34 | 3.20 | n.a. | 0.42 | 5.43 | Br | 0.59 | 0.41 | 3.29 | 5.74 | 2.13 | 1.40 | 1.19 | 1.78 | 0.67 | 14.56 |
| As | n.a. | n.a. | n.a. | n.a. | 0.21 | 2.00 | n.a. | 0.21 | 1.19 | Ag | 46.50 | 44.35 | 43.84 | 50.85 | 28.89 | 24.49 | 54.76 | 26.03 | 46.53 | 43.12 |
| Br | n.a. | n.a. | n.a. | n.a. | 2.46 | 3.42 | n.a. | 1.41 | 1.37 | Sb | n.a. | n.a. | 1.79 | 1.00 | 0.66 | 0.50 | n.a. | n.a. | n.a. | n.a. |
| Ag | 58.53 | 51.17 | 7.60 | 63.59 | 52.67 | 33.00 | 64.88 | 46.43 | 35.74 | Au | 30.24 | 31.54 | 43.72 | 28.44 | 54.30 | 59.77 | 22.42 | 53.02 | 29.16 | 30.90 |
| Sb | 2.60 | 1.60 | 17.23 | 2.30 | 1.87 | 0.76 | 1.68 | 0.85 | 0.88 | C | n.a. | n.a. | n.a. | n.a. | 3.64 | n.a. | n.a. | n.a. | n.a. | n.a. |
| Au | 27.84 | 35.32 | 21.97 | 23.76 | 33.46 | 47.97 | 25.73 | 40.35 | 45.17 | Total | 87.12 | 85.90 | 94.33 | 87.23 | 91.22 | 87.73 | 89.44 | 86.45 | 88.17 | 89.20 |
| Pb | n.a. | n.a. | 3.28 | n.a. | n.a. | n.a. | n.a. | n.a. | n.a. | | | | | | | | | | | |
| Total | 88.97 | 88.09 | 71.29 * | 89.65 | 91.50 | 91.19 | 92.29 | 90.46 | 90.66 | | | | | | | | | | | |

| | | | | | | Atomic quantities (%) | | | | | | | | | | | | | | |
|---|---|---|---|---|---|---|---|---|---|---|---|---|---|---|---|---|---|---|---|---|
| Point # | 1 | 2 | 3 | 4 | 5 | 6 | 7 | 8 | 9 | Point # | 10 | 11 | 12 | 13 | 14 | 15 | 16 | 17 | 18 | 19 |
| Mg | n.a. | n.a. | 1.0 | n.a. | n.a. | n.a. | n.a. | n.a. | n.a. | S | 1.5 | 28.3 | 5.6 | 4.6 | 4.3 | 3.8 | 29.0 | 20.3 | 1.6 | 2.2 |
| Si | n.a. | n.a. | 5.5 | n.a. | n.a. | n.a. | n.a. | n.a. | n.a. | Cl | 30.0 | 0.3 | 1.4 | 0.5 | 1.0 | 0.6 | 0.4 | 0.4 | 34.7 | 0.3 |
| S | n.a. | n.a. | 4.1 | n.a. | 1.9 | 3.3 | n.a. | 2.7 | 3.2 | Fe | 0.3 | 1.3 | n.a. | n.a. | n.a. | 1.4 | 2.5 | 1.8 | n.a. | n.a. |
| Cl | n.a. | n.a. | n.a. | n.a. | 0.2 | 0.3 | n.a. | 0.7 | 0.5 | As | n.a. | 1.9 | n.a. | n.a. | n.a. | 0.6 | 1.1 | 0.6 | n.a. | n.a. |
| Fe | n.a. | n.a. | 44.2 | n.a. | 0.8 | 8.1 | n.a. | 1.1 | 13.4 | Br | 0.9 | 0.6 | 5.6 | 9.8 | 2.9 | 3.0 | 1.6 | 3.2 | 0.9 | 24.0 |
| As | n.a. | n.a. | n.a. | n.a. | 0.4 | 3.8 | n.a. | 0.4 | 2.2 | Ag | 49.7 | 48.6 | 55.2 | 64.3 | 28.9 | 38.5 | 53.4 | 34.8 | 46.8 | 52.7 |
| Br | n.a. | n.a. | n.a. | n.a. | 4.2 | 6.0 | n.a. | 2.5 | 2.4 | Sb | n.a. | n.a. | 2.0 | 1.1 | 0.6 | 0.7 | n.a. | n.a. | n.a. | n.a. |
| Ag | 76.9 | 71.1 | 9.3 | 80.9 | 67.0 | 43.2 | 80.6 | 62.0 | 45.7 | Au | 17.7 | 18.9 | 30.2 | 19.7 | 29.7 | 51.4 | 12.0 | 38.8 | 16.1 | 20.7 |
| Sb | 3.0 | 2.0 | 18.6 | 2.6 | 2.1 | 0.9 | 1.8 | 1.0 | 1.0 | C | n.a. | n.a. | n.a. | n.a. | 32.7 | n.a. | n.a. | n.a. | n.a. | n.a. |
| Au | 20.0 | 26.9 | 14.7 | 16.5 | 23.3 | 34.4 | 17.5 | 29.5 | 31.6 | | | | | | | | | | | |
| Pb | n.a. | n.a. | 2.1 | n.a. | n.a. | n.a. | n.a. | n.a. | n.a. | | | | | | | | | | | |

Note: *—includes Al 0.10 wt. %; n.a.—not analyzed.

**Table 2.** Electron microprobe data in weight percent for the mustard gold.

| Point # | 20 | 21 | 22 | 23 | 24 | 25 | 26 | 27 | 28 | 29 | 30 |
|---|---|---|---|---|---|---|---|---|---|---|---|
| S | n.a. | n.a. | n.a. | n.a. | 0.27 | 0.24 | 0.22 | 0.21 | 0.23 | 0.04 | 11.15 |
| Cl | n.a. | n.a. | n.a. | n.a. | 11.16 | 11.44 | 11.49 | 11.48 | 11.61 | 10.72 | 0.00 |
| Fe | 13.5 | 11.16 | 1.71 | 2.08 | 0.04 | 0.10 | 0.14 | 0.00 | 0.05 | 0.00 | 0.15 |
| Al | 0.19 | 0.21 | n.a. | n.a. | n.a. | n.a. | n.a. | n.a. | n.a. | n.a. | n.a. |
| Cu | n.a. | n.a. | 0.25 | 0.24 | n.a. | n.a. | n.a. | n.a. | n.a. | n.a. | n.a. |
| As | 1.16 | 1.26 | 0.61 | 0.74 | 0.13 | 0.00 | 0.00 | 0.00 | 0.00 | 0.00 | 0.00 |
| Ag | 0.81 | 0.52 | 12.22 | 10.38 | 49.90 | 47.38 | 49.15 | 48.11 | 47.37 | 81.49 | 59.86 |
| Sb | 0.11 | 0.18 | 0.42 | 0.32 | n.a. | n.a. | n.a. | n.a. | n.a. | n.a. | n.a. |
| Au | 66.27 | 72.33 | 73.19 | 65.88 | 38.51 | 36.45 | 35.72 | 37.39 | 36.25 | 1.20 | 26.88 |
| Tl | n.a. | n.a. | 5.02 | 6.34 | n.a. | n.a. | n.a. | n.a. | n.a. | n.a. | n.a. |
| Total | 82.04 | 85.66 | 93.43 | 85.97 | 100.00 | 95.60 | 96.72 | 97.20 | 95.51 | 93.45 | 98.04 |
| Atomic quantities (%) | | | | | | | | | | | |
| S | n.a. | n.a. | n.a. | n.a. | 0.8 | 0.8 | 6.7 | 0.7 | 0.8 | 0.1 | 33.4 |
| Cl | n.a. | n.a. | n.a. | n.a. | 32.0 | 33.7 | 31.4 | 33.5 | 34.2 | 28.4 | 0.0 |
| Fe | 40.2 | 25.3 | 5.5 | 7.2 | 0.1 | 0.2 | 0.2 | 0.0 | 0.1 | 0.0 | 0.3 |
| Al | 1.2 | 1.0 | n.a. | n.a. | n.a. | n.a. | n.a. | n.a. | n.a. | n.a. | n.a. |
| Cu | n.a. | n.a. | 0.7 | 0.7 | n.a. | n.a. | n.a. | n.a. | n.a. | n.a. | n.a. |
| As | 2.6 | 25.3 | 1.5 | 1.9 | 0.2 | 0.0 | 0.0 | 0.0 | 0.0 | 0.0 | 0.0 |
| Ag | 1.2 | 2.1 | 20.4 | 18.7 | 47.0 | 45.9 | 44.1 | 46.2 | 45.8 | 70.9 | 53.3 |
| Sb | 0.2 | 0.6 | 0.6 | 0.5 | n.a. | n.a. | n.a. | n.a. | n.a. | n.a. | n.a. |
| Au | 55.9 | 46.6 | 66.9 | 64.9 | 19.9 | 19.4 | 17.5 | 19.7 | 19.2 | 0.6 | 13.1 |
| Tl | n.a. | n.a. | 4.4 | 6.0 | n.a. | n.a. | n.a. | n.a. | n.a. | n.a. | n.a. |

Note: n.a.—not analyzed.

## 4. Results

Mustard gold was found in quartz-rich metasomatic rocks with rich Au–Ag–Sb–Pb mineralization in lens #2 (Figure 1) and in the overlaying glacial deposits. The first grain of mustard gold forms an intergrowth with küstelite (Au 28–35 wt%); the mustard gold comprises the central part of the grain, but is overgrown by küstelite, and forms a thin rim around the küstelite (Figure 2).

The mustard gold is brown-gray under reflected light, strongly anisotropic, with a yellow color effect under crossed polars. The mosaic structure of this mustard gold can be easily seen even with an optical microscope (Figure 2A), and with an SEM we see that the blocks differ in size of pores, which varies from the nanoscale to 8 μm, and in the direction of the gold filaments (Figure 2C–E). This mustard gold contains Au, Fe, Sb, Ag, Pb, and O (17.0 wt%) as the main components, and minor S, Si, Al, Mg (Table 1, #3). X-ray analysis of the mustard gold shows the presence of Ag–Au alloy, but Fe and Sb oxides are not identified.

Another grain of mustard gold ~0.1 mm in size was noted in association with electrum (wt%: Au 61.2%, Ag 34.6%, Sb 0.5%, total 96.3%). The grain is brick red, strongly anisotropic, with a bright yellow color effect in reflected light under crossed polars. This grain exhibits block structure, the size of the particles is 0.1–0.3 μm. Microprobe analysis shows that the main elements are Au and Fe, and minor elements (<1 wt%) are As, Ag, Sb, and Al (Table 2, #20 and 21). The low "Total" values are due to porosity and oxides. This mustard gold grain is coated with a thin rim of sulfur-rich simplesite (wt%: Fe 30.62%, As 15.11%, S 4.24%, Sb 0.10%) 5–10 μm thick.

One of the studied grains of mustard gold contains 5–6 wt% of Tl (Table 2, #22, and #23). This grain has brick red color, and its appearance is very similar to native copper (Figure 3). The backscatter electron image shows fine intergrowth of different mineral phases. Except for Au, Ag, and Tl, the substance contains Fe (~2%), minor As, Cu, Sb (Table 2). Zonation in the grain reflects the change in gold, silver, iron, and thallium content. The low "Total" values are due to oxides and porosity.

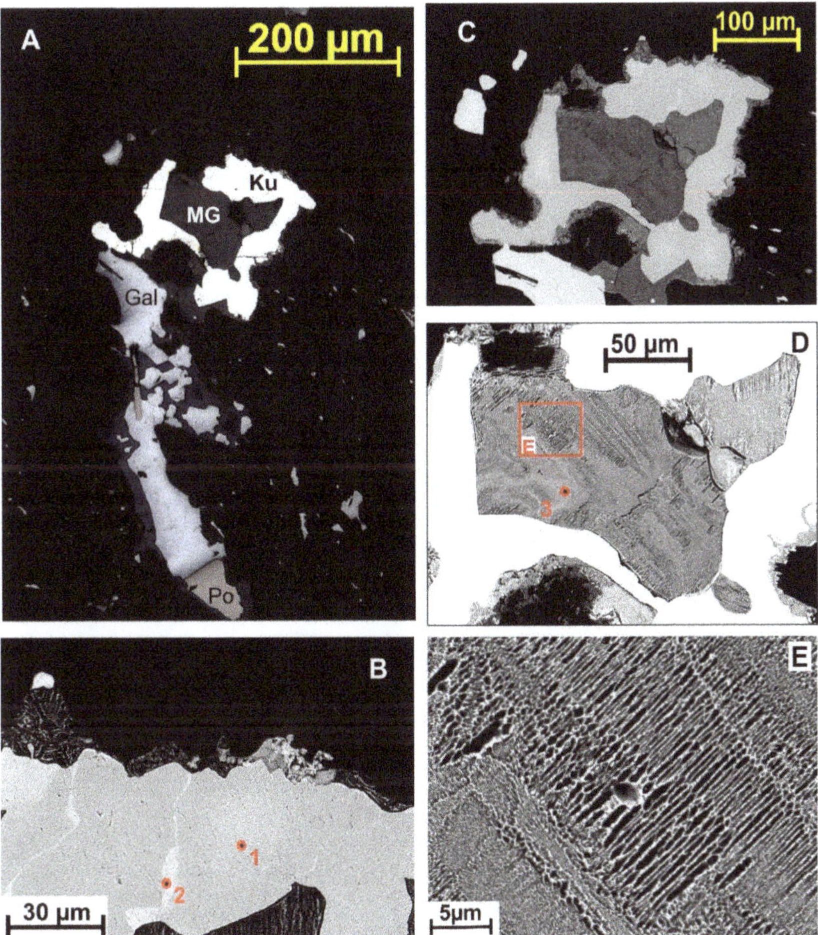

**Figure 2.** Mustard gold with küstelite. (**A**) Polished section of the mineralized quartz metasomatic rock under plane polarized light: Ku—küstelite; Gal—galena; MG—mustard gold; Po—pyrrhotite; (**B–E**) back scattered electron (BSE) images. (**B**) Heterogeneity of küstelite with channels of migration of electrum across küstelite; (**C–E**) mustard gold grain, images at different scales. Red figures (and also in Figures 3–8) are the points of analyses presented in Tables 1 and 2.

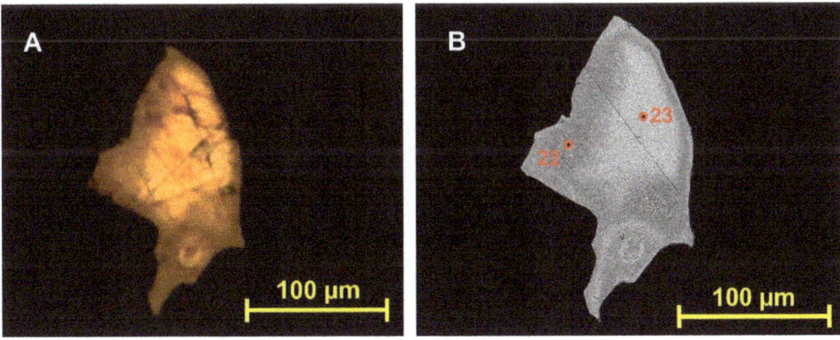

**Figure 3.** A grain of mustard gold with thallium. (**A**) Polished section under plane polarized light; (**B**) BSE-image with the points of microprobe analysis.

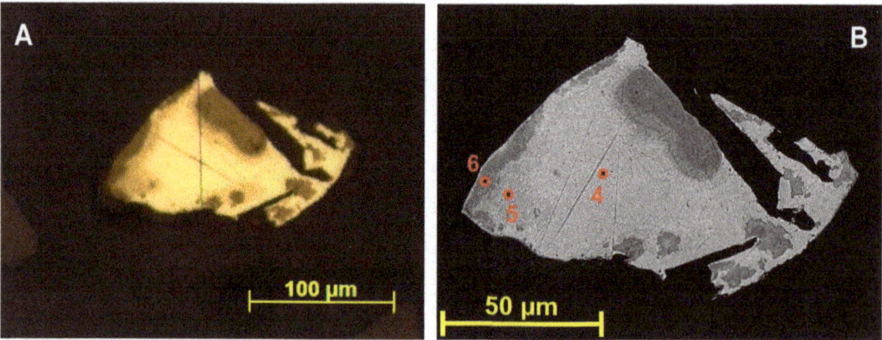

**Figure 4.** A grain of küstelite, partly replaced by mustard gold with silver chlorides and bromides. (**A**) Polished section under plane polarized light; (**B**) BSE-image of the same grains. The darkest parts of the mustard gold grains contain >3 wt% Fe.

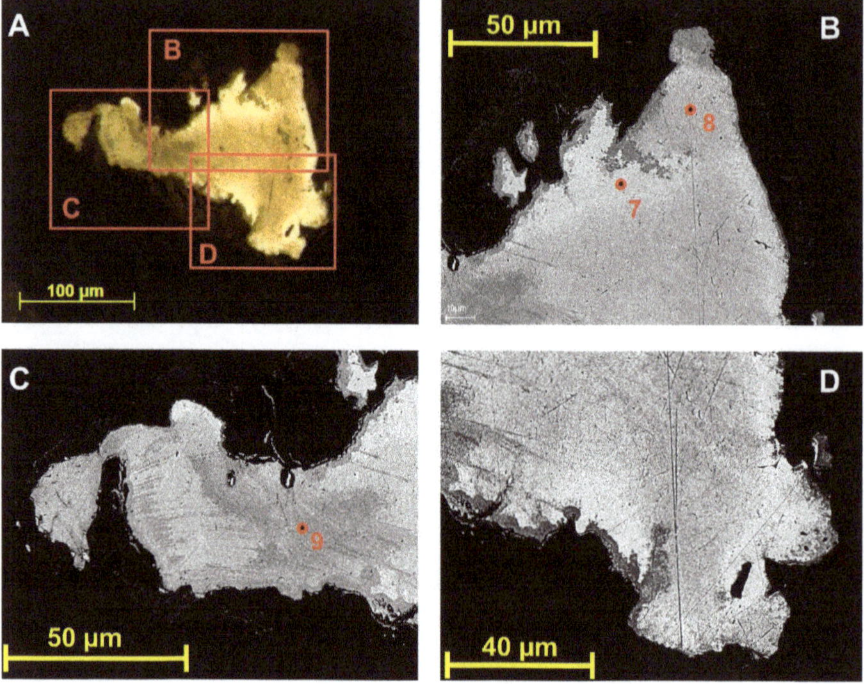

**Figure 5.** A grain of küstelite, partly replaced by mustard gold with silver chlorides and bromides. (**A**) Polished section under plane polarized light; (**B–D**) BSE-images of different parts of the grain, illustrating fine intergrowths of different mineral phases.

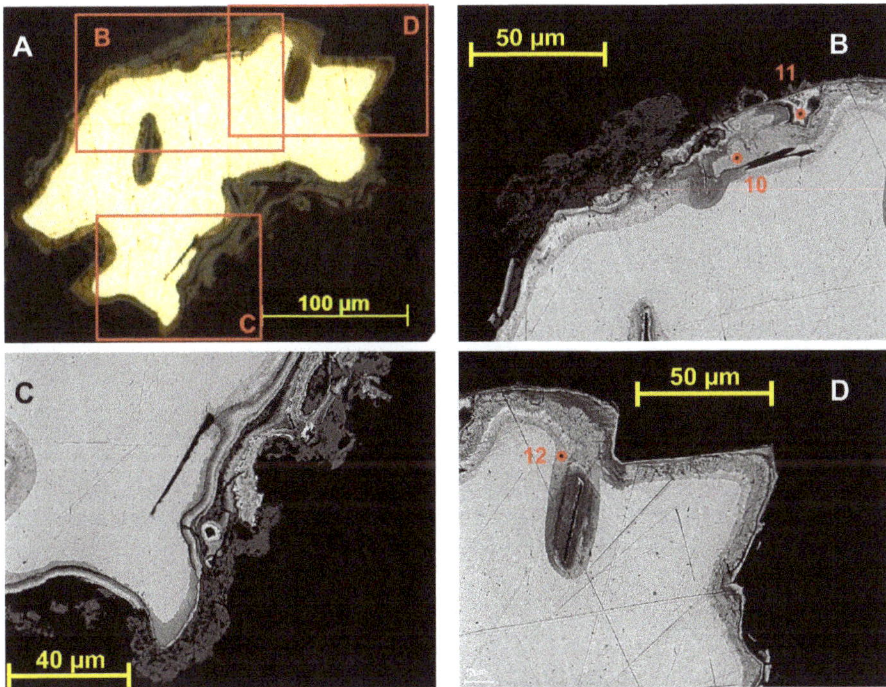

**Figure 6.** A grain of küstelite, with a rim of mustard gold with silver chlorides, bromides, and sulfides. (**A**) Polished section under plane polarized light; (**B–D**) BSE-images of different parts of the grain. Spongy dark-gray mineral, overgrowing the grain, is chlorargyrite.

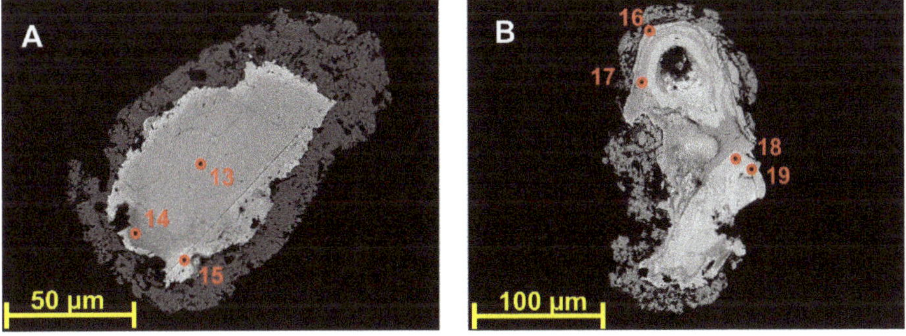

**Figure 7.** (**A**) Mustard gold with Br, Cl, Sb, C, and S in different ratios; the grain is overgrown with chlorargyrite; BSE-image; (**B**) zoning in the grain of mustard gold, connected with zonal distribution of Au and Ag, Cl and Br, halogens and S; the grain is overgrown with chlorargyrite; BSE-image.

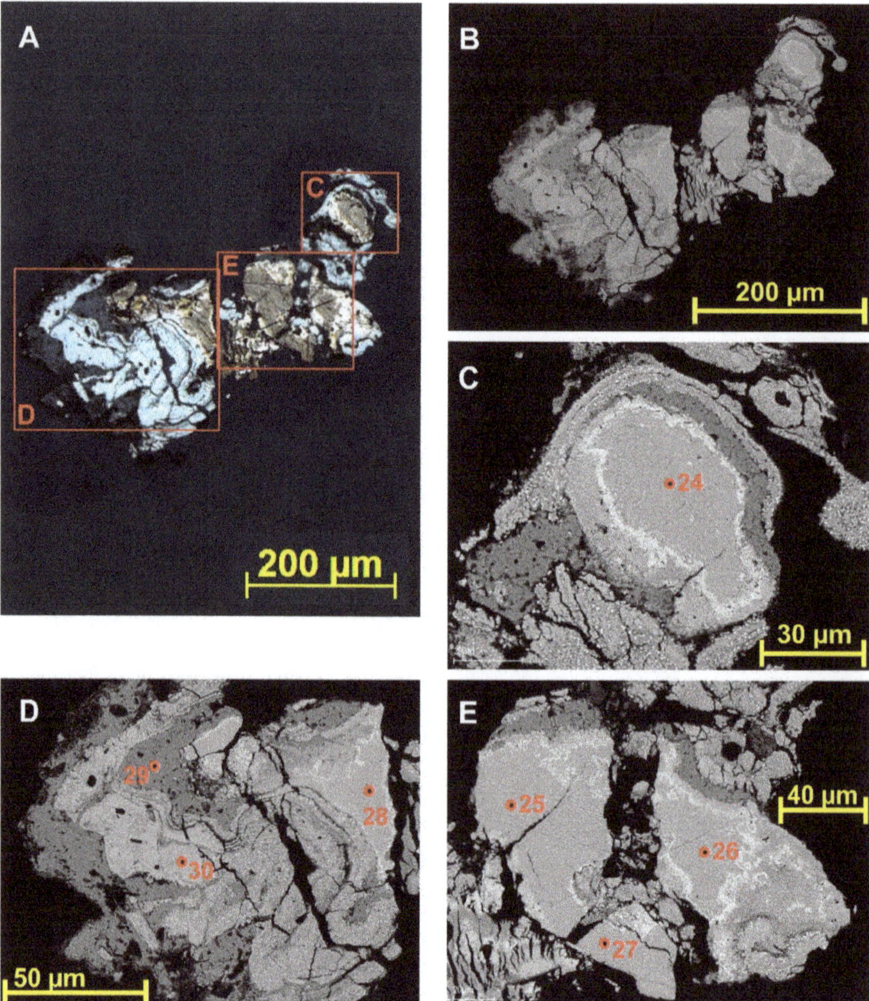

**Figure 8.** Zonal grain of mustard gold with gold and silver chlorides and sulfides. (**A**) Polished section under plane polarized light; (**B–E**) BSE-images of different parts of the grain with points of analysis.

Heavy mineral concentrates from crushed mineralized samples and from the glacial deposits overlaying the mineralized rocks contain küstelite grains, which are partly altered to mustard gold with S, Cl, and Br in its composition. The "primary" küstelite contains ~25 wt% Au, ~65 wt% Ag, and ~2 wt% Sb (Table 1, #4 and #7). Altered parts of the grains are of yellow-brown color, with dull luster (Figures 4–6). Contents of Au increase, Ag decrease, and Br, Cl, S, Fe, As (and C in one case) appear in different proportions in the altered parts of the grains (Table 1, #4–12). The substance is heterogeneous, and we see in the BSE-images (Figures 4 and 5) that it is a fine intergrowth of Au–Ag alloy with other phases.

Many grains of mustard gold are zonal (Figures 4–7) due to zoning in the distribution of Ag and Au, Br and Cl, halogens, and S. Transition from one zone to another is more often sharp, but in some grains it is gradual. Zones with chlorides and bromides of Ag and Au are yellow–brown, and those rich in S are blue-gray. Some grains are coated with chlorargyrite or goethite.

Some grains do not preserve relics of "primary" küstelite, and their cores consist of creamy-colored material (Figures 7B and 8), containing Au, Ag, Cl, and rarely, some Br and S (Table 1, #17, Table 2, #24–#28). This mustard gold is a fine intergrowth of Ag–Au alloy and chlorargyrite, the presence of both phases was verified with X-ray analysis. Conversion of the results of microprobe analysis from weight percent to atomic quantities (Table 1, #18, Table 2, #24, #25) shows that (Ag + Au)/Cl ratio is close to 2, i.e., the content of Ag–Au alloy and chlorargyrite in the intergrowth is near equal.

Mustard gold grains (intergrowths of Ag–Au alloy and chlorargyrite) are fringed with a strip, enriched in very fine (<1 μm) re-deposited electrum grains (Figure 8A,C,E).

The grains of mustard gold are surrounded by a zone of silver-rich minerals (gray in Figure 8A). This zone is of banded texture, up to 0.2 mm thick. It consists of strips of gray material with a dull luster and light blue-gray with metallic luster (Figure 8A,D). The gray, with the dull luster material contains Ag, Cl, and a little (1.2 wt%) Au, (Ag + Au)/Cl ratio is 2 (Table 2, #29). This is a very fine intergrowth of metallic Ag and chlorargyrite in equal quantities. The presence of both mineral phases was confirmed with X-ray analysis.

The composition of the light blue-gray with metallic luster mineral (Table 2, #30) corresponds to formulae $Ag_{3.2}Au_{0.8}S_2$, i.e., this is probably silver–gold sulfide yutenbogaardtite.

Chlorargyrite makes the outermost zone of loose material with a dark-gray color (better seen in Figure 7B). The thickness of this zone is up to 20 μm.

## 5. Discussion

The mustard gold forms in a process of decomposition of Au–Te and Au–Sb compounds [6,8,10,11] mainly in the zone of oxidation and disintegration of Au–Te and Au–Sb ore, and in residual placers, not actively reworked by alluvial processes [3]. Findings of mustard gold are numerous in the zones of hypergenesis of epithermal low-depth Au–Ag–Te and Au–Ag–Sb deposits in the Mesozoic and Cenozoic volcanic belts [2–14], which got to the surface and are actively denudated at present.

The Oleninskoe gold deposit differs from all other gold deposits in the Fennoscandian Shield in its geochemical and mineralogical characteristics: It is rich in Sb-sulfosalts of silver, lead, and copper, contains Au–Sb minerals gold-bearing dyscrasite and aurostibite, gold is low-grade (küstelite makes three of four generations of Ag–Au alloys), and contains 1–2 wt% Sb [22]. These specific features of the Oleninskoe deposit, similar to those in Au–Ag–Sb epithermal deposits, favored the formation of mustard gold after küstelite and Au–Ag–Sb minerals in the zone of ore oxidation.

A hydrothermal origin of mustard gold was considered to be possible along with hypergene genesis after the findings of mustard gold in the deep horizons of Au–Sb deposits in Sakha–Yakutia in aggregates with non-altered arsenopyrite, pyrite, chalcostibite, and fahlore [2,5]. The possibility of the hydrothermal genesis of mustard gold is confirmed by the results of the experiment with the decomposition of gold telluride (calaverite) in fluids with alkaline and normal pH, and the formation of mustard gold in the temperature interval 140–220 °C [25].

All samples with mustard gold in the Oleninskoe deposit were taken from the surface of bedrocks in the trenches or from the bottom of glacial deposits, overlaying the mineralized rocks. This indicates that the mustard gold formed in the zone of hypergenesis. Only one mustard gold grain in the aggregate with küstelite, non-altered galena, and pyrrhotite (see Figure 2) looks like a mineral of hydrothermal genesis.

Mustard gold is usually a product of decomposition of gold tellurides and antimonides, but it can form after Ag-rich Au–Ag alloys as well [2,3,5]. Replacement of Ag–Au alloys with impurities of Sb by mustard gold was described in one of Au–Sb epithermal deposit in Sakha–Yakutia [5]. In the Oleninskoe deposit, Ag-rich Au–Ag alloy küstelite with 1–3 wt% of Sb is the main mineral, which is replaced by mustard gold. Another mineral, which is supposed to produce mustard gold, is gold-bearing dyscrasite (up to 17 wt% of Au). Many dyscrasite grains in the Oleninskoe deposit have the porous structure of decomposition of the mineral (Figure 9), very similar to that one in the mustard gold.

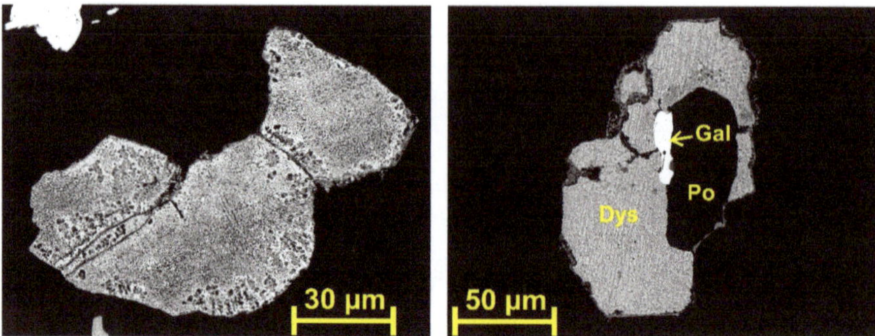



**Figure 9.** Altered Au-bearing dyscrasite with a porous structure, BSE-images. Po—pyrrhotite, Gal—galena, Dys—dyscrasite.

As is shown in [26,27], silver is more mobile than gold in low-temperature processes, and acid or neutral fluids can easily mobilize silver in complexes with Cl⁻ or Br⁻ from the Ag–Au alloys. This process is realized during the process of the formation of mustard gold in the Oleninskoe deposit. A "gold sponge" in the mustard gold grains forms through mobilization and partial removal of Ag with a corresponding increase in gold content in the residual material: The composition of the "primary" alloy corresponds mainly to küstelite with Au/Ag ratio ~1/3, but in the mustard gold Au/Ag ratio is more than 1, i.e., close to that one in electrum. The mobilized silver is partially removed and can be re-deposited in the outer parts of the mustard gold grains in the form of chloride and sulfide minerals.

The substance filling the pores in the "gold sponge", is of different composition. As a rule, it contains some Fe oxides (or hydroxides)—probably it is goethite with impurities of As, Sb, Al, Si. Generally, Fe-oxide/hydroxide is the most common substance to fill the pores [2,9], and it determines the brown color of the mustard gold grains. Except for the Fe oxide, we found high content of Sb (>17 wt%) in one of the studied grains. This gave us a reason to suppose that the grain was a product of decomposition of Au-bearing dyscrasite, not küstelite.

Thallium is not mentioned in the list of elements, typical for mustard gold [8,9], and it is the first finding of mustard gold with Tl. Thallium was reported earlier as an impurity in late sulfides and oxides, which form thin (1–3 µm) rims around Ag–Au grains and, rarely, around pyrite and arsenopyrite [28] in the Oleninskoe deposit. Thallium-bearing minerals were crystallized, by all indications, at the very late stages of hydrothermal ore alteration [28].

The mustard gold with silver chlorides and bromides filling the pores has not been described earlier. In the zone of ore oxidation in the Oleninskoe deposit, we can see the gradual replacement of küstelite by mustard gold with the formation of zonal grains. Silver, mobilized from küstelite during its decomposition, filled the pores in the form of chlorargyrite and bromargyrite, and it was partially removed and re-deposited in the outer parts of the mustard gold grains.

Silver is a metal less stable than gold under the conditions of ore oxidation [26,27,29]. Hence, it hardly can form structures similar to mustard gold. Nevertheless, native silver can form aggregates of very thin filaments, as was described for the process of removal of extra silver from freibergite [30]. Fine intergrowth of native silver with chlorargyrite, structurally similar to mustard gold, was found in the Oleninskoe (Figure 8, outer zone of the grain). But mechanism of formation of this intergrowth differs from that one of mustard gold: The native silver–chlorargyrite intergrowth formed in the outer parts of the mustard gold grains of the elements, mobilized from the "primary" küstelite, whereas the mustard gold formed in situ of the residual substance of the "primary" grain.

## 6. Conclusions

Mustard gold is a rare mineral formation, and finding it in the Oleninskoe deposit is the first finding in the Precambrian Fennoscandian Shield. In the Oleninskoe deposit, the mustard gold formed as a product of decomposition of the Ag–Au–Sb minerals—küstelite with an impurity of Sb and, probably, gold-bearing dyscrasite.

The studied mustard gold is composed of a net of very fine gold filaments ("gold sponge") with pores filled with Fe and Sb oxides or Ag chlorides. Mustard gold grains are non-homogenous. Some of them are a mosaic structure, connected with change in pore size and orientation of the gold filaments in different microblocks. Many mustard gold grains are zonal, and zonation formed due to partial removal of Ag and corresponding enrichment in gold of the residual material and due to uneven distribution of the substance filling the pores.

Thallium oxides, silver chlorides, bromides, and sulfides have not been mentioned earlier as a substance to fill the pore in mustard gold. Tl, Cl, Br, and S extend the list of elements found in the mustard gold.

Native silver–chlorargyrite fine intergrowths, structurally similar to mustard gold, were found in the outer parts of the mustard gold grains. These intergrowths probably formed of the elements, mobilized from altered küstelite during the formation of mustard gold.

**Author Contributions:** Methodology, A.A.K.; Conceptualization, A.A.K.; Original Draft Writing, A.A.K.; Electron Microscope Study and Microprobe Analysis, Y.E.S.; X-ray Analyses, E.A.S.; Editing and Review, A.A.K., Y.E.S., E.A.S.

**Funding:** The work was carried out under Project 0226-2019-0053 of the Russian Academy of Sciences.

**Acknowledgments:** The authors thank Valentina Basalaeva, Anna Kalachyova, and Mikhail Sidorov (GI KSC RAS) for the preparation of heavy mineral concentrates and polished sections. The authors are grateful to the reviewers for their comments, which helped to improve the manuscript.

**Conflicts of Interest:** The authors declare no conflict of interest.

## References

1. Lindgren, W. *Mineral Deposits*; McGraw-Hill Book Company, Inc.: New York, NY, USA; London, UK, 1933; 930p.
2. Nekrasov, I.Y. *Geochemistry, Mineralogy, and Genesis of Gold Deposits*; Routledge: Moscow, Russia, 1991; 302p. (In Russian)
3. Litvinenko, I.S.; Shilina, L.A. Hypergene new formations of gold in placer deposits in the Nizne–Myakitsky placer-ore field, north-East Russia. *Ores Met.* **2017**, *1*, 75–90. (In Russian)
4. Petersen, S.B.; Makovicky, E.; Juiling, L.; Rose-Hansen, J. Mustard gold from the Dongping Au–Te deposit, Hebei province, People's Republic of China. *N. Jb. Miner. Mh.* **1999**, *8*, 337–357.
5. Gamyanin, G.N.; Zhdanov, Y.Y.; Nekrasov, T.Y.; Leskova, N.V. "Mustard" gold from gold–antimony ores of Eastern Yakutia. *New Data Miner.* **1987**, *34*, 13–20. (In Russian)
6. Okrugin, V.M.; Skilskaya, E.; Etschmann, B.; Pring, A.; Li, K.; Zhao, J.; Griffiths, G.; Lumpkin, G.R.; Triani, G.; Brugger, J. Microporous gold: Comparison of textures from nature and experiments. *Am. Mineral.* **2014**, *99*, 1171–1174. [CrossRef]
7. Tolstykh, N.D.; Vymazalova, A.; Tuhý, M.; Shapovalova, M. Conditions of formation of Au–Se–Temineralization in the Gaching ore occurrence (Maletoyvayam ore field), Kamchatka, Russia. *Mineral. Mag.* **2018**, *82*, 1–44. [CrossRef]
8. Tolstykh, N.D.; Palyanova, G.A.; Bobrova, O.V.; Sidorov, E.G. Mustard gold of the Gaching Ore Deposit (Maletoyvayam Ore Field, Kamchatka, Russia). *Minerals* **2019**, *9*, 489. [CrossRef]
9. Kudaeva, A.L.; Andreeva, E.D. Mustard gold: Characteristics, types, and chemical composition. In *'Natural Environment of Kamchatka', Proceedings of the XIII Regional Youth Scientific Conference, Petropavlovsk, Russia, 15 April 2014*; Institute of volcanology and seismology, Far-East Branch of RAS: Petropavlovsk, Russia, 2014; pp. 17–30. (In Russian)
10. Parfyonov, M.I. Spongy and 'Mustard' Gold in the Tumannoe Gold–Antimony Deposit (Eastern Chukotka, Russia). Available online: http://www.minsoc.ru/viewreports.php?id=1&cid=1383&rid=1740 (accessed on 24 October 2019). (In Russian).

11. Amuzinsky, V.A.; Anisimova, G.S.; Zhdanov, Y.U.A.; Ivanov, G.S.; Koksharsky, M.G.; Nedosekin, Y.D.; Polyansky, P.M. *Sarylahskoe and Sentachanskoe Gold-Antimony Deposits: Geology, Mineralogy, and Geochemistry*; MAIK 'Nauka/Interperiodika': Moscow, Russia, 2001; 218p. (In Russian)

12. Zhao, J.; Pring, A. Mineral Transformations in Gold–(Silver) Tellurides in the Presence of Fluids: Nature and Experiment. *Minerals* **2019**, *9*, 167. [CrossRef]

13. Makovicky EMartin Chovan, M.; Bakos, F. The stibian mustard gold from the Kriváň Au deposit, Tatry Mts., Slovak Republic. *N. Jahrb. Miner. Abh.* **2007**, *184*, 207–215.

14. Dill, H.G.; Weiser, T.; Benhardt, I.R.; Kilibarda, R. The composite gold—Antimony vein deposit at Kharma (Bolivia). *Econ. Geol.* **1995**, *90*, 51–66. [CrossRef]

15. Kalinin, A.A. *Gold in Metamorphic Complexes of the North-Eastern Part of the Fennoscandian Shield*; Federal Research Centre 'Kola Science Centre': Apatity, Russia, 2018; 250p. (In Russian)

16. Kalinin, A.A.; Kazanov, O.V.; Bezrukov, V.I.; Prokofiev, V.Y.u. Gold Prospects in the Western Segment of the Russian Arctic: Regional Metallogeny and Distribution of Mineralization. *Minerals* **2019**, *9*, 137. [CrossRef]

17. Volkov, A.V.; Novikov, I.A. The Oleninskoe gold sulfide deposit (Kola peninsula, Russia). *Geol. Ore Depos.* **2002**, *44*, 361–372.

18. Galkin, N.N. Geology and Mineralogy of Gold Occurrences in the Pellaphk–Oleninsky Ore Unit. Ph.D. Thesis, Kola Sci. Centre, Apatity, Russia, 2006; 22p. (In Russian).

19. Kudryashov, N.M.; Lyalina, L.M.; Apanasevich, E.A. Age of rare metal pegmatites from the Vasin Myl'k deposit (Kola region): Evidence from of U–Pb geochronology of microlite. *Dokl. Earth Sci.* **2015**, *461*, 321–325. [CrossRef]

20. Nemčok, M.; Henk, A.; Gayer, R.A.; Vandycke, P.; Hathaway, T.M. Strike-slip fault bridge fluid pumping mechanism: Insights from field-based palaeostress analysis and numerical modeling. *J. Struct. Geol.* **2002**, *24*, 1885–1901. [CrossRef]

21. Kalinin, A.A.; Savchenko, Y.E.; Selivanova, E.A. Minerals of precious metals in the Oleninskoe gold deposit (Kola peninsula). *Zap. RMO* **2017**, *146*, 43–58. (In Russian)

22. Kalinin, A.A. Composition of arsenopyrite as an indicator of conditions of formation of mineralization in metasomatic rocks in the Voron'ya tundra area. In *Metamorphism and Metamorphic Ore Genesis in the Early Precambrian*; KFAN SSSR: Apatity, Russia, 1984; pp. 54–57. (In Russian)

23. Ivashchenko, V.I.; Golubev, A.I. *Gold and Platinum of Karelia: Genetic Types of Mineralization and Prospects*; Runkvist, D.V., Ed.; Karelian Sci. Centre RAS: Petrozavodsk, Russia, 2011; 369p. (In Russian)

24. Kalinin, A.A.; Prokofiev, V.Y.; Telezhkin, A.A. Composition of hydrothermal fluids in the gold occurrences of the Kola region. In *Physico-Chemical Parameters of Petro- and Ore Genesis, Proceedings of the All-Russian Conference, Moscow, Russia, 7–9 October 2019*, Electronic Edition ed; IGEM RAS: Moscow, Russia, 2019; pp. 74–76. (In Russian)

25. Zhao, J.; Bruger, J.; Gundler, P.V.; Xia, F.; Chen, G.; Pring, A. Mechanism and kinetics of a mineral transformation under hydrothermal conditions: Calaverite to metallic gold. *Am. Mineral.* **2009**, *94*, 1541–1555. [CrossRef]

26. Pal'yanova, G.A.; Kolonin, G.R. Geochemical mobility of Au and Ag during hydrothermal transfer and precipitation: Thermodynamic simulation. *Geochem. Int.* **2007**, *45*, 744–757. [CrossRef]

27. Pal'yanova, G.A. *Physico-Chemical Features of Behavior of Gold and Silver in the Processes of Hydrothermal Ore Genesis*; Siberian Branch of Russian Academy of Science: Novosibirsk, Russia, 2008; 221p. (In Russian)

28. Kalinin, A.A.; Savchenko, Y.E. Thallium mineralization in the Oleninskoe gold prospect, Kolmozero–Voronya belt. *Proc. Fersman Sci. Sess.* **2019**, *16*, 240–244. [CrossRef]

29. Korobushkin, I.M.; Glotov, A.M.; Vasilkova, N.A. Character of oxidation of gold–sulfide deposits and methods of processing of the oxidized ores. *Izv. Tomsk. Polytech. Inst. Probl. Geol. Gold Depos.* **1977**, *239*, 239–248. (In Russian)

30. Savva, N.E. *Mineralogy of Silver in the North-East of Russia*; Sidorov, A.A., Ed.; Triumph: North–East Interdisciplinary Scientific Research Institute: Magadan, Russia, 2018; 544p. (In Russian)

*Article*

# Characteristics of Supergene Gold of Karst Cavities of the Khokhoy Gold Ore Field (Aldan Shield, East Russia)

**Galina S. Anisimova *, Larisa A. Kondratieva and Veronika N. Kardashevskaia**

Diamond and Precious Metal Geology Institute, SB RAS, 677000 Yakutsk, Russia; lkon12@yandex.ru (L.A.K.); kardashevskaya92@mail.ru (V.N.K.)
* Correspondence: gsanisimova@diamond.ysn.ru; Tel.: +7-4112-33-58-72

Received: 20 December 2019; Accepted: 3 February 2020; Published: 6 February 2020

**Abstract:** Typomorphic features of supergene gold in karst cavities were studied in the recently discovered Au–Te–Sb–Tl deposit within the Khokhoy gold ore field of the Aldan-Stanovoy auriferous province (Aldan shield, East Russia). Two morphological types of supergene gold, massive and porous, are recognized there. The first type is represented by gold crystals and irregular mass, with the fineness ranging from 835 to 1000‰. They are closely associated with goethite, siderite, unnamed Fe, Te, and Tl carbonates, Tl tellurites/tellurates and antimonates, as well as avicennite with a Te impurity. The second type is represented by mustard gold of two types with different internal structure: microporous and dendritic. The supergene gold is characterized by persistently high fineness. Along with Ag, it invariably contains Hg (up to 5.78 wt%) and Bi, and, rarely, Pb, Cu, and Fe. The supergene gold is chemically homogeneous, and its particles are all marked by high fineness, without any rims or margins. The obtained characteristics made it possible to prove the existence of two genetic types of supergene gold. Mustard microporous gold is the result of the decomposition of the associated minerals—goethite, Tl oxides, tellurium, Fe, Mn and Tl carbonates and antimonates, containing microinclusions of gold. Massive gold and dendrites are newly formed. The decomposition, remobilization, and reprecipitation of residual gold nanoparticles and their aggregation led to the formation of dendrites, and with further crystal growth and filling of pores, to gold of massive morphology. In terms of morphology, internal structure, fineness, and trace element composition, supergene gold of the Khokhoy gold ore field is comparable to gold from the Kuranakh deposit (Russia) and the Carlin-type gold deposits. It also is similar to spungy and mustard gold from Au–Te and Au–Sb deposits, weathering crusts, and placers. Its main characteristic feature is a close paragenesis with Tl minerals.

**Keywords:** gold ore field; karst cavities; monolithic and porous gold; Tl oxides (avicennite); Tl carbonates; Tl tellurates and antimonates

---

## 1. Introduction

The importance of gold-bearing supergene zones increased significantly in relation to the discovery and commercial exploitation of the Carlin-type gold deposits (USA) [1–5]. In Russia, assigned to this type are deposits localized in karst cavities such as the Kuranakh deposit within the Central Aldan district of Yakutia and the Vorontsovskoe deposit in the Urals [6–9]. The question of native gold formation in supergene conditions, including in karst cavities, remains are as yet poorly investigated. Supergene gold in karst cavities has specific features, exhibits various textural and chemical characteristics that can help in elucidating the process of formation of supergenic gold.

That's why studies on the typomorphic features of the recently discovered supergene gold from the Khokhoy gold ore field within the Aldan-Stanovoy auriferous province are quite topical now.

*Brief Geological Characteristics of the Khokhoy Ore Field*

The Khokhoy gold ore field is located in the Verkhneamginskaya auriferous zone, in the up-stream basin of the river Khokhoy, a right tributary of the river Amga [10]. The territory is a part of the Aldan-Stanovoy shield, in the basemen subsidence and sedimentary cover enlargement areas (Aldan shield, East Russia) (Figure 1). The majority of the sediments are those of Lower Cambrian terrigenous-carbonate sediments (predominantly dolomitic ones) with stratigraphic mismatches of closed Lower Jurassic terrigenous sandstone sediments. Mesozoic magmatism resulted in little stratified intrusions, monzonite lakkolithes and syenite-porphyries as well as in alkaline gabbroides. The ore field is structured by a vast faulted area with the North-Eastern strike. The faults' kinematics implies them being normal faults and slip faults with the amplitude of more than 100 m with echelon faults of the North-Western strike forming a ladder-shaped block structure of the ore field. Mineralization takes place where the echelon strippings intersect each other within the tectonized contact of Cambric and Jurassic sediments. Ores are accumulated in the karst cavities forming a vast zone with the North-Western strike of more than 10 km crossing the ore field from the South to the North. Almost the whole anomalous zone consists of karst holes of various size and mineralization extent, open as well as closed with sandstones. The latitude of the karst cavities opened during the mine working ranges from 5–15 to 50 more than 50 m, the depth is 15–45 m or more. Karst cavities are formed of intensive limonitized argillaceous-sandy fulvous matter with primary ores and enclosing rocks fragments of various size. According to X-ray phase and thermal analyses, karst include quartz, potassium feldspars, muscovite, goethite, and less often calcite, hematite, and fluorite. Clinochlor, Fe-clinochlor, lepidocrocite, jarosite and kaolinite also attested. Primary ores exist as fragments of jasperoids, pyrite-adular-quartz metasomatites (Figure 2). The metasomatites contain 93.53–94.45% of $SiO_2$, 1.41–1.91% of $Al_2O_3$ and 0.88–1.56% of $K_2O$. The primary ores almost exclusively consist of fine-grain quartz with some adular in the form of little, frequently idiomorphic interpositions with late druse quartz lenses, often with chalcedony rim.

The geochemical association of the Khokhoy gold ore field elements is the following Au, Sb, Te and Tl. The main minerals of karst cavities are of fine grain quartz, chalcedony, opal, adular, sericite, calcite, barite, fluorite, goethite, limonite, hematite. Unnamed Fe, Mn, Te, and Tl carbonates, galena, weissbergite, berthierite, arsenopyrite, chalcocite, unnamed sulfide Re and W, avicennite, hollandite, chalcopyrite, acanthite, chlorargyrite, fine grain native gold and silver are rarely. Gold mineralization of the Khokhoy gold ore field has a hypogene-supergene nature. The loose gold-bearing rocks here are secondary formations, which originated from oxidation, disintegration, and redeposition in the karst cavities of primary ores such as pyrite–adular–quartz metasomatites that resulted from silicic-potassic metasomatism of carbonate rocks. Strong Tl enrichment occurs during potassic metasomatism, in the fault zones, as is well seen in the Khokhoy gold ore field. The Au–Tl–As–Sb–Te–Ba geochemical profile of the mineralization and its low-temperature formative conditions suggest the epithermal origin of the ores. With regard to structural-morphological and mineralogical-geochemical parameters, gold mineralization of the Khokhoy gold ore field is comparable to the Kuranakh-type gold deposits of the Central Aldan district of Yakutia, representing unique supergene, shallow-depth, friable ores with free gold localized in karst cavities. They are characterized by large reserves with a relatively low gold grade.

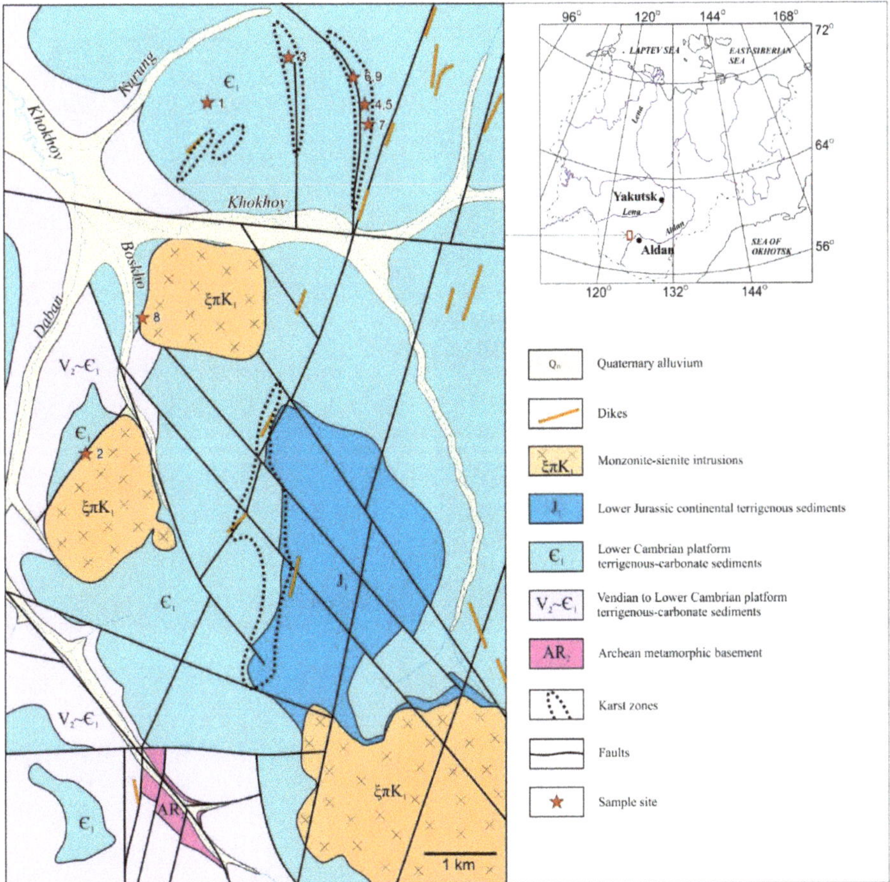


**Figure 1.** Structure of the Khokhoy gold ore field (Aldan shield, East Russia).

**Figure 2.** Ore bearing karst, (**a**) general look, (**b**) pyrite-adular-quartz jasperoids, (**c**) section of **b**.

## 2. Materials and Methods

We used grab samples and friable materials of argillaceous and sandy fragments, taken from the surface mines and the core drilling holes of the Khokhoy gold ore field karst (Table 1). The grab samples of 0.3–0.5 kg taken from the karst fragments were used for making polished slides. The friable materials of 3 kg, on the other hand, were hydroseparated and to get the heavy residue. All samples were investigated with the use of binocular microscope, then we detected individual gold lumps and the associated minerals, which we forced together under pressure, made solid with epoxide and polished. All the polished samples were investigated with the use of Jenavert microscope in the reflected light, photographed and made ready for microprobe analysis. In order to investigate the chemical composition of the native gold and the associating minerals (ore, vein, and supergene minerals) as well as to detect unknown minerals we used microprobe analysis. The samples were analyzed with Camebax microanalyzer (Cameca, Courbevoie, France). We investigated the composition of gold making 3–5 probes in the center and on the edges of the gold lumps (the analysis was performed by N. Khristoforova). The majority of the samples were analyzed with the use of scanning electron microscope JEOL JSM-6480LV and energy spectrometer by Oxford (JEOL, Tokyo, Japan), which was used for taking the picture (the analysis was performed by S. Popova and S. Karpova). We made the quantitative XPP analyses using OXFORD INCA ENERGY 350 (Oxford Instruments, UK). The analysis conditions are as follows, the accelerating voltage of 20 kV, the measuring current flow is 1.08 nA, the measurement time is 10 s. The photographing conditions are as follows, voltage is 20 kV, the current flow is 17 nA. The analytic lines are Cu, Fe, Zn—K$\alpha$; Sb, S—L$\alpha$. The standards are gold 750‰—Au, Ag; HgTe (coloradoite)—Hg, Te; CuSbS$_2$ (chalcostibite)—Cu, Sb, S; ZnS (sphalerite)—Zn; CuFeS$_2$ (chalcopyrite)—Fe; PbS (galena)—Pb; FeAsS (arsenopyrite)—As; BaSO$_4$ (barite)—Ba; ZrSiO$_4$ (zircon)—Zr; manganese 100%—Mn.

**Table 1.** Details of samples studied.

| No. | Sample Label | Grid | | No. Grains | Primary Gold | Secondary Gold | |
| --- | --- | --- | --- | --- | --- | --- | --- |
| | | Easting | Northing | | | Massive | Porous |
| 1 | 4004 | 6555710 | 21514040 | 2 | 2 | 0 | 0 |
| 2 | 3237 | 6550789 | 21510138 | 1 | 1 | 0 | 0 |
| 3 | K-21 | 6556300 | 21513025 | 1 | 1 | 0 | 0 |
| 4 | 12014 | 6556295 | 21501333 | 1 | 2 | 0 | 0 |
| 5 | 5-15-53, 54, 57-60, 64, 65, 67-74, 76-78, 80-82, 84-87, 90, 94-95, 97-98, 100-103, 105-108, 110-113, 115-118, 6-15 | 6555705 | 21514030 | 82 | 0 | 72 | 10 |
| 6 | 1-17-1-7 | 6556046 | 21513843 | 7 | 0 | 0 | 7 |
| 7 | 1-17-8-14 | 6555603 | 21514221 | 7 | 0 | 0 | 7 |
| 8 | 1-17-15 | 6552872 | 21511211 | 1 | 0 | 0 | 1 |
| 9 | 1-17-16-21 | 6556022 | 21513960 | 6 | 0 | 0 | 6 |

Limits of element detection (wt%) X-ray spectral microprobe analysis: Au 0.117, Ag 0.061, Hg 0.083, Cu 0.031, Fe 0.019, Pb 0.066, Bi 0.095. Limits of element detection (wt%) scanning electron microscope equipped with energy spectrometer: Au 1.84, Ag 0.96, Hg 1.6, Cu 1.22, Fe 1.04, Pb 1.78, Bi 2.7.

## 3. Results

### 3.1. Supergene Gold of the Khokhoy Gold Ore Field

Primary Native Gold

Visually native gold in primary ores is extremely rare [10]. It is assumed that there is invisible finely divided gold, as evidenced by the gold content in the ore. Native gold is largely attested in porous oxidized pyrite relics or in quartz associated with barite and galena as submicroscopic cloddy effusions of the size up to 5 μm, which is why we could not make conditional analysis. The native gold on the whole barely contains any impurities (970–999‰), which is consistent with the ratio of Au/Ag in ores 100/1. Very rarely, grains with up to 15 wt% of silver are attested.

### 3.2. Supergene Gold of the Karst Cavities

The physicochemical processes that have taken place during the development of the karst have influenced the typomorphic characteristics of gold. The karst cavities gold is most frequently attested in loose condition. At the same time, it is bigger in its size than that of primary ores. The Khokhoy gold ore field gold is represented by two types: massive and porous.

#### 3.2.1. Massive Gold

Massive gold is represented by crystals and irregular mass (Figures 3 and 4). Crystals are rarely attested in karst holes and make 1–2% of the total mass. They are represented by individuals of dodecahedral (Figure 3a) and octahedral (Figure 3b) shapes with clear cut facets, and rarely with smooth edges. On the surface of the facets some dints and dents indicating sliding processes are attested. The size of the crystals and irregular mass of gold is reaches 0.2 mm. The fineness of the gold is very high, namely 950–1000‰ (Table 2; Nos. 1 and 2).

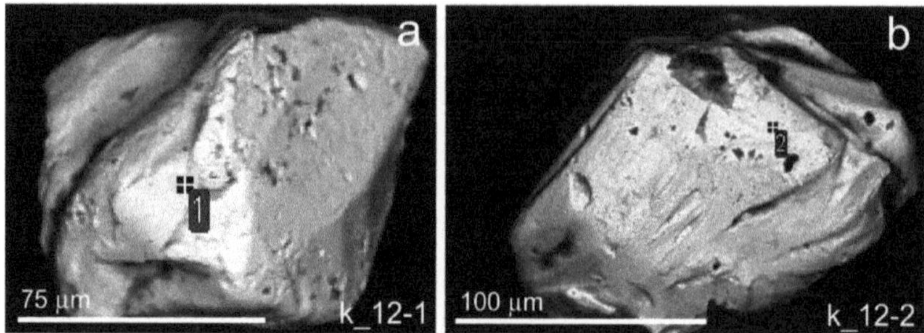

Figure 3. The dodecahedral (**a**) and octahedral (**b**) crystals of Khokhoy ore field. Mounted polished sections, scanning electronic microscope in back-scattering electron mode. The numbers in microphotographs are analitical spots, here and in other figures.

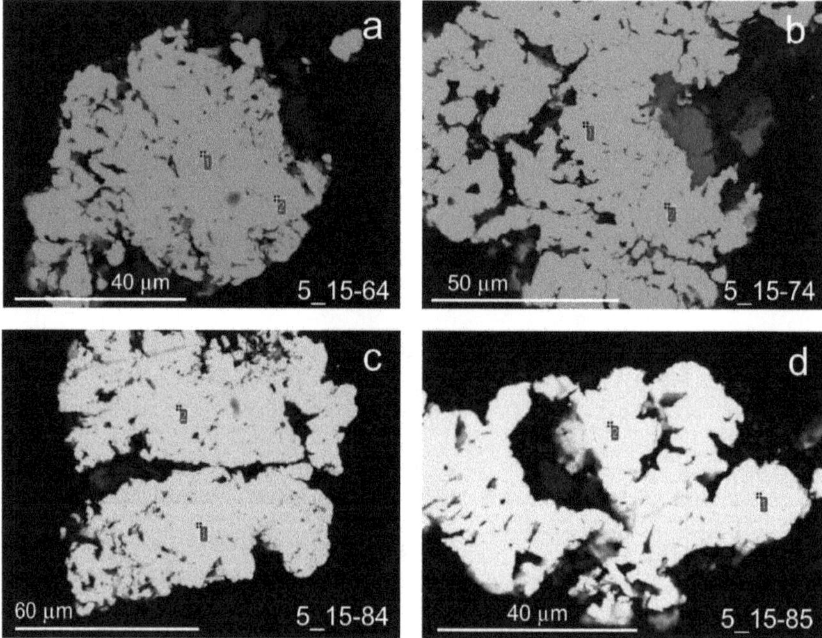

Figure 4. Massive gold. (**a**) gold with Hg alloy, (**b,c**) pure gold, (**d**) gold with Ag alloy. Mounted polished sections, scanning electronic microscope in back-scattering electron mode.

**Table 2.** The composition (in wt%) and fineness (in ‰) of the supergene gold shown in Figures 3, 4, 8 and 10.

| No. | Sample * | Sp. | Ag | Au | Hg | Total | Fineness |
|---|---|---|---|---|---|---|---|
| 1 | K_12 | 1 | - | 98.31 | - | 98.31 | 1000 |
| 2 | | 2 | 4.73 | 94.28 | - | 99.01 | 952 |
| 3 | 5_15-64 | 1 | - | 94.01 | 3.61 | 97.62 | 963 |
| 4 | | 2 | - | 95.90 | 2.92 | 98.82 | 970 |
| 5 | 5_15-74 | 1 | - | 100.48 | - | 100.48 | 1000 |
| 6 | | 2 | 0.64 | 99.11 | - | 99.75 | 993 |
| 7 | 5_15-84 | 1 | - | 98.16 | - | 98.16 | 1000 |
| 8 | | 2 | - | 98.39 | - | 98.39 | 1000 |
| 9 | 5_15-85 | 1 | 4.42 | 95.13 | - | 99.55 | 956 |
| 10 | | 2 | - | 101.40 | - | 101.40 | 1000 |
| 11 | 1_17-5 | 1 | - | 101.78 | - | 101.78 | 1000 |
| 12 | | 2 | - | 98.71 | - | 98.71 | 1000 |
| 13 | | 3 | - | 99.08 | - | 99.08 | 1000 |
| 14 | | 4 | - | 99.35 | - | 99.35 | 1000 |
| 15 | 1_17-15 | 1 | - | 95.56 | 4.57 | 100.13 | 954 |
| 16 | | 3 | 0.28 | 96.64 | 3.54 | 100.47 | 962 |
| 17 | | 4 | - | 100.29 | - | 100.29 | 1000 |
| 18 | | 5 | - | 100.91 | - | 100.91 | 1000 |
| 19 | | 6 | 0.37 | 96.37 | 2.85 | 99.59 | 967 |
| 20 | | 7 | - | 95.50 | 4.74 | 100.24 | 953 |
| 21 | | 8 | - | 96.75 | 3.08 | 99.83 | 969 |
| 22 | | 9 | - | 98.85 | 1.56 | 100.41 | 984 |
| 23 | | 10 | - | 93.91 | 4.33 | 98.24 | 956 |
| 24 | 1_17-13 | 1 | - | 98.45 | - | 98.45 | 1000 |
| 25 | | 4 | - | 100.57 | - | 100.57 | 1000 |
| 26 | | 5 | - | 99.77 | - | 99.77 | 1000 |
| 27 | | 6 | - | 97.90 | - | 97.90 | 1000 |
| 28 | | 7 | - | 97.25 | - | 97.25 | 1000 |
| 29 | | 8 | - | 101.06 | - | 101.06 | 1000 |
| 30 | | 9 | - | 99.11 | - | 99.11 | 1000 |
| 31 | 1_17-18 | 1 | - | 99.01 | - | 99.01 | 1000 |
| 32 | | 2 | 5.98 | 92.21 | - | 98.19 | 939 |
| 33 | | 3 | - | 100.12 | - | 100.12 | 1000 |
| 34 | | 4 | 2.77 | 95.59 | - | 98.36 | 972 |
| 35 | | 5 | - | 97.49 | - | 97.49 | 1000 |
| 36 | | 6 | 6.08 | 92.56 | - | 98.64 | 938 |
| 37 | 1_17-7 | 2 | 0.59 | 95.41 | 2.57 | 98.57 | 968 |
| 38 | | 6 | 1.25 | 98.76 | 0.61 | 100.62 | 982 |
| 39 | | 8 | 0.95 | 97.18 | 2.68 | 100.81 | 964 |
| 40 | | 9 | 2.03 | 95.51 | 3.04 | 100.58 | 949 |
| 41 | 1_17-14 | 1 | 1.97 | 97.67 | 1.54 | 100.18 | 975 |
| 42 | | 2 | - | 100.08 | - | 100.08 | 1000 |
| 43 | | 3 | - | 99.78 | - | 99.78 | 1000 |
| 44 | | 4 | 3.14 | 92.32 | 3.01 | 98.48 | 937 |
| 45 | | 5 | | 100.20 | - | 100.20 | 1000 |
| 46 | | 7 | 2.95 | 95.42 | 1.17 | 99.54 | 958 |
| 47 | | 8 | - | 95.05 | 4.33 | 99.38 | 956 |
| 48 | 1_17-20 | 2 | 5.40 | 93.43 | - | 98.83 | 945 |
| 49 | | 4 | 3.99 | 97.98 | - | 101.97 | 961 |
| 50 | | 5 | 3.13 | 96.10 | - | 99.24 | 968 |
| 51 | | 6 | 2.18 | 98.61 | - | 100.78 | 978 |
| 52 | | 7 | 2.54 | 96.86 | - | 99.40 | 974 |
| 53 | 1_17-6 | 3 | - | 98.02 | - | 98.02 | 1000 |
| 54 | | 6 | - | 98.74 | - | 98.74 | 1000 |
| 55 | | 7 | - | 98.78 | - | 98.78 | 1000 |
| 56 | | 8 | - | 99.82 | - | 99.82 | 1000 |

Notes: * Nos. 1–10—massive gold; 1–2—crystals; 3–10—irregular mass gold; 11–55—porous gold. Sp.—analysis spot in figures. The analyzes were performed on a scanning electron microscope equipped with energy spectrometer.

The most widespread are cloddy gold particles of massive structure (Figure 4). In terms of morphology, they belong to the irregularly-shaped type, and associate closely with various minerals formed from the weathering of primary ores (Figure 5). The supergene gold often occurs in association with unnamed Fe, Te, and Tl carbonates (Figure 5a, Table 3; Nos. 1–8). Aggregates of intimately intergrown unnamed tellurates of Tl and native gold are occasionally observed along cracks in the gold particles (Figure 5b, Table 3; Nos. 9–15). In the interstices between massive gold there are found dendritic gold particles closely interwoven with goethite (Figure 5c, Table 3, Nos. 16–23). The most intricate forms of gold are seen in a very rare mineral avicennite ($Tl_2O_3$), an oxide of thallium with a Te impurity (Figure 5d, Table 3; Nos. 24–39). Sometimes massive gold occurs in assemblage with siderite (Figure 5e,f, Table 3; Nos. 40–43) and goethite (Figure 6, Table 3, Nos. 14–15). At a higher magnification, micron gold inclusions are visible in associating minerals (Figure 5f, Table 3, Nos. 44–55; Figure 6, Nos. 9–13). At a higher magnification, micron-sized gold grains are seen in siderite (Figure 5f, Table 3; Nos. 44–55). Fineness of the monolithic gold varies from 835 to 1000‰, with the high-fineness particles prevailing (Tables 2 and 3). Along with Ag, trace elements include mercury (up to 3.61 wt%), bismuth, and, more rarely Fe, Cu, Zn, and Pb (Table 4).

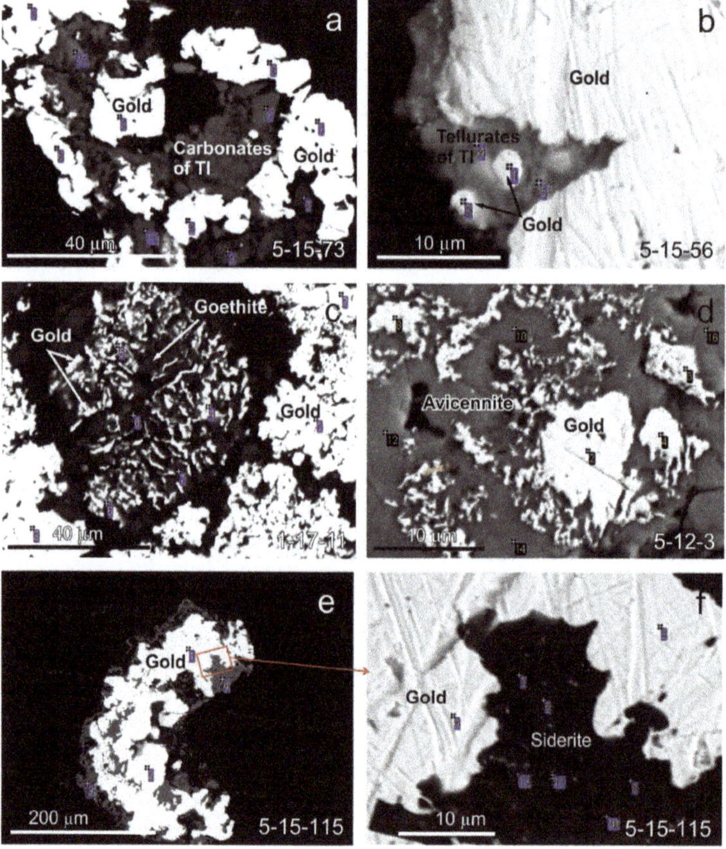


**Figure 5.** Massive gold in association with various supergene minerals. (**a**) massive gold in association with unnamed carbonates Te, Fe, Tl, (**b**) massive gold with inclusions of unnamed tellurates Tl, (**c**) honeycomb gold at the edges, sponge gold in the center in close association with goethite, (**d**) inclusion of massive and sponge gold in avicennite, (**e**) massive gold in association with siderite, (**f**) enlarged fragment of e. Mounted sections, scanning electron microscope, image in the back-scattering electron mode.

**Table 3.** The composition of the massive gold and associated minerals (in wt%) shown in Figures 5 and 6.

| No. | Sample | Sp. | Au | Ag | Hg | Te | Fe | Tl | Si | O | Total |
|---|---|---|---|---|---|---|---|---|---|---|---|
| 1 | 5_15_73 | 1 | 99.61 | - | 1.48 | - | - | - | - | - | 101.09 |
| 2 | 5_15_73 | 2 | 101.61 | - | - | - | - | - | - | - | 101.61 |
| 3 | 5_15_73 | 6 | 96.87 | 1.38 | 2.6 | - | - | - | - | - | 100.85 |
| 4 | 5_15_73 | 7 | - | - | - | 13.73 | 11.23 | 22.78 | 1.53 | 17.78 | 67.05 |
| 5 | 5_15_73 | 8 | - | - | - | 10.97 | 12.29 | 15.58 | 0.92 | 15.1 | 54.86 |
| 6 | 5_15_73 | 9 | - | - | - | 13.13 | 12.44 | 20.99 | 2.51 | 17.26 | 66.33 |
| 7 | 5_15_73 | 10 | - | - | - | 16.08 | 14.58 | 24.42 | | 16.08 | 71.16 |
| 8 | 5_15_73 | 11 | - | - | - | 14.93 | 11.82 | 20.9 | 1.39 | 14.51 | 63.55 |
| 9 | 5_15_101 | 1 | 50.27 | 3.62 | - | - | 23.15 | - | - | 23.06 | 100.1 |
| 10 | 5_15_101 | 2 | 50.45 | 2.56 | - | - | 20.81 | - | - | 25.94 | 99.76 |
| 11 | 5_15_101 | 3 | 52.45 | 2.6 | - | - | 20.57 | - | - | 23.26 | 98.88 |
| 12 | 5_15_101 | 4 | 53.02 | 3.95 | - | - | 21.93 | - | - | 19.83 | 98.73 |
| 13 | 5_15_101 | 5 | 51.43 | 1.64 | - | - | 24.96 | - | - | 30.32 | 108.35 |
| 14 | 5_15_101 | 6 | 92.07 | 7.39 | - | - | - | - | - | - | 99.46 |
| 15 | 5_15_101 | 7 | 93.01 | 6.74 | - | - | - | - | - | - | 99.75 |
| 16 | 1_17_11 | 1 | 93.41 | 5.64 | - | - | - | - | - | - | 99.05 |
| 17 | 1_17_11 | 2 | 98.17 | - | - | - | - | - | - | - | 98.17 |
| 18 | 1_17_11 | 3 | 94.18 | 6.00 | - | - | - | - | - | - | 100.18 |
| 19 | 1_17_11 | 4 | 93.28 | - | - | - | 1.68 | - | - | 4.01 | 98.97 |
| 20 | 1_17_11 | 5 | 98.79 | - | - | - | - | - | - | - | 98.79 |
| 21 | 1_17_11 | 6 | 97.45 | 2.33 | - | - | - | - | - | - | 99.78 |
| 22 | 1_17_11 | 7 | - | - | - | 4.67 | 60.42 | - | - | 35.45 | 100.54 |
| 23 | 1_17_11 | 8 | - | - | - | 5.48 | 58.33 | - | - | 35.23 | 99.04 |
| 24 | 5_15_3 | 1 | 99.39 | - | - | - | - | - | - | - | 99.39 |
| 25 | 5_15_3 | 2 | 99.39 | - | - | - | - | - | - | - | 99.39 |
| 26 | 5_15_3 | 3 | 94.79 | 6.28 | - | - | - | - | - | - | 101.07 |
| 27 | 5_15_3 | 4 | 93.91 | 6.42 | - | - | - | - | - | - | 100.33 |
| 28 | 5_15_3 | 5 | 93.39 | 4.44 | - | - | - | - | - | - | 97.83 |
| 29 | 5_15_3 | 6 | 91.78 | 5.77 | - | - | - | - | - | - | 97.55 |
| 30 | 5_15_3 | 7 | 93.87 | 5.91 | - | - | - | - | - | - | 99.78 |
| 31 | 5_15_3 | 8 | 93.13 | 4.56 | - | - | - | - | - | - | 97.69 |
| 32 | 5_15_3 | 9 | - | - | - | 4.93 | - | 75.65 | - | 17.25 | 97.83 |
| 33 | 5_15_3 | 10 | - | - | - | 3.96 | - | 81.6 | - | 13.65 | 99.21 |
| 34 | 5_15_3 | 11 | - | - | - | 7.5 | - | 79.55 | - | 13.49 | 100.54 |
| 35 | 5_15_3 | 12 | - | - | - | 5.15 | - | 83.25 | - | 13.37 | 101.77 |
| 36 | 5_15_3 | 13 | - | - | - | 5.53 | - | 78.79 | - | 13.41 | 97.73 |
| 37 | 5_15_3 | 14 | - | - | - | 3.85 | - | 81.09 | - | 13.04 | 97.98 |
| 38 | 5_15_3 | 15 | - | - | - | 7.74 | - | 76.21 | - | 15.31 | 99.26 |
| 39 | 5_15_3 | 16 | - | - | - | 4.87 | - | 81.91 | - | 16.16 | 102.94 |
| 40 | 5_15-115 | 1 | 93.99 | 5.07 | - | - | - | - | - | - | 99.06 |
| 41 | 5_15-115 | 2 | 92.51 | 6.95 | - | - | - | - | - | - | 99.46 |
| 42 | 5_15-115 | 3 | - | - | - | - | 45.22 | - | 3.20 | 30.85 | 79.27 |
| 43 | 5_15-115 | 4 | - | - | - | - | 45.28 | - | 2.23 | 30.24 | 77.75 |
| 44 | 5_15-115f | 1 | 90.63 | 7.02 | - | - | - | - | - | - | 97.65 |
| 45 | 5_15-115f | 2 | 91.13 | 6.80 | - | - | - | - | - | - | 97.93 |
| 46 | 5_15-115f | 3 | 92.50 | 6.65 | - | - | - | - | - | - | 99.15 |
| 47 | 5_15-115f | 4 | 90.57 | 7.56 | - | - | - | - | - | - | 98.13 |
| 48 | 5_15-115f | 5 | - | - | - | - | 48.28 | - | 2.24 | 28.48 | 79.00 |
| 49 | 5_15-115f | 6 | - | - | - | - | 47.20 | - | 2.10 | 27.50 | 76.80 |
| 50 | 5_15-115f | 7 | - | - | - | - | 46.93 | - | 2.27 | 30.94 | 80.58 |
| 51 | 5_15-115f | 8 | - | - | - | - | 48.23 | - | 2.32 | 30.48 | 81.03 |
| 52 | 5_15-115f | 9 | - | - | - | - | 47.95 | - | 2.42 | 28.68 | 79.06 |
| 53 | 5_15-115f | 10 | - | - | - | - | 48.55 | - | 2.07 | 30.76 | 81.38 |
| 54 | 5_15-115f | 11 | 15.33 | 2.87 | - | - | 38.46 | - | 1.89 | 37.28 | 95.84 |
| 55 | 5_15-115f | 12 | 15.66 | - | - | - | 41.85 | - | 2.23 | 36.90 | 96.64 |

Note: The analyzes were performed on a scanning electron microscope.

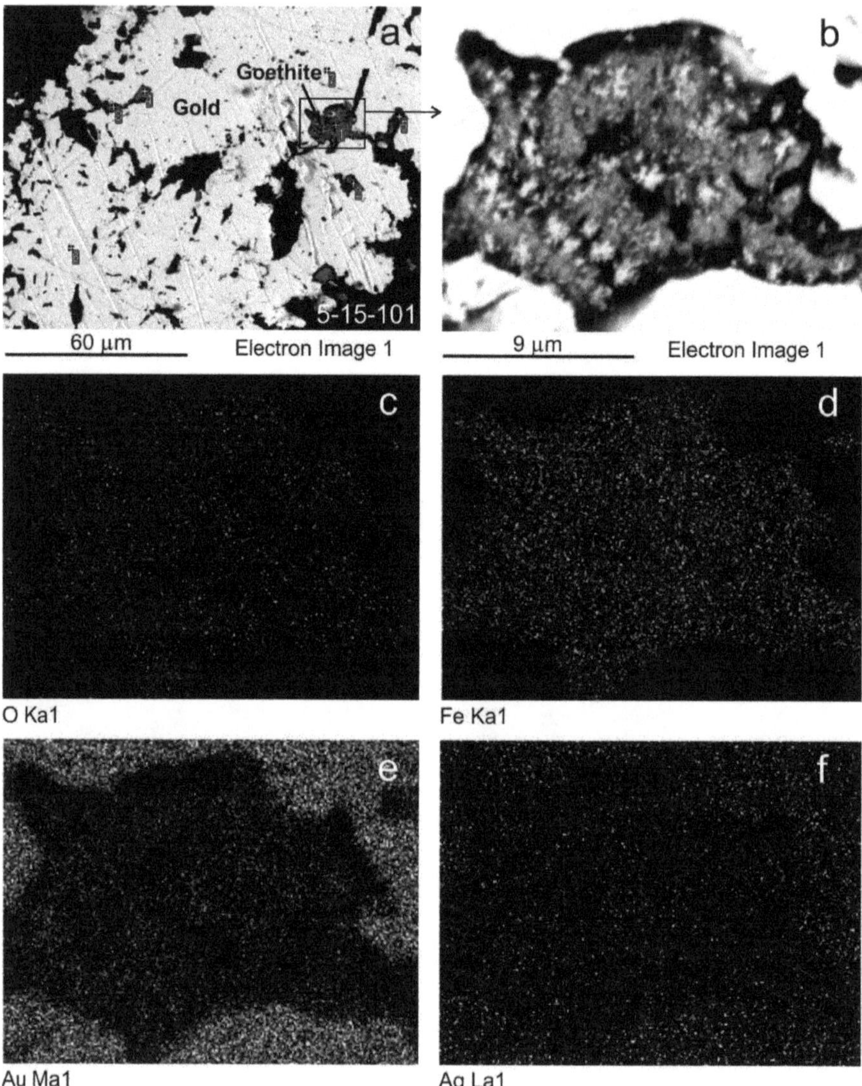

**Figure 6.** Massive gold with goethite: (**a**) general view, (**b**) enlarged goethite fragment with microinclusions of gold in backscattered electrons, (**c–f**) in X-rays of O, Fe, Au and Ag.

**Table 4.** The composition (in wt%) and fineness (in ‰) of the massive gold (in wt%) shown in Figure 11.

| No. | Sample | Fe | Cu | Pt | Ag | Au | Hg | Pb | Bi | Total | Fineness |
|---|---|---|---|---|---|---|---|---|---|---|---|
| 1 | 5-15-53 | 0.01 | - | - | 1.29 | 100.15 | 0.01 | 0.10 | 0.17 | 101.73 | 984 |
| 2 | 5-15-54 | 0.05 | - | - | 3.18 | 94.70 | 0.10 | 0.03 | - | 98.06 | 965 |
| 3 | 5-15-57 | 0.06 | 0.00 | 0.11 | 0.28 | 97.43 | 0.08 | 0.21 | 0.19 | 98.35 | 990 |
| 4 | 5-15-58 | 0.04 | - | 0.07 | 1.56 | 97.17 | 0.10 | 0.08 | 0.12 | 99.14 | 980 |
| 5 | 5-15-59 | 0.03 | - | 0.08 | 4.71 | 93.17 | - | 0.10 | 0.18 | 98.27 | 947 |
| 6 | 5-15-60 | 0.02 | - | 0.01 | 1.59 | 98.26 | 0.02 | 0.07 | 0.08 | 100.05 | 981 |
| 7 | 5-15-64 | 0.01 | - | 0.01 | 1.40 | 96.88 | 0.02 | 0.07 | 0.18 | 98.57 | 982 |
| 8 | 5-15-65 | - | - | - | 0.93 | 99.64 | 0.10 | 0.10 | 0.13 | 100.89 | 987 |
| 9 | 5-15-67 | 0.01 | - | 0.04 | 1.58 | 97.67 | 0.07 | 0.02 | 0.06 | 99.45 | 981 |
| 10 | 5-15-68 | - | - | - | 1.13 | 97.37 | 0.14 | 0.10 | 0.23 | 98.97 | 983 |
| 11 | 5-15-69 | 0.01 | 0.00 | 0.09 | 0.83 | 96.84 | 0.14 | 0.10 | 0.18 | 98.19 | 985 |
| 12 | 5-15-70 | 0.01 | 0.02 | - | 1.48 | 99.47 | 0.08 | 0.20 | 0.22 | 101.48 | 980 |
| 13 | 5-15-71 | - | 0.01 | 0.16 | 1.11 | 99.05 | - | 0.11 | 0.21 | 100.66 | 984 |
| 14 | 5-15-72 | 0.01 | - | 0.01 | 1.02 | 100.11 | 0.08 | 0.06 | 0.12 | 101.39 | 987 |
| 15 | 5-15-73 | - | 0.01 | 0.05 | 0.73 | 99.35 | 0.03 | 0.05 | 0.15 | 100.38 | 990 |
| 16 | 5-15-74 | 0.01 | - | - | 0.60 | 100.57 | 0.01 | 0.06 | 0.12 | 101.38 | 992 |
| 17 | 5-15-76 | - | 0.00 | 0.14 | 1.33 | 96.65 | 0.05 | 0.03 | 0.10 | 98.30 | 982 |
| 18 | 5-15-77 | - | - | - | 1.46 | 97.81 | 0.13 | 0.03 | 0.06 | 99.48 | 983 |
| 19 | 5-15-78 | 0.00 | 0.01 | - | 0.71 | 99.74 | - | 0.12 | - | 100.58 | 992 |
| 20 | 5-15-80 | 0.01 | 0.01 | 0.13 | 0.92 | 97.97 | 0.11 | 0.05 | 0.19 | 99.38 | 986 |
| 21 | 5-15-81 | 0.02 | 0.02 | 0.13 | 0.90 | 97.48 | 0.10 | - | 0.14 | 98.78 | 986 |
| 22 | 5-15-82 | 0.04 | 0.02 | 0.05 | 0.91 | 98.35 | 0.03 | 0.12 | 0.17 | 99.68 | 987 |
| 23 | 5-15-84 | 0.00 | 0.02 | 0.15 | 0.88 | 98.13 | 0.01 | - | 0.19 | 99.38 | 987 |
| 24 | 5-15-85 | - | 0.01 | 0.01 | 7.46 | 90.65 | 0.04 | - | 0.03 | 98.20 | 923 |
| 25 | 5-15-86 | 0.07 | 0.02 | 0.05 | 5.13 | 92.93 | 0.26 | 0.06 | 0.03 | 98.55 | 942 |
| 26 | 5-15-87 | 0.01 | 0.01 | 0.04 | 1.00 | 98.02 | 0.19 | 0.13 | 0.09 | 99.47 | 985 |
| 27 | 5-15-90 | - | 0.01 | 0.02 | 1.45 | 96.79 | 0.02 | 0.05 | 0.19 | 98.53 | 982 |
| 28 | 5-15-94 | - | 0.01 | - | 1.29 | 99.27 | 0.12 | 0.08 | 0.25 | 101.02 | 983 |
| 29 | 5-15-95 | 0.12 | - | 0.05 | 8.14 | 92.21 | - | 0.05 | 0.02 | 100.60 | 917 |
| 30 | 5-15-97 | 0.01 | 0.11 | 0.08 | 2.99 | 97.31 | 0.09 | 0.03 | 0.10 | 100.79 | 965 |
| 31 | 5-15-98 | - | 0.13 | 0.06 | 2.62 | 98.32 | - | 0.00 | - | 101.20 | 972 |
| 32 | 5-15-100 | - | 0.09 | 0.10 | 3.20 | 94.78 | 0.04 | 0.02 | 0.06 | 98.30 | 964 |
| 33 | 5-15-101 | - | 0.08 | 0.08 | 7.27 | 90.71 | 0.11 | - | 0.05 | 98.35 | 922 |
| 34 | 5-15-102 | 0.00 | 0.03 | 0.01 | 7.41 | 90.18 | 0.01 | 0.07 | 0.04 | 97.88 | 921 |
| 35 | 5-15-103 | 0.00 | 0.08 | 0.04 | 4.12 | 93.65 | 0.05 | 0.07 | 0.10 | 98.13 | 954 |
| 36 | 5-15-105 | - | 0.07 | 0.07 | 7.39 | 91.03 | - | 0.02 | 0.23 | 98.84 | 921 |
| 37 | 5-15-106 | - | 0.06 | 0.09 | 7.16 | 91.45 | - | - | - | 98.79 | 926 |
| 38 | 5-15-107 | 0.15 | 0.01 | 0.08 | 15.95 | 81.64 | - | 0.05 | 0.01 | 97.93 | 834 |
| 39 | 5-15-108 | 0.00 | 0.09 | 0.14 | 3.77 | 94.80 | - | 0.04 | 0.17 | 99.09 | 957 |
| 40 | 5-15-110 | 0.11 | 0.04 | 0.17 | 8.49 | 87.04 | 0.16 | 0.77 | 0.17 | 96.96 | 898 |
| 41 | 5-15-111 | - | 0.04 | 0.06 | 7.17 | 91.49 | 0.06 | 0.04 | 0.20 | 99.08 | 923 |
| 42 | 5-15-112 | 0.00 | 0.05 | 0.01 | 6.88 | 92.73 | - | 0.02 | 0.14 | 99.82 | 929 |
| 43 | 5-15-113 | 0.01 | 0.01 | - | 7.36 | 92.62 | 0.06 | 0.06 | 0.15 | 100.28 | 924 |
| 44 | 5-15-115 | - | 0.08 | 0.05 | 3.68 | 97.45 | - | 0.01 | 0.15 | 101.42 | 961 |
| 45 | 5-15-116 | - | 0.08 | 0.03 | 3.37 | 97.24 | - | 0.04 | 0.09 | 100.86 | 964 |
| 46 | 5-15-117 | 0.00 | 0.07 | 0.07 | 7.09 | 91.55 | - | - | 0.14 | 98.96 | 925 |
| 47 | 5-15-118 | 0.11 | 0.01 | 0.05 | 6.47 | 93.29 | 0.06 | - | - | 100.04 | 933 |

Note: Analyzes performed on an X-ray microanalyzer.

## 3.2.2. Porous Gold

Besides the massive gold, porous gold particles can be found. Prevalent are irregular mass, with lesser flattened forms. Both varieties are characterized by microporosity and a mustard color, indicating a wide occurrence of mustard gold in the Khokhoy gold ore field (Figure 7). The term "mustard gold" was got in by W. Lindgren [11]. Typical features of mustard gold are low reflectivity, porous or colloidal texture and rusty, reddish, orange-red and brown-yellow colors in reflected light. It is characteristic of gold-telluride and gold-antimony deposits.

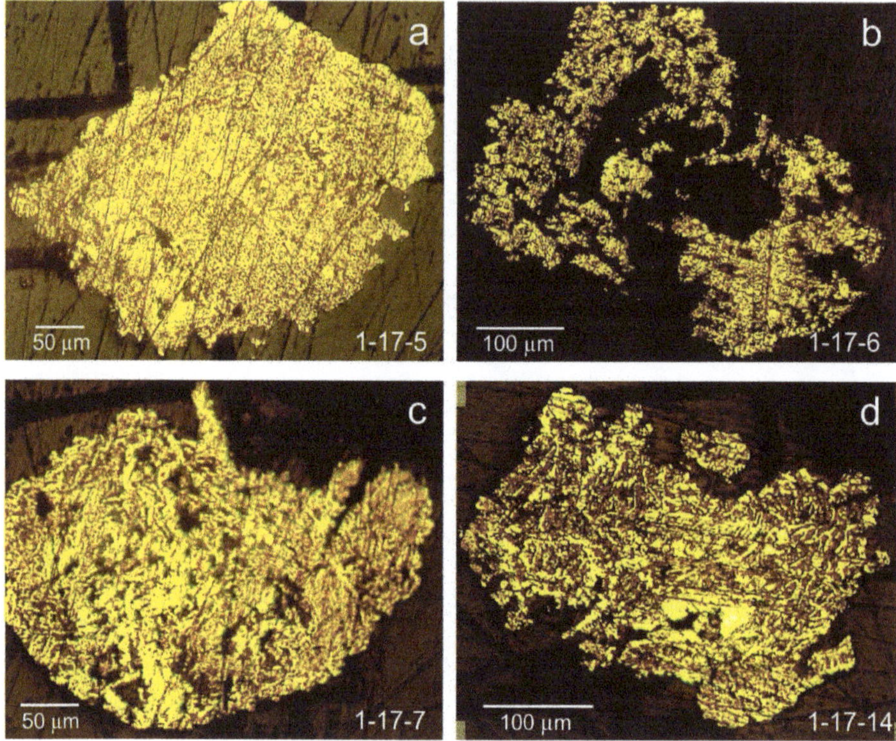

**Figure 7.** Native gold of the Khokhoy gold ore field, (**a**) microporous spongy-mustard gold, (**b**) framework gold with massive gold in the center, (**c,d**) dendritic mustard gold. Mounted sections, ore microscope, image in reflected light.

"Mustard" gold in the ore field according to the internal structure is microporous and dendritic. Porous irregular mass and flattened gold are friable aggregates of fulvous color. The porosity of gold is clearly reflected in the investigation of gold in a scanning electron microscope. The closeness of mustard and sponge gold is often observed (Figure 8a,b). Sometimes, massive gold grains occur with the development of hollow spongy and mustard gold closer to the center (Figure 8c). Less commonly observed is microporous mustard gold with elongated massive gold fibers in the central part (Figure 8d). The mustard gold micropores can be hollow or filled with various chemical elements Fe, Te, Cu, Mn, Sb, and Tl.

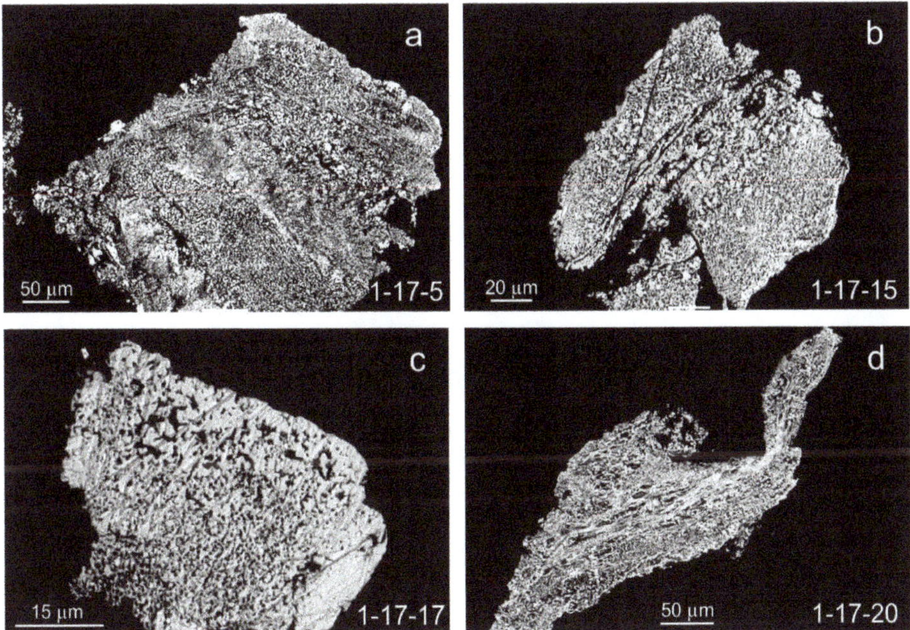

**Figure 8.** Mustard and sponge gold of microporous structure: (**a**,**b**) closely located mustard and sponge gold, (**c**) massive gold with spongy and mustard gold close the center, (**d**) microporous mustard gold with elongated fibers of massive gold in the center. Mounted sections, scanning electron microscope, image in the backscattering electron mode.

Mustard gold of dendritic structure is more widespread (Figure 9). Dendrites are represented in the center by dendritic-branched lumps of gold surrounded by spongy mustard gold (Figure 9a) in close association with iron hydroxides. Sometimes there are looped massive gold grains interspersed with mustard-sponge gold (Figure 9b). Varieties of mustard twiggy gold prevail, the pores of which are mostly hollow (Figure 9c,d).

According to microprobe analyses, in individual grains one can assume the presence of iron oxides (hydroxides) in the pores. Figure 10 shows the internal structure of the enlarged fragments of dendritic mustard gold.

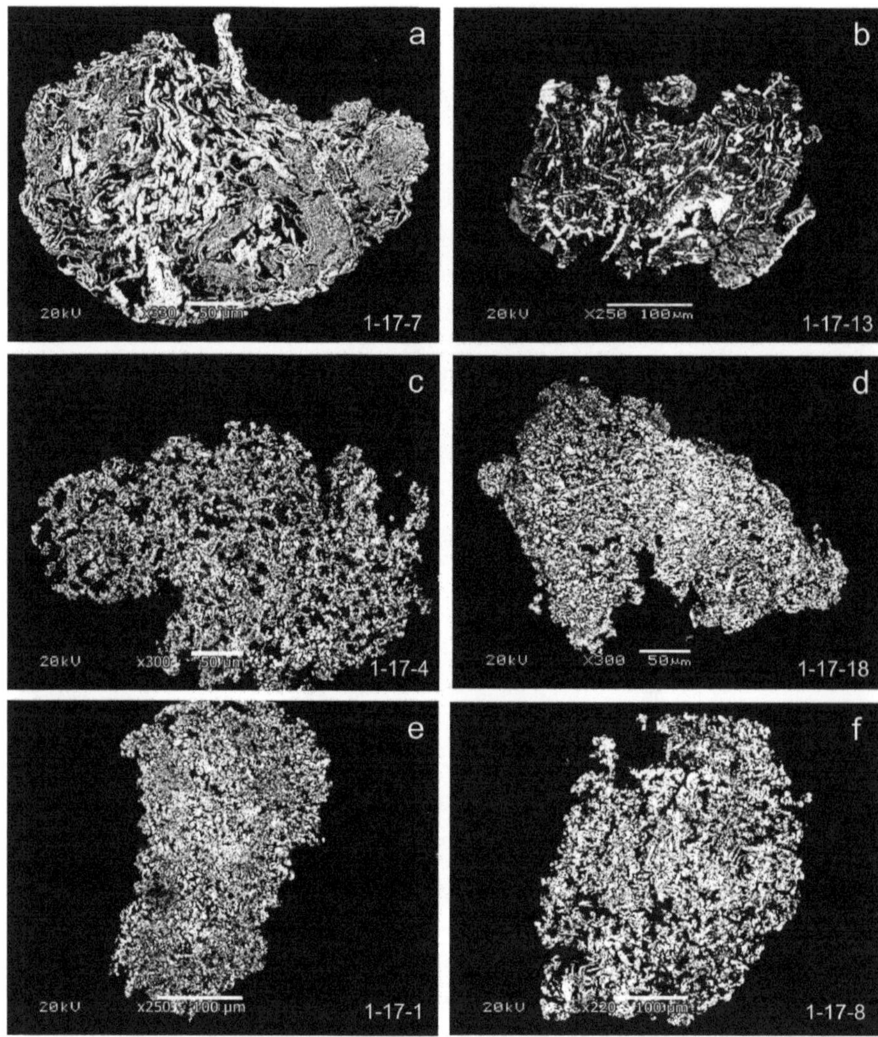

**Figure 9.** Mustard gold of dendritic structure, (**a**,**b**) dendritic gold in the center, meandering massive golds, mustard porous gold along the edges in close association with iron hydroxides, (**c**–**f**) varieties of twiggy gold. Mounted sections, scanning electron microscope, image in the backscattering electron mode.

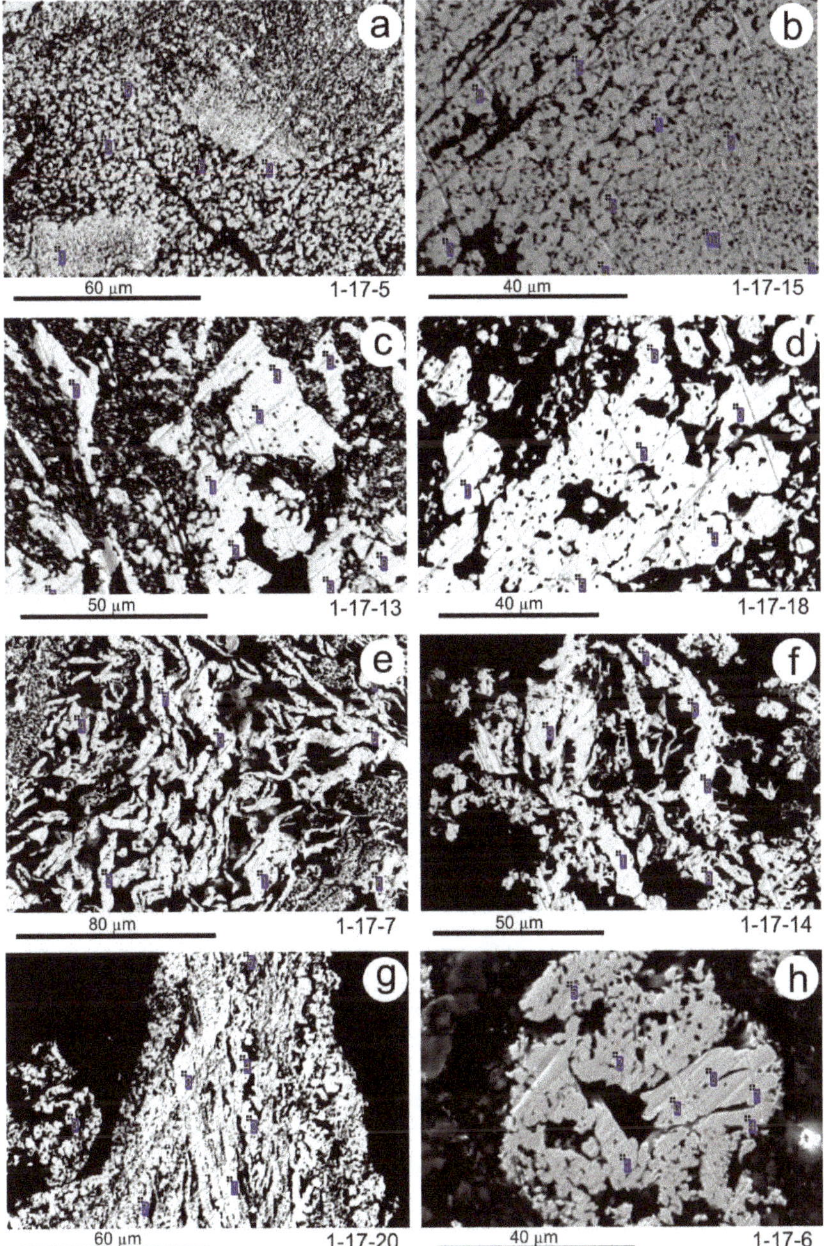

**Figure 10.** The internal structure of mustard-spongy (**a,b,h**) and dendritic mustard (**c–g**) gold. Fragments of the gold shown in Figures 6–8. (**a,b**) brain structure; (**c,d**) massive golds are cemented with microporous mustard gold; (**e,f**) tortuous gold in the cement of microporous mustard-sponge gold; (**g**) streaky-meandering golds in the center are bordered with brain-shaped golds; (**h**) a lump of gold of the brain-structure. Mounted sections, scanning electron microscope, back-scattering electron mode.

Sponge and mustard gold fineness is above >900‰. Among impurities, in addition to silver, Hg (up to 5.48 wt%) and Bi (up to 0.42 wt%) are constantly present, the remaining elements Fe, Zn, Pb, Cu, and Pt are found sporadically (Table 2, Nos. 11–56, Table 5).

**Table 5.** The composition (in wt%) and fineness (in ‰) of the porous gold shown in Figure 11.

| No. | Sample | Sp. | Fe | Cu | Zn | Ag | Au | Hg | Pb | Bi | Total | Fineness |
|---|---|---|---|---|---|---|---|---|---|---|---|---|
| 1 | 1-17-1 | 1 | - | 0.02 | - | 4.79 | 94.29 | 0.15 | - | 0.15 | 99.40 | 949 |
| 2 | 1-17-1 | 2 | 0.06 | - | 0.01 | 1.01 | 96.85 | 0.07 | 0.08 | 0.10 | 98.17 | 986 |
| 3 | 1-17-2 | 1 | 0.05 | 0.02 | - | 8.86 | 89.44 | - | 0.08 | 0.05 | 98.50 | 908 |
| 4 | 1-17-6 | 1 | 0.05 | 0.02 | - | 0.852 | 97.14 | 0.02 | 0.01 | 0.16 | 98.25 | 989 |
| 5 | 1-17-6 | 2 | 0.03 | 0.02 | 0.02 | 0.717 | 99.28 | 0.17 | - | 0.22 | 100.46 | 988 |
| 6 | 1-17-6 | 3 | 0.03 | 0.01 | - | 2.52 | 94.86 | 0.22 | 0.01 | 0.08 | 97.73 | 971 |
| 7 | 1-17-7 | 1 | - | 0.04 | - | 0.752 | 98.68 | 0.09 | 0.03 | 0.15 | 99.74 | 989 |
| 8 | 1-17-7 | 2 | - | 0.02 | - | 0.092 | 99.74 | 0.25 | 0.08 | 0.16 | 100.34 | 994 |
| 9 | 1-17-7 | 3 | 0.05 | 0.01 | 0.01 | 0.194 | 99.6 | 0.26 | 0.04 | 0.26 | 100.42 | 992 |
| 10 | 1-17-7 | 4 | 0.06 | 0.02 | - | 0.627 | 97.98 | 0.14 | - | 0.12 | 98.945 | 990 |
| 11 | 1-17-8 | 1 | - | - | - | 5.17 | 95.24 | - | 0.02 | 0.10 | 100.54 | 947 |
| 12 | 1-17-8 | 2 | - | - | - | 5.67 | 95.96 | 0.07 | 0.04 | 0.09 | 101.84 | 942 |
| 13 | 1-17-8 | 3 | - | 0.02 | - | 6.07 | 94.35 | 0.27 | - | 0.07 | 100.77 | 936 |
| 14 | 1-17-8 | 4 | - | 0.02 | 0.01 | 5.09 | 91.64 | 0.18 | 0.04 | 0.12 | 97.1 | 944 |
| 15 | 1-17-8 | 5 | - | - | - | 4.89 | 95.79 | 0.29 | 0.02 | 0.14 | 101.13 | 947 |
| 16 | 1-17-9 | 1 | - | - | - | 6.04 | 92.1 | 0.27 | - | 0.01 | 98.42 | 936 |
| 17 | 1-17-9 | 2 | 0.01 | 0.01 | - | 3.28 | 96.79 | 0.17 | - | 0.14 | 100.40 | 964 |
| 18 | 1-17-10 | 1 | - | 0.01 | - | 1.07 | 98.75 | 0.02 | 0.11 | 0.16 | 100.11 | 986 |
| 19 | 1-17-10 | 2 | - | 0.01 | - | 1.16 | 99.94 | 0.32 | 0.09 | 0.13 | 101.66 | 983 |
| 20 | 1-17-10 | 3 | - | - | - | 1.19 | 99.93 | 0.10 | - | 0.28 | 101.50 | 985 |
| 21 | 1-17-11 | 1 | - | - | - | 1.42 | 98.56 | 0.24 | 0.02 | 0.25 | 100.49 | 981 |
| 22 | 1-17-11 | 2 | - | 0.01 | 0.012 | 3.99 | 96.46 | 0.19 | - | 0.17 | 100.83 | 957 |
| 23 | 1-17-11 | 3 | - | 0.01 | - | 3.95 | 96.14 | 0.35 | 0.03 | 0.18 | 100.66 | 955 |
| 24 | 1-17-12 | 1 | - | - | - | 2.56 | 97.78 | 0.20 | 0.07 | 0.02 | 100.63 | 972 |
| 25 | 1-17-13 | 1 | - | 0.03 | - | - | 99.62 | - | - | 0.24 | 99.89 | 997 |
| 26 | 1-17-14 | 1 | 0.07 | - | - | 0.79 | 98.82 | 0.1 | 0.05 | 0.22 | 100.04 | 988 |
| 27 | 1-17-15 | 1 | 0.03 | - | 0.03 | 0.30 | 96.62 | 0.30 | 0.10 | 0.24 | 97.62 | 990 |
| 28 | 1-17-17 | 1 | - | - | 0.02 | 0.29 | 99.72 | 0.27 | 0.01 | 0.24 | 100.56 | 992 |
| 29 | 1-17-18 | 1 | - | - | - | 3.87 | 90.87 | 4.63 | - | 0.18 | 99.55 | 913 |
| 30 | 1-17-19 | 1 | 0.04 | - | - | 6.78 | 91.49 | 2.97 | - | 0.066 | 101.35 | 903 |
| 31 | 1-17-21 | 1 | - | - | - | 1.88 | 97.46 | 0.22 | 0.03 | 0.42 | 100.01 | 975 |
| 32 | 1-17-21 | 2 | 0.02 | 0.02 | 0.01 | 4.00 | 93.29 | 0.04 | - | 0.18 | 97.56 | 956 |

Note: Analyzes performed on an X-ray microanalyzer.

All gold varieties are chemically homogeneous and characterized by high fineness of the gold particles without any rims or margins. The distribution patterns of fineness in both types of native gold massive and porous, do not differ essentially. The high-fineness gold (>950‰) makes up 64%, the amount of particles of lower fineness (800–950‰) comes to only 36% of the total. Along with Ag, the porous native gold invariably contains mercury (up to 5.78 wt%) and bismuth, and, less frequently, lead, copper, and iron.

Cumulative diagrams clearly show high concentrations of Pb and Cu in the massive gold, and of Hg in the porous one (Figure 11). The amount of Fe and Bi impurities in both types of supergene gold is similar.

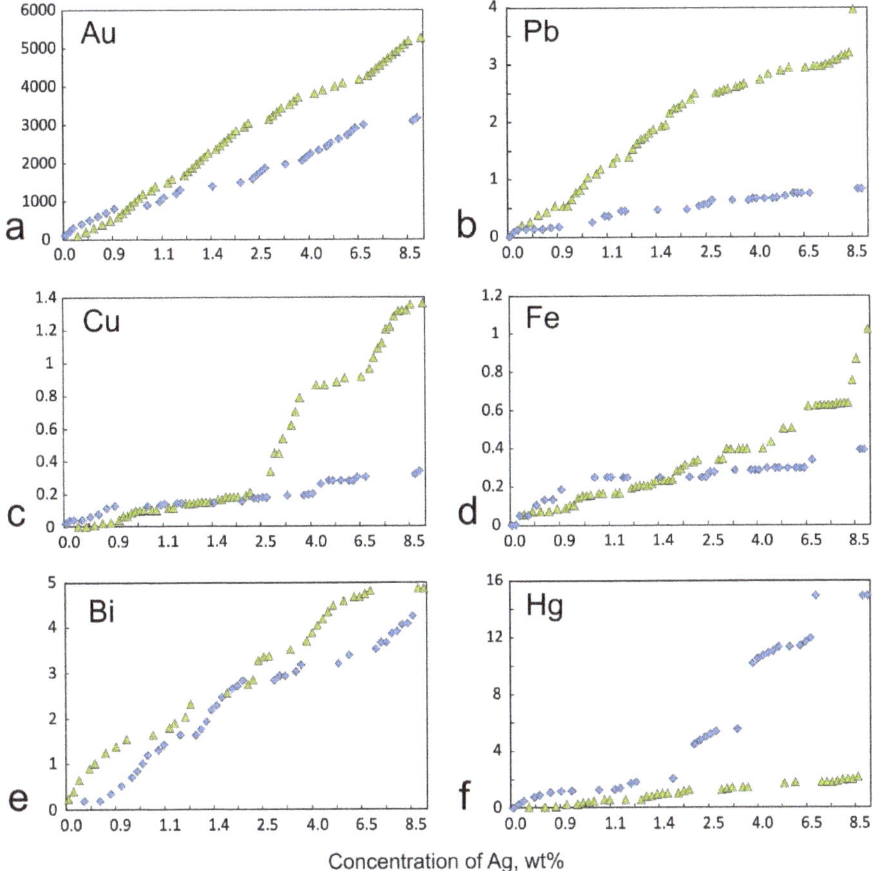



**Figure 11.** Cumulative plots of trace element ((**a**) Au, (**b**) Pb, (**c**) Cu, (**d**) Fe, (**e**) Bi, (**f**) Hg) vs. silver concentrations in massive (green triangle) and porouse (blue rombus) gold of the Khokhoy gold ore field.

## 4. Discussion

Two genetic types of supergene gold, massive and porous, are recognized there.

The first type is represented by gold crystals and irregular mass, with the fineness ranging from 835 to 1000‰. They are closely associated with goethite, siderite, unnamed Fe, Te, and Tl carbonates, Tl tellurites/tellurates and antimonates, as well as avicennite with a Te impurity.

The second type is represented by mustard gold with different internal structure: microporous and dendritic.

Evidentially, mustard gold is characteristic of gold-telluride [12–15] and gold-antimony [16–18] deposits, weathering crust [19–21] and placers [22,23]. Recently, mustard gold, sometimes with an admixture of Tl (up to 6.34 wt%), was found in the Oleninsky Au–Ag deposit (Kola Peninsula, Russia) [24]. In all the cases, mustard gold was developed in the oxidation zone as a result of the decomposition of tellurides, antimonides, sulfides, bismuthides Au (Ag) and low-fineness gold [14]. In the Kuranakh deposit, mustard gold is the product of the decomposition of tellurates [25]. Researchers consider the occurrence of mustard gold a result of hypogenous as well as supergenous processes. The proponents of its hypogenic origin believe that mustard gold, Au tellurates/antimonates, and complex gold oxides were formed from hydrothermal low-temperature solutions with a high oxidation potential [16]. Supergene processes explain the origin of the secondary mustard gold of high fineness of placers in the North-East of Russia [23].

The formation of mustard gold at the Dongping mines (Hebei province, China) has been related to decomposition of calaverite by selective leaching of tellurium while leaving the gold alloy in the cavity formed by the alteration reaction [12,13]. This type of pseudomorphic alteration was also documented by Palache et al. [26]. The occurrence of microporous gold has also been observed under cold climatic conditions, such as at the Aginskoe low-sulfidation ephithermal deposit in Central Kamchatka, Russia [27]. In this deposit, calaverite is the main Au telluride mineral and it has been partially replaced by porous gold. By comparing the textures of microporous gold from this natural occurrence with those obtained experimentally via the dealloying of gold–(silver) tellurides [28–30], Okrugin et al. [31] confirmed that natural microporous gold can form via the replacement of telluride minerals and assessed the role that hydrothermal fluids may play in the formation of microporous gold. The formation of the secondary high-fineness mustard gold from placers in northeast Russia is interpreted to be due to supergene processes. Although no antimonides, nor tellurides, or gold bismuthides were found in the ores of the Khokhoy gold ore field, their presence is assumed by the minerals associated with supergenic gold and the microchemical composition. They are represented by Fe, Mn and Tl tellurates and carbonates, as well as avicennite, Tl antimonates, goethite, limonite, hydrogoethite, and siderite of which the breakdown contributes to the formation of microporous gold particles of spungy habit.

So, what is the formation mechanism of dendrites? Interesting data on the formation of native gold dendrites in epithermal ores are presented by Saunders [32–34]. Author showed a scanning electron microscope (SEM) image of electrum dendrites that appeared to have formed from aggregation of smaller colloid-sized particles. Metallic nanoparticles appear to form from supersaturated solutions and can form dendrites by the self-assembly and aggregation of the nanoparticles. These dendrites are typically an intermediary stage to more traditional crystal formation as the infilling of branches of the dendrites occurs. In epithermal ores, the dendrites of electrum appear to be preserved due to the infilling of other nanoparticles between the branches, such as silica nanoparticles. The 'fractal' electrum dendrites have been observed in many the Tertiary bonanza epithermal ores in northern Nevada, and have been interpreted to be evidence of nanoparticle nucleation and aggregation in ore formation. More recently, similar textures and genetic interpretations have been made from ores from the southeastern USA [33] and Bulgaria [35]. Finally, disseminated electrum nanoparticles have recently been discovered in the epithermal Round Mountain (Nevada) deposit and have been proposed to be precursors for coarser electrum crystals there [36,37].

We believe that this is a more suitable mechanism of formation for the supergene gold of the Khokhoy gold ore field. The colloidal gold transfer is thought to be possible not only in hydrothermal conditions but in supergene zones too [38]. In the oxidation zone of low-sulfidation ores, to which group the Khokhoy ores belong, the role of gold colloids could be significant. Nanoparticles or colloids of gold with nanoparticles of silica (opal, chalcedony) could form dendrites by the self-assembly and aggregation of the nanoparticles. These dendrites were an intermediary stage to the formation of massive gold particles and crystals by the infilling of the dendrite branches. The massive gold type occurs in the most hypsometrically high levels of karst formations of the Khokhoy gold ore field.

The style of mineralization is among the main factors defining the microchemical composition of native gold, particularly the placer gold [39]. In the Khokhoy gold ore field, the main ore minerals associated with primary gold are pyrite, hematite, galena, and chalcosine. Supergene gold occurs in paragenesis with goethite, siderite, oxides, carbonates, tellurates, and antimonates of thallium. The set of trace elements (Cu, Pb, Fe), both in the massive and porous gold, corresponds to main elements of the associated minerals. The constant presence of Bi in both gold types suggests an intrusive source for them. The preferential accumulation of Hg in porous gold may be explained by its friable texture.

The lack of gold placers in the Khokhoy gold ore field may be due to that microaggregates of mustard gold are instable in supergene process because of friable texture, thus having a weak potential for placer formation.

The fact that native gold is unvisible and fine-grained in primary ores and visible in the karst cavities indicates that it grew larger in size in the oxidation zone of karst formations.

## 5. Conclusions

Relationships of massive gold with unnamed tellurates and carbonates of thallium and with avicennite ($Te_2O_3$) are first described for the Khokhoy gold ore field. Along with massive gold closely associated with Tl minerals, there are abundant porous particles, the so called spungy and mustard gold.

The research has shown that the Khokhoy gold ore field, according to its mineralogical and geochemical features, should be classified as a deposit of Au–Te–Sb–Tl mineral composition localized in karst cavities. A typical gold ore deposit in karst cavities is the Kuranakh deposit of Central Aldan district of South Yakutia [6], with which our data is being compared. It should be noted that in the Kuranakh deposit one of the typomorphic geochemical elements is thallium, but its mineral form has not been identified. In the Khokhoy ore field, a diverse spectrum of thallium minerals is attested. This fact makes closer to Carlin-type gold deposits in the West of the USA [1–5], Alshar epithermal Au–As–Hg–Tl deposit in Macedonia [40,41], Vorontsovskoe deposit in the Urals, Russia [7–9,42] etc. The main difference between the Khokhoy gold ore field and these deposits at this stage of research is the absence of As and Hg minerals—realgar, auripigment, cinnabar. In addition to thallium minerals, Sb and Te minerals (weissbergite, antimonite, berthierite, and unnamed antimonates and tellurates of thallium) are widespread in the ores.

Supergene gold in karst cavities has specific features: (1) morphology—massive (gold crystals and cloddy particles of monolithic structure) and porous (microporous and dendritic); (2) associated minerals—goethite, limonite, avicennite, hydrogoethite, and siderite, Fe, Mn and Tl tellurates and carbonates, Tl antimonates; (3) persistently high fineness and chemically homogeneous; (4) microchemical elements Hg, Bi, Fe. The obtained characteristics made it possible to prove the existence of two genetic types of supergene gold. Mustard microporous gold is the result of the decomposition of the associated goethite minerals, Tl oxides, tellurates, carbonates and antimonates Fe, Mn and Tl containing microinclusions of gold. Massive gold and dendrites are newly formed. The decomposition, remobilization, and reprecipitation of residual gold nanoparticles and their aggregation led to the formation of dendrites, and with further crystal growth and filling of pores, to gold of massive morphology.

The supergene gold of the Khokhoy gold ore field is comparable in its typomorphic characteristics to that of the Kuranakh deposit, as well as with to that of Carlin-type gold deposits in the West of the USA, to that of the Alshar epithermal deposit in Macedonia, to that of the Vorontsovskoe deposit in the Urals, Russia, and also have similarities with those of sponge and mustard gold of gold-telluride and gold-antimony deposits, weathering crusts and placers.

However, at the same time, the gold has certain uniqueness, namely the paragenesis with thallium minerals. To date, no such relationship has been cited in the literature.

**Author Contributions:** The idea of the research belongs to G.S.A., who has then designed it, analyzed the results and wrote the paper. L.A.K. was responsible for the fieldwork in the auriferous zone; she also supported the research with valuable ideas. V.N.K. was also responsible for the fieldwork in the auriferous zone; moreover she made the mineralgraphic description and took the photographs of the polished sections. All authors have read and agreed to the published version of the manuscript.

**Funding:** The research was carried out under the governmental project of Diamond and Precious Metal Geology Institute, SB RAS, which was financially supported by the Ministry of science and higher education of the Russian Federation, the project number is 0381-2019-0004 and Russian Foundation for Basic Reseach, the project number is 18-45-140045.

**Acknowledgments:** We would like to thank N. Khristoforova, S. Popova and S. Karpova for their help with an enormous part of this research, analyzing the samples with scanning electronic microscope and performing X-ray spectral analysis (Diamond and Precious Metal Geology Institute, SB RAS). We are also very grateful to E. Sokolov for the organization of the fieldwork at the auriferous field (JSC 'Yakutskgeologiya'). Authors thanks anonymous reviewers for their comments and constructive suggestions.

**Conflicts of Interest:** We declare no conflict of interest.

## References


1. Hofstra, A.H.; Cline, J.S. Characteristics and models for Carlin-type gold deposits. *Econ. Geol.* **2000**, *13*, 163–220.
2. Cline, J.S.; Hofstra, A.H.; Muntean, J.L.; Tosdal, R.M.; Kenneth, K.A. Carlin-Type Gold Deposits in Nevada: Critical Geologic Characteristics and Viable Models. *Econ. Geol.* **2005**, *100*, 451–484.
3. Volkov, A.V.; Sidorov, A.A. The geological-genetic model of Carlin type gold deposits. *Litosphere* **2016**, *6*, 145–165. (In Russian)
4. Zhong, H.R.; Chao, S.W.; Wu, B.X.; Zhi, T.G.; Hofstra, A.H. Geology and geochemistry of Carlin-type gold deposits in China. *Miner. Depos.* **2002**, *37*, 378–392. [CrossRef]
5. Xia, Y.; Su, W.; Zhang, Z.; Liu, J. Geochemistry and Metallogenic Model of Carlin-Type Gold Deposits in Southwest Guizhou Province, China. In *Geochemistry Earth's System Processes*; INTECH: London, UK, 2012.
6. Rodionov, S.M.; Fredericksen, R.S.; Berdnikov, N.V.; Yakubchuk, A.S. The Kuranakh epithermal gold deposit (Aldan Shield, East Russia). *Ore Geol. Rev.* **2014**, *59*, 55–65. [CrossRef]
7. Sazonov, V.N.; Murzin, V.V.; Grigor'ev, N.A. Vorontsovsk gold deposit: An example of carlin-type mineralization in the Urals, Russia. *Geol. Ore Depos.* **1998**, *40*, 139–151.
8. Murzin, V.V.; Naumov, E.A.; Azovskova, O.B.; Varlamov, D.A.; Rovnushkin, M.Y.; Pirajno, F. The Vorontsovskoe Au–Hg–As ore deposit (Northern Urals, Russia): Geological setting, ore mineralogy, geochemistry, geochronology and genetic model. *Ore Geol. Rev.* **2017**, *85*, 271–298. [CrossRef]
9. Vikentyev, I.V.; Tyukova, E.E.; Vikent'eva, O.V.; Chugaev, A.V.; Dubinina, E.O.; Prokofiev, V.Y.; Murzin, V.V. Vorontsovka Carlin-style gold deposit in the North Urals: Mineralogy, fluid inclusion and isotope data for genetic model. *Chem. Geol.* **2019**, *508*, 144–166. [CrossRef]
10. Anisimova, G.S.; Kondratieva, L.A.; Sokolov, E.P.; Kardashevskaya, V.N. Gold mineralization of the Lebedinsky and Kuranakh types in the Verkhneamginsky district (South Yakutia). *Otech. Geol.* **2018**, *5*, 3–13. (In Russian)
11. Lindgren, W. *Mineral Deposits*; McGraw-Hill Book Company, Inc.: New York, NY, USA; London, UK, 1933; p. 930.
12. Petersen, S.B.; Makovicky, E.; Li, J.L.; Rose-Hansen, J. Mustard gold from the Dongping Au–Te deposit, Hebei province, People's Republic of China. *Neues Jahrb. Mineral. Mh.* **1999**, *8*, 337–357.
13. Li, J.; Makovicky, E. New studies on mustard gold from the Dongping Mines, Hebei Province, China: The tellurian, plumbian, manganoan and mixed varieties. *Neues Jahrb. Mineral. Abh.* **2001**, *176*, 269–297.
14. Tolstykh, N.D.; Palyanova, G.A.; Bobrova, O.V.; Sidorov, E.G. Mustard Gold of the Gaching Ore Deposit (Maletoyvayam Ore Field, Kamchatka, Russia). *Minerals* **2019**, *9*, 489–506. [CrossRef]
15. Tolstykh, N.; Vymazalova, A.; Tuhy, M.; Shapovalova, M. Conditions of formation of Au–Se–Te mineralization in the Gaching ore occurrence (Maletoyvayam ore field), Kamchatka, Russia. *Mineral. Mag.* **2018**, *82*, 649–674. [CrossRef]
16. Gamyanin, G.N.; Nekrasov, I.; Zhdanov, J.J.; Leskova, I.V. Auroantimonate—A new natural compound of gold. *Dokl. Akad. Nauk SSSR* **1988**, *301*, 947–950. (In Russian)
17. Amuzinsky, V.A.; Anisimova, G.S.; Zhdanov, Y.Y. *Sarylakhskoe and Sentachanskoe Gold Antimony Deposits*; Nauka/Interperiodika: Moscow, Russia, 2001; 218p. (In Russian)
18. Zacharias, J.; Nemec, M. Gold to aurostibite transformation and formation of Au–Ag–Sb phases: The Krásná Hora deposit, Czech Republic. *Mineral. Mag.* **2017**, *81*, 987–999. [CrossRef]
19. Kalinin, Y.A.; Kovalev, K.R.; Naumov, E.A.; Kirillov, M.V. Gold in the weathering crust the Suzdal' deposit (Kazakhstan). *Russ. Geol. Geophys.* **2009**, *50*, 174–187. [CrossRef]
20. Kalinin, Y.A.; Palyanova, G.A.; Bortnikov, N.S.; Naumov, E.A.; Kovalev, K.R. Aggregation and Differentiation of Gold and Silver during the Formation of the Gold-Bearing Weathering Crusts (on the Example of Kazakhstan Deposits). *Dokl. Earth Sci.* **2018**, *482*, 1193–1198. [CrossRef]

21. Kalinin, Y.A.; Palyanova, G.A.; Naumov, E.A.; Kovalev, K.R.; Pirajno, F. Supergene remobilization of Au in Au-bearing regolith related to orogenic deposits: A case study from Kazakhstan. *Ore Geol. Rev.* **2019**, *109*, 358–369. [CrossRef]

22. Reith, F.; Stewart, L.; Wakelin, S.A. Supergene Au transformation: Secondary and nano-particulate Au from southern New Zealand. *Chem. Geol.* **2012**, *320–321*, 32–45. [CrossRef]

23. Litvinenko, I.S.; Shilina, L.A. Hypergene gold neomineralization in placer deposits of Nizhne-Myakitsky ore-placer field, North-East Russia. *J. Ores Met.* **2017**, *1*, 75–90. (In Russian)

24. Kalinin, A.A.; Savchenko, Y.E.; Selivanova, E.A. Mustard Gold in the Oleninskoe Gold Deposit, Kolmozero–Voronya Greenstone Belt, Kola Peninsula, Russia. *Minerals* **2019**, *9*, 786–800. [CrossRef]

25. Yablokova, S. About a new morphological variety of gold and its origin. *Dokl. Akad. Nauk SSSR* **1972**, *205*, 936–939. (In Russian)

26. Palache, C.; Berman, H.; Frondel, C. *Dana's System of Mineralogy I*; Wiley: New York, NY, USA, 1944.

27. Andreeva, E.D.; Matsueda, H.; Okrugin, V.M.; Takahashi, R.; Ono, S. Au–Ag–Te mineralization of the low-sulfidation epithermal Aginskoe deposit, Central Kamchatka, Russia. *Resour. Geol.* **2013**, *63*, 337–349. [CrossRef]

28. Xu, W.; Zhao, J.; Brugger, J.; Chen, G.; Pring, A. Mechanism of mineral transformations in krennerite, Au3AgTe8, under hydrothermal conditions. *Am. Miner.* **2013**, *98*, 2086–2095. [CrossRef]

29. Zhao, J.; Xia, F.; Pring, A.; Brugger, J.; Grundler, P.V.; Chen, G. A novel pre-treatment of calaverite by hydrothermal mineral replacement reactions. *Miner. Eng.* **2010**, *23*, 451–453. [CrossRef]

30. Zhao, J.; Brugger, J.; Xia, F.; Ngothai, Y.; Chen, G.; Pring, A. Dissolution-reprecipitation vs. solid-state diffusion: Mechanism of mineral transformations in sylvanite, $(AuAg)_2Te_4$, under hydrothermal conditions. *Am. Miner.* **2013**, *98*, 19–32. [CrossRef]

31. Okrugin, V.M.; Andreeva, E.; Etschmann, B.; Pring, A.; Li, K.; Zhao, J.; Griffiths, G.; Lumpkin, G.R.; Triani, G.; Brugger, J. Microporous gold: Comparison of textures from Nature and experiments. *Am. Miner.* **2014**, *99*, 1171–1174. [CrossRef]

32. Saunders, J.A.; Schoenly, P.A. Boiling, colloid nucleation and aggregation, and the genesis of bonanza gold mineralization at the Sleeper Deposit, Nevada. *Miner. Depos.* **1995**, *30*, 199–211. [CrossRef]

33. Saunders, J.A. Textural evidence of episodic introduction of metallic nanoparticles into bonanza epithermal ores. *Minerals* **2012**, *2*, 228–243. [CrossRef]

34. Saunders, J.A.; Burke, M. Formation and aggregation of gold (electrum) nanoparticles in epithermal ores. *Minerals* **2017**, *7*, 163–173. [CrossRef]

35. Marinova, I.; Ganev, V.; Titorenkova, R. Colloidal origin of colloform-banded textures in the Paleogene low-sulfidation Khan Krum gold deposit, SE Bulgaria. *Miner. Depos.* **2014**, *49*, 49–74. [CrossRef]

36. Burke, M. An Electron Microscopy Investigation of Gold and Associated Minerals from Round Mountain, Nevada. Master's Thesis, Miami University, Oxford, OH, USA, 2017.

37. Burke, M.; Rakovan, J.; Krekeler, M.P.S. A study by electron microscopy of gold and associated minerals from Round Mountain, Nevada. *Ore Geol. Rev.* **2017**, *91*, 708–717. [CrossRef]

38. Petrovskaya, N.V. *Native Gold*; Nauka: Moscow, Russian, 1973; 347p. (In Russian)

39. Chapman, R.J.; Mortensen, J.K.; LeBarge, W.P. Styles of lode gold mineralization contributing to the placers of the Indian River and Black Hills Creek, Yukon Territory, Canada as deduced from microchemical characterization of placer gold grains. *Miner. Depos.* **2011**, *46*, 881–903. [CrossRef]

40. Volkov, A.V.; Serafimovski, T.; Kochneva, N.T.; Tomson, I.N.; Tasev, G. The Alshar epithermal Au–As–Sb–Tl deposit, southern Macedonia. *Geol. Ore Depos.* **2006**, *48*, 175–192. [CrossRef]

41. Palinkaš, S.S.; Hofstra, A.H.; Percival, T.J.; Šoštari'c, S.B.; Palinkaš, L.; Bermanec, V.; Pecskay, Z.; Boev, B. Comparison of the Allchar Au–As–Sb–Tl Deposit, Republic of Macedonia, with Carlin-Type Gold Deposits. *Rev. Econ. Geol.* **2018**, *20*, 335–363.

42. Kasatkin, A.V.; Makovicky, E.; Plášil, J.; Škoda, R.; Agakhanov, A.A.; Karpenko, V.Y.; Nestola, F. Tsygankoite, $Mn_8Tl_8Hg_2(Sb_{21}Pb_2Tl)_{\Sigma24}S_{48}$, a New Sulfosalt from the Vorontsovskoe Gold Deposit, Northern Urals, Russia. *Minerals* **2018**, *8*, 218. [CrossRef]

Article

# Formation of Au-Bearing Antigorite Serpentinites and Magnetite Ores at the Massif of Ophiolite Ultramafic Rocks: Thermodynamic Modeling

Valery Murzin [1], Konstantin Chudnenko [2], Galina Palyanova [3,4,]*and Dmitry Varlamov [5]

[1]  A.N. Zavaritsky Institute of Geology and Geochemistry, Ural Branch of Russian Academy of Sciences, Akademika Vonsovskogo str., 15, 620016 Ekaterinburg, Russia; murzin@igg.uran.ru

[2]  A.P. Vinogradov Institute of Geochemistry, Siberian Branch of Russian Academy of Sciences, Favorskogo str., 1a, 664033 Irkutsk, Russia; chud@igc.irk.ru

[3]  V.S. Sobolev Institute of Geology and Mineralogy, Siberian Branch of Russian Academy of Sciences, Akademika Koptyuga pr., 3, 630090 Novosibirsk, Russia

[4]  Department of Geology and Geophysics, Novosibirsk State University, Pirogova str., 2, 630090 Novosibirsk, Russia

[5]  Institute of Experimental Mineralogy, Russian Academy of Sciences, 142432 Chernogolovka, Moscow Region, Russia; dima@iem.ac.ru

*  Correspondence: palyan@igm.nsc.ru

Received: 10 October 2019; Accepted: 2 December 2019; Published: 5 December 2019

**Abstract:** We constructed thermodynamic models of the formation of two types of gold-ore mineralization at the Kagan ultramafic massif in the Southern Urals (Russia). The first type of gold-mineralization is widely spread at the massif in the tectonic zones of schistose serpentinites containing typically $\leq 0.1$ ppm Au. The second type of gold-ore mineralization is represented by veined massive, streaky and impregnated magnetite ores in contact with serpentinites. It contains to 5 vol.% sulfides and 0.2–1.2 ppm Au. Our thermodynamic calculations explain the formation of two types of gold-ore mineralization in the bedrocks of ultramafic massifs. Metamorphic water, which is the result of the dehydration of early serpentinites (middle Riphean) during high-temperature regional metamorphism (700 °C, 10 kbar) (late Precambrian), is considered as the source of ore-bearing fluid in the models. The metasomatic interaction of metamorphic fluid with serpentinites is responsible for the gold-poor mineralization of the 1st type at T = 450–250 °C and P = 2.5–0.5 kbar. The hydrothermal gold-rich mineralization of the 2nd type was formed during mixing of metamorphic and meteoric fluids at T = 500–400 °C and P = 2–3 kbar and discharge of mixed fluid in the open space of cracks in serpentinites. The model calculations showed that the dominant forms of gold transport in fluids with pH = 3–5 are $AuCl_2{}^-$ complexes ($\geq$450 °C) and, as the temperature decreases, $AuHS^0$, or $AuOH^0$. Mineral associations obtained in model calculations are in general similar to the observed natural types of gold mineralization.

**Keywords:** Kagan ultramafic massif; Southern Urals; antigorite serpentinites; magnetite veins; gold mineralization; native gold; thermodynamic modeling; fluid regime

---

## 1. Introduction

The Urals is one of the most important gold provinces in the Russia and in the world [1,2]. By the end of the 20th century, the Urals was one of the five largest regions in the world in gold mining. By that time, about 500 gold deposits of various scales were known [3]. Nowadays, the main proportion of gold is extracted as an accessory component from volcanogenic massive sulfide deposits localized in the Tagilo–Magnitogorskian zone (Figure 1A). These deposits are associated with the upper mantle basaltoid magmatism. The largest gold-ore deposits (Kochkarskoe, Berezovskoe, Svetlinskoe, etc.)

belong to the quartz veined and sulfide disseminated types, occuring within the East Uralian zone, and are related to the products of crustal granitoid magmatism [4]. Haloes of gold-bearing placers are widespread there owing to primary deposits. Massifs of ophiolite mafic-ultramafic rock complexes occur in the Main Uralian Fault zone (MUF). The long-term occurrence of tectonic, metamorphic, and metasomatic processes resulted in a deep transformation of these complexes and the origin of various types of gold-bearing rocks: antigorite serpentinites, rodingites, magnetite-chlorite-carbonate, and other varieties [5–7].

Deposits and ore occurrences in ultramafic massifs (Karabash, Verh-Neivinsk, Talovsk, Kagan, etc.) are scarce. They contain small amounts of the gold, but industrial placers of these occuring metals are widespread throughout the massif. Native gold in these placers is represented by Au–Cu intermetallides, as well as Au–Ag–Cu, Au–Ag, and Au–Ag–Hg solid solutions. Placer gold is larger in size and is frequently intergrown with magnetite, chlorite, pyroxene, and serpentine. The genetic aspects of formation of gold mineralization in ultramafic rocks are problematic and their discussion has started quiet recently [8–13]. This work continues the discussion.

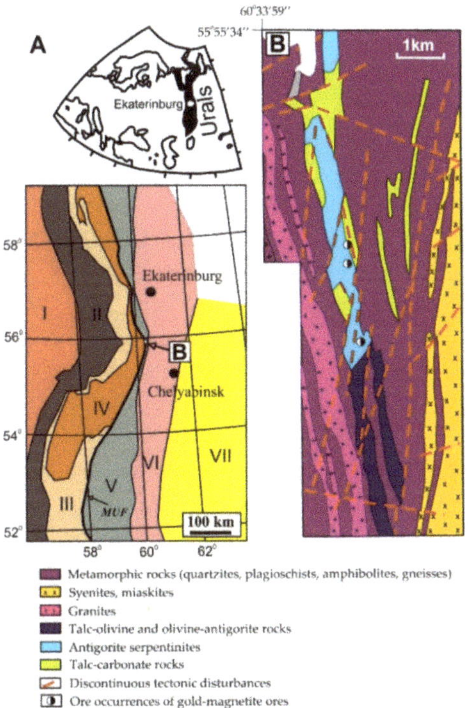


**Figure 1.** Location of the Kagan massif on the scheme for tectonic structure of the Urals from [4] (**A**); and its geologic structure from [14] (**B**). A—Tectonic zones of the Urals; I—East European platform; II—Preuralian foredeep; III—West Uralian zone; IV—Central Uralian zone; V—Tagilo–Magnitogorskian zone; VI—East Uralian zone; VII—Transuralian zone. MUF—Main Uralian Fault.

The object of our study is gold-bearing serpentinites and their sulfide–magnetite ores at the Kagan ultramafic massif in the Southern Urals (Russia) (Figure 1A). The massif occurs in the Main fault zone within the Sysert–Vishnevogorsk metamorphic complex. Gold-bearing ultramafic rocks and sulfide–magnetite ores of the Kagan massif occur in antigorite serpentinites and were referred to a gold-productive serpentinite (antigorite) metasomatic formation [7]. Gold-magnetite ore occurrences were exploited in the middle of the 20th century by two small mines.

Based on the results of the previous study [9] of altered rocks from the Kagan massif and gold–sulfide–magnetite ores, a suggestion was made that the metamorphic fluid, which mobilized metals from ultramafic rocks, could be the source of gold and other metals (Fe, Cu, Ag). The aim of this work is to develop physicochemical models of the formation of gold mineralization in the processes of metamorphogenic–metasomatic transformation of ultramafic substance, using computer thermodynamic modeling with the help of the Selektor-C software and available geological, geochemical, mineralogical, and other data on the object under study. Further, in the following section we provide a review of literature data on the studied model object, including published and unpublished results obtained by us. These data were used to construct a thermodynamic model of the object under study.

## 2. Geological and Metallogenic Overview

### 2.1. Geological Setting and Types of Gold Mineralization of the Kagan Massif

Premetamorphic ultramafic rocks in the Vishnevogorsk and other metamorphic complexes of the Middle (Sysert) and South (Ilmenogorsk) Urals are distinquished as part of the Riphean ophiolite association developed on the Archean–early Proterozoic basement [14]. Ultramafic and volcano–sedimentary rocks of ophiolite association were metamorphosed under the conditions of subgranulite to upper green–schist facies. A several stages of metamorphic and metasomatic alterations of ultramafic rocks are identified (Table 1). Gold and anthophyllite mineralization occur with them [9,15,16].

**Table 1.** Stages of metamorphism and metasomatism of ultramafic rocks in the Sysert–Ilmenogorsk metamorphic complex [14] and the types of mineralization associated with them [9,15,16].

| Stages of Metamorphism and Metasomatism, Age | Composition of Metamorphic and Metasomatic Rocks (From High to Low Temperature) | Type of Mineralization |
|---|---|---|
| Regional zonal dynamothermal metamorphism, Late Precambrian | Enstatite–olivine<br>Talc–olivine<br>Antigorite–olivine<br><br>Antigorite | Gold–sulfide–magnetite, Gold–antigorite |
| Regional silicic acid metasomatism, $O_2$–$S_1$<br>Local silicic acid metasomatism, $P_3$–T | Tremolite–chlorite<br>Enstatite<br>Anthophyllite, tremolite–anthophyllite<br>enstatite–anthophyllite, talc–carbonate–anthophyllite<br>chlorite-biotite, chlorite–actinolite | Gold–sulfide<br>Anthophyllite–asbest |

The Kagan massif is a concordant tabular body 400–600 m thick and 16 km long. It occurs in the plagioschists and amphibolites of the middle Riphean in the northwestern wing of the Vishnevogorsk anticline (Figure 1B). To the west of the massif, among metamorphic rocks, there are dikes of plagioclase and two-feldspar granites. In the east, metamorphic rocks make contact with the Vishnevogorsk syenite–miaskite intrusion in the Ilmenogorsk alkaline complex. The metamorphism of the rocks of the Kagan massif is associated with the late Precambrian stage and is manifested in the zoning of the massif. Talc–olivine and olivine–antigorite rocks are developed in the southern part of the massif; to the north they are replaced by antigorite serpentinite with the sites of olivine–antigorite rocks and talc–carbonate rocks (Figure 1B). Silica–acid metasomatism in the Kagan massif is weakly expressed and is manifested in the development of anthophyllite in talc–olivine rocks and talc in olivine–antigorite and antigorite rocks. The periphery parts of the Kagan massif in its narrow zones are transformed into tremolite–anthophyllite rocks [14].

Gold-ore mineralization occurs among antigorite serpentinites in the central and northern parts of the Kagan massif. Its two types are distinguished.

The first type of gold-ore mineralization is widely spread at the massif in the tectonic zones of schistose serpentinites. During the prochispecting and mapping works for gold in 1991, it was revealed that schistose serpentinites of the Kagan massif, tested on drill holes, pits, and ditches, have low contents of gold, typically ≤ 0.1 ppm. Nevertheless, crushed samples frequently contain grains of native gold with a size of 0.05–0.1 mm, less often 0.2–0.35 mm. The mechanisms of enlargement of native gold particles in natural migration processes (dissolution, transfer, and deposition) of noble metal are interpreted in many works [17–21].

The second type of gold-ore mineralization is represented by veined massive, streaky and impregnated magnetite ores (Figure 2) containing to 5 vol.% sulfides. Gold–sulfide–magnetite ores occur in the tectonic zone stretching along the eastern contact of the northern part of the massif for 2 km (Figure 1B). Magnetite lenses are to 5–6 m long and to 0.2 m thick and arranged in chains along the tectonic zone. Assay analysis shows that Au content in such ores is 0.2–1.2 ppm, and it drastically increases in the areas with visible gold. Chemical–spectral analysis also showed the presence of platinum group elements (PGE) (in ppm): to 0.77 Pd, to 0.02 Pt, and to 0.01–0.02 Rh, Ir, Os, and Ru [22]. PGE anomalies are common for many serpentinites [23].

**Figure 2.** Magnetite ores from Kagan deposit: (**A**) banded ore composed of magnetite and serpentinite bands (brown); (**B**) disseminated ore with oxidation products of copper sulfides (green); (**C**) banded massive ore with bands of serpentinite; (**D**) massive ore (black) in contact with serpentinite.

## 2.2. Mineralogy of Gold-Bearing Antigorite Serpentinites and Gold–Sulfide–Magnetite Ores

Serpentinites containing native gold are composed of a lamellar aggregate of antigorite with a small amount of talc, chlorite, tremolite, metamorphic olivin (Mg/(Mg + Fe) = 0.96–0.98) as well as relict olivine (Mg/(Mg + Fe) = 0.92–0.93), clinopyroxene, and loop-shaped serpentine. Mg/(Mg + Fe) of relic olivine corresponds to the typical composition of accessory olivine in dunites and harzburgite. Antigorite contains 6.4–12.3 wt % FeO. Its laths are zoned, with its edges richer in iron than the central parts. Talc and chlorite replace antigorite. In serpentinites one can observe scattered grains of Cr–spinel up to 2 mm in size, tiny crystals of magnetite, and rare particles of pentlandite (to 0.5 mm). Cr–spinel is replaced by Cr–magnetite and Cr–clinochlore. Magnetite contains 0.9–1.5 wt % MgO and 0.4–2.0 wt % $Cr_2O_3$. It is typical for most other serpentinites [24–26].

Tremolite (Mg/(Mg + Fe) = 0.95–0.96) is likely synchronous to antigorite. It develops after clinopyroxene and is actively replaced through the net of thin veinlets by late serpentinite and chlorite of clinochlore–pennine series.

The grains of native gold from antigorite serpentinites have flattened, less often, isometric shapes and sizes up to 0.35 mm. Their chemical composition is given in Table 2 and in Figure 3A. Native gold is characterized by content up to 22.23 wt % Ag and small amounts of 0.2–0.4 wt % Cu. Individual grains contain 0.29–0.78 wt % Cu (average 0.42), one grain 6.5 wt % Cu and 0–1.7 (average 0.17) wt % Hg. The fineness of native gold varies from 772 to 996‰.

**Table 2.** Chemical composition and fineness of native gold from zones of schistose serpentinites of Kagan massif.

| No. Sample/ Grain | Au wt % | Ag wt % | Cu wt % | Hg wt % | Total, wt % | Fineness, ‰ * |
|---|---|---|---|---|---|---|
| K-10/1 | 99.06 | 0.04 | 0.63 | 0.0 | 99.73 | 993 |
| K-10/2 | 98.98 | 0.09 | 0.59 | 0.0 | 99.66 | 993 |
| K-10/10 | 92.21 | 0.43 | 6.52 | 0.0 | 99.16 | 930 |
| K-5/11 | 85.86 | 13.93 | 0.34 | 0.0 | 100.13 | 857 |
| K-6/12 | 77.41 | 22.23 | 0.60 | 0.0 | 100.24 | 772 |
| C-437/13 | 92.55 | 5.44 | 0.21 | 1.72 | 99.92 | 926 |
| Sp-207/14 | 92.17 | 8.13 | 0.26 | 0.11 | 100.67 | 916 |
| Sp-207/15 | 95.10 | 4.23 | 0.36 | 0.03 | 99.72 | 954 |
| Sp-207/16 | 99.13 | 0.92 | 0.31 | 0.0 | 100.36 | 988 |
| Sp-207/17 | 97.70 | 2.43 | 0.78 | 0.0 | 100.91 | 968 |
| Sp-201/18 | 90.74 | 9.66 | 0.29 | 0.0 | 100.69 | 901 |
| Sp-212/19 | 99.90 | 0.16 | 0.47 | 0.0 | 100.53 | 994 |
| Sp-212/20 | 89.03 | 9.17 | 0.33 | 0.53 | 99.06 | 899 |
| Sp-212/21 | 99.14 | 0.0 | 0.41 | 0.0 | 99.55 | 996 |
| Sp-212/22 | 91.96 | 8.18 | 0.29 | 0.0 | 100.43 | 916 |
| Sp-212/23 | 97.72 | 1.74 | 0.78 | 0.0 | 100.24 | 975 |
| Sp-328/1 | 94.05 | 5.52 | 0.42 | 0.0 | 99.99 | 941 |
| Sp-328/25 | 92.87 | 6.79 | 0.41 | 0.0 | 100.07 | 928 |
| Sp-328/26 | 95.70 | 4.11 | 0.52 | 0.0 | 100.33 | 954 |

Note. * $Au \times 1000/(Au + Ag + Cu + Hg)$. Composition was determined in the Institute of Geology and Geochemistry, Ural Branch of RAS on the electron-probe micro analyzer JXA-5a (analyst V.A. Vilisov). Analytical conditions: U = 20 kV, standards are pure metals and HgTe. The average measurement accuracy is 0.3 wt % for Au, Ag, Cu, Pd, and 1 wt % for Hg (which is associated with the possible migration of mercury from the analysis zone).

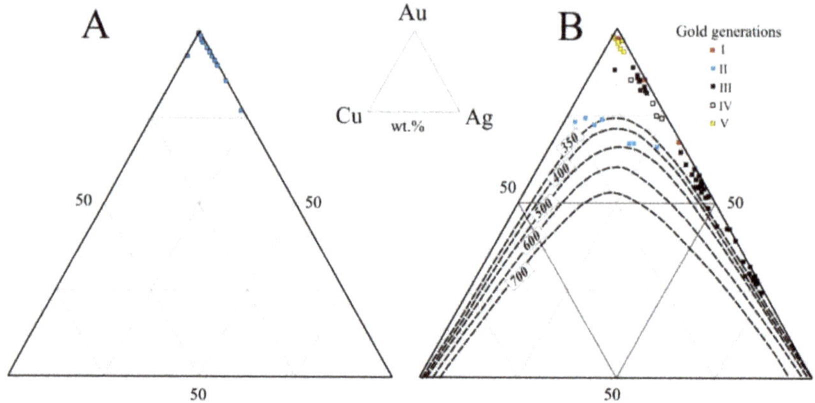


**Figure 3.** Composition of native gold from Kagan massif on the Au–Ag–Cu diagram: (**A**) from schistose antigorite serpentinites; (**B**) various generations of native gold from magnetite ores. The diagram contains the isotherms (°C) of ternary solid solution from experimental data [27].

Disseminated and massive ores are composed of polygonally grained aggregates of magnetite with grains smaller than 1 mm and a low content of sulfides (no less than 2–5 vol.%). Accumulations of magnetite are cemented with antigorite and minor talc (Mg/(Mg + Fe) = 0.97–0.98), chlorite (Mg/(Mg + Fe) = 0.91–0.95), and tremolite (Mg/(Mg + Fe) = 0.95–0.96). Talc is developed after serpentine (Figure 4A).

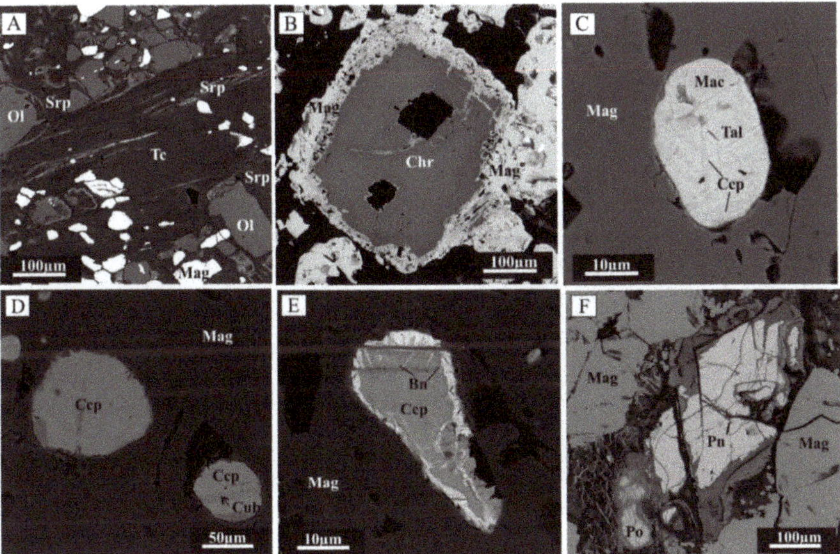


**Figure 4.** Relationships of minerals in gold–sulfide–magnetite ores: (**A**) development of talc (Tc) after relict olivine (Ol) and serpentine (Srp); (**B**) rim of magnetite (Mag) replacing a grain of relict Cr–spinel (Chr); (**C,D**) sulfides of early paragenesis: (**C**) grain of talnakhite (Tal) with plates of chalcopyrite (Ccp) and mackinawite (Mac) inclusion in magnetite, (**D**) chalcopyrite inclusions (one of them with plates of cubanite (Cub)) in magnetite; (**E,F**) sulfides of late paragenesis: (**E**) inclusion composed of chalcopyrite and bornite (Bn) in magnetite; (**F**) pentlandite (Pn) and pyrrhotite (Po) in magnetite. Here and in Figure 5 are electronic-microscopic images in the regime of back-scattered electrons (BSE).

Among aggregates of antigorite one can also observe Cr–spinel (Cr/(Cr + Al) = 0.56–0.67), olivine (Mg/(Mg + Fe) = 0.92–0.98) and loop-shaped serpentine.

The whole mass of magnetite ore is penetrated by a net of thin veinlets of chrysotile (Mg/(Mg + Fe) = 0.94–0.99), chlorite (Mg/(Mg + Fe) = 0.93 to –0.94) and dolomite.

The groundmass of magnetite from massive and disseminated ores contains 1.0–2.4 wt % MgO. The grains of Cr–spinel present in magnetite ore are replaced by Cr–magnetite and Cr–clinochlore (Figure 4B). The composition of Cr–spinel and Cr–magnetite is uncommon and contains a high amount of MnO (0.4–5.2 wt %) and ZnO (0.1–3.5 wt %).

Sulfides are represented by early and late parageneses. Sulfides of early paragenesis are chalcopyrite $CuFeS_2$, pyrrhotite $Fe_{1-x}S$, talnakhite $Cu_9(Fe,Ni)_8S_{16}$, cubanite $CuFe_2S_3$, and Cu-, Co-bearing mackinawite $(Fe,Ni)_9S_8$ (Figure 4C,D). These sulfides form rounded inclusions less than 50 µm in size in magnetite. Sulfides of late paragenesis are chalcopyrite $CuFeS_2$, bornite $Cu_5FeS_4$, pyrrhotite $Fe_{1-x}S$, Co-pentlandite $(Fe,Ni)_9S_8$, and Fe-sphalerite ZnS. They are localized in the intergranular space among aggregates of magnetite and form segregations to 2–3 mm in size (Figure 4E,F). Sulfides of both parageneses contain inclusions of native gold.

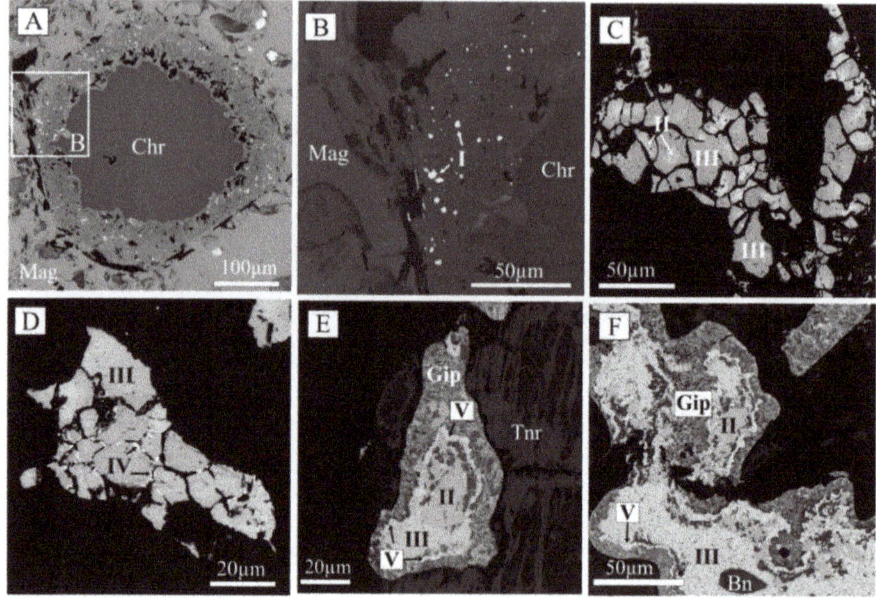

**Figure 5.** Native gold of generations I–V from magnetite ores of Kagan massif: (**A**) inclusions of native gold of generation I and copper sulfides in magnetite (Mag) replacing Cr–spinel (Chr); (**B**) fragments on "a"; (**C**) aggregates of zoned grains of gold of generations II and III; (**D**) thin gold veinlets of generation IV in grained aggregate of generation III; (**E,F**) zoned grain of gold of generations II and III with a gold rim of generation V and fine-grained aggregate of hypergenic, tenorite, and carbonate (Gip); (**F**) bornite inclusions (Bn) in gold of generation III.

We distinguish five generations of native gold [28]. Native gold of generation I occurs together with early paragenesis sulfides and forms 1–10 µm inclusions in magnetite, including that replacing the grains of Cr-spinel (Figure 5A,B). Its fineness varies in a wide range (580–960‰) even when gold particles are within a single cluster. It contains 0.4–1 wt % Pd and 0.1–1.3 wt % Cu (Table 3, Figure 3B).

Native gold of II–IV generations was deposited in the intergranular space of magnetite aggregates of late paragenesis. Gold grains attain 1–2 mm and are composed of polygrained aggregate of its various generations. Gold of generation II makes up the central parts of individual grains in such aggregates (Figure 5C) and is represented by Au–Ag–Cu–(Hg) solid solutions. Gold of this generation is characterized by considerable variations in the contents of Ag (2.8–26.7 wt %), Cu (6.6–24.2 wt %) and Hg (0–2.0 wt %) and rather stable fineness (648–744‰) (Table 2). Gold of generation III makes up the major volume of polygrained aggregates (Figure 5C,D). Its composition corresponds to Cu-bearing (to 2.9 wt % Cu) kustelite and electrum (fineness 280–514‰) or Au–Ag solid solution (fineness 810–853‰).

Gold of generation IV makes up a small part in the total mass of gold. It includes veinlet-like zones no more than 5–10 µm in thickness, developed in the intergranular space of polygrained gold aggregate (Figure 5D). The fineness of gold of generation IV is 740–853‰ and it contains Cu (1.4–3.8 wt %) and Hg (to 1.4 wt %) (Table 2).

Gold of generation V has the appearance typical of hypergenic gold. It replaces gold of earlier generations and has the highest fineness (933–976‰). Gold of generation V is also present in the fine-grained aggregate of minerals of hypergenic origin (iron hydroxides, tenorite, Cu–Mg carbonate, etc.) together with oxidized copper sulfides (Figure 5E,F). Its chemical composition corresponds to Au–Ag solid solution containing to 5 wt % Ag and 2.2 wt % Cu.

**Table 3.** Typical chemical composition and fineness of native gold of different generations from magnetite ores.

| No. Sample/Grain | Generation | Au wt % | Ag wt % | Cu wt % | Hg wt % | Pd wt % | Total wt % | Fineness ‰ |
|---|---|---|---|---|---|---|---|---|
| 888/23 | | 93.99 | 2.00 | 1.33 | 0.00 | 0.44 | 97.76 | 961 |
| 888/24 | I | 81.58 | 13.79 | 0.20 | 0.00 | 0.75 | 96.33 | 847 |
| 888/25 | | 55.24 | 40.11 | 0.12 | 0.00 | 0.97 | 95.44 | 579 |
| KAG-3/16 | | 67.51 | 19.70 | 13.48 | 0.26 | 0.00 | 100.95 | 669 |
| KAG-3/20 | | 74.18 | 2.82 | 24.19 | 0.04 | 0.00 | 101.23 | 733 |
| KAG-3/31 | | 73.81 | 4.84 | 20.57 | 0.00 | 0.00 | 99.22 | 744 |
| KAG-3/33 | II | 73.51 | 9.10 | 16.68 | 0.32 | 0.00 | 99.61 | 738 |
| KAG-3/35 | | 71.94 | 8.02 | 19.41 | 0.00 | 0.00 | 99.37 | 724 |
| 890b/42 | | 64.86 | 26.70 | 6.58 | 2.00 | 0.00 | 100.14 | 648 |
| KAG-3/5 | | 65.82 | 20.36 | 11.91 | 0.00 | 0.00 | 98.09 | 671 |
| 890s/1 | | 31.73 | 67.93 | 0.08 | 0.49 | 0.00 | 100.23 | 317 |
| 890s/6 | | 28.22 | 72.28 | 0.56 | 0.00 | 0.00 | 101.06 | 279 |
| 890s/7 | III | 43.51 | 53.86 | 2.89 | 0.32 | 0.00 | 100.58 | 433 |
| KAG-3/27 | | 84.66 | 13.19 | 1.94 | 0.00 | 0.00 | 99.79 | 848 |
| KAG-3/38 | | 51.45 | 46.82 | 0.98 | 0.82 | 0.00 | 100.07 | 514 |
| 890/13 | | 33.35 | 65.60 | 1.08 | 0.00 | 0.00 | 100.03 | 333 |
| 890s/9 | | 78.43 | 19.88 | 1.51 | 1.38 | 0.00 | 101.20 | 775 |
| 890s/4 | IV | 74.11 | 22.35 | 2.31 | 1.43 | 0.00 | 100.20 | 740 |
| 890/7 | | 84.88 | 10.85 | 3.79 | 0.00 | 0.00 | 99.52 | 853 |
| 890/14 | | 74.09 | 24.26 | 1.38 | 0.00 | 0.00 | 99.73 | 743 |
| KAG-3/29 | | 95.60 | 2.99 | 0.45 | 0.00 | 0.00 | 99.04 | 965 |
| KAG-3/45 | V | 96.46 | 0.20 | 2.21 | 0.00 | 0.00 | 98.87 | 976 |
| KAG-3/47 | | 92.24 | 4.99 | 1.62 | 0.00 | 0.00 | 98.85 | 933 |

Note. Microanalysis was performed in the IEM RAS (Institute of Experimental Mineralogy, Russian Academy of Sciences, Chernogolovka, Moscow Region, Russia), using electron scanning microscope CamScan MV2300 with energy-dispersive X-ray microanalyzer Link INCA Energy 300 (analyst D.A. Varlamov). Values of element concentration below 2σ (mean square error of analysis) are shown in italics. The reduced sum of elements in sample 888 is due to the small sizes of gold particles and partial entrapment of host magnetite (iron is excluded from the sum). The average measurement accuracy is 0.3 wt % for Au, Ag, Cu, Pd, and 1 wt % for Hg (which is associated with the possible migration of mercury from the analysis zone).

### 2.3. Physico-Chemical Conditions, Sources of Substance and Fluid and the Sequence of Formation of Gold Mineralization

Data on the research and estimation of P–T–X-parameters, sources of ore substance and fluid during the formation of gold mineralization veins at the Kagan ultramafic massif are reported in [7,9]. The isotopic composition of antigorite serpentinites is characterized by highly lightweight hydrogen -$\delta^{18}O$ = +5.2‰, $\delta D$ = −127.7‰ [9]. This isotopic composition corresponds to continental serpentinites of lizardite–chrysotile type [29,30]. It is obvious that the studied antigorite serpentinites inherited the isotopic composition of earlier loop-shaped serpentinites. The lightweight isotopic composition of hydrogen includes talc from magnetite ores ($\delta^{18}O$ = +7.4‰, $\delta D$ = −132.7‰) and amphibole from amphibole–pyroxene ($\delta^{18}O$ = +9.3‰, $\delta D$ = −124.6‰) and pyroxene-amphibole rocks ($\delta^{18}O$ = +7.1‰, $\delta D$ = −126.2‰) from the Kagan massif. All these facts suggest a metamorphic origin of the fluid that takes part in the formation of gold mineralization. Such a fluid could result from the release of water (deserpentinization) during progressive metamorphism of early continental serpentinites and their replacement by olivine, enstatite, anthophyllite, and talc. Ultramafic rocks composed of these minerals are widespread at the Kagan massif, particularly in its southern part.

Study of the gas composition of fluid released during the thermal decrepitation (method gas chromatography) of primary inclusions at 300–600 °C in the minerals of antigorite serpentinite and magnetite ore showed that the composition of fluid has a $X_{H2O}$ = 0.975–0.978 and fraction of carbon dioxide $X_{CO2}$ = 0.019–0.020. To estimate the redox conditions of gold mineralization, we used the indicator of the oxidation degree of volatiles $CO_2/(CO_2 + CO + H_2 + CH_4)$, which takes into account the ratio of oxidized and reduced gasses in the composition of inclusions [31]. This indicator was 0.84 during antigorization and 0.86 during the deposition of magnetite, which is the evidence that the gold mineralization of the studied types was formed under oxidizing conditions.

On the basis of available information on this object we propose the following scheme of the formation sequence of gold mineralization:

1. Upward movement of ultramafic rocks (harzburgites, dunites) to the surface, their serpentinization with the formation of early loop-shaped lizardites and/or chrysotile serpentinites with disseminated magnetite. Lizardite serpentinites at the Kagan massif only occur as small relics, whereas at the other ultramafic massifs of the Vishnevogorsk metamorphic complex (Nyashevskii, Ishkelskii, Borzovskii, etc.) they are widespread [14]. The presence of relicts of early loop-shaped serpentine (lizardite, chrysotile) in association with magnetite indicates that their formation temperatures were 100–300 °C.

2. Formation of the initial ore-forming fluid as a result of zonal metamorphism of early serpentinites and, probably, host rocks under the conditions of subgranulite-amphibolites facies (T = 60–700 °C and P = 5–10 kbar). Fluid is the result of release of water during deserpentinization of early serpentinites and transition of some ore components, such as Cu, Ag, Au, Zn, Mn, and S to solution.

3. Manifestation of fault tectonics within the massif and movement of the ore-bearing initial fluid along faults from the high-temperature zone to the lower-temperature zone of stability of antigorite and its discharge. In this zone the early loop-shaped serpentinite transforms into antigorite serpentinite with the growth of magnetite grains and formation of zoned magnetite rims on accessory Cr–spinel. Simultaneously, gold-ore mineralization of the first type is deposited along the thin cracks in schistose antigorite serpentinites. The formation of antigorite serpentinites took place under the upper degrees of green schist facies at the temperature of antigorite stability of 350–500 °C.

4. Continuation of the arrival of initial fluid into the open space of cracks and its mixing with meteoritic waters. The subsequent discharge of mixed fluid leads to the formation of the second type magnetite ores which carry sulfides and native gold of generation I of early paragenesis. Results of study of the composition of native gold with impurity of copper (II and III generations), combined with the diagram of the Au–Ag–Cu system [27], show that the temperature of gold deposition in magnetite ores could reach 350–450 °C (Figure 3B).

5. Cooling of the massif to conditions when antigorite becomes unstable and is replaced by chrysotile, chlorite, and talc. Recrystallization of rocks and gold-magnetite ores results in larger particles of copper sulfides and native gold of II–IV generations of late paragenesis. The temperature deposition conditions of minerals of late paragenesis were estimated using a chlorite geothermometer and were 310–210 °C [28].

6. In the conditions of hypergenesis zone, oxidation of copper and iron sulfides takes place with the formation of fine-grained aggregates of iron hydroxide (FeO(OH)), tenorite (CuO), Cu–Mg carbonate, and native gold of V generation.

## 3. Research Methods

### 3.1. Software and Thermodynamic Dataset for Modeling

Thermodynamic modeling of the first and second types of gold-ore mineralization was carried out using the Selektor-C software package (version 3.01, A.P. Vinogradov Institute of Geochemistry, Irkutsk, Russia), which implements the Gibbs energy minimization method based on the convex programming approach [32,33]. A detailed description of computational algorithms was given in [34]. The models used thermodynamic databases in the range of specified T–P parameters: for minerals [35–42], gases [43], and components of aqueous solution [44–52].

This method was widely used in the modeling of the formation of native gold: hypogene and hypergene models of the Kyuchyus Au–Sb–Hg deposit (Yakutia, Russia), including mercuric gold [53]; models of quaternary solid solutions in hydrothermal conditions at the Aitik Au–Ag–Cu porphyry deposit (Sweden) and the hydrothermal-hypergene model of formation of Au–Cu intermetallics at the Wheaton Creek placer deposit (Canada) [54]; the model of the Au–Ag, Ag–Pb, and Ag–Au–Pb mineralization at the Rogovik epithermal gold–silver deposit in the Omsukchan ore district (northeastern Russia) [55].

In the study we used the method of multi-reservoir dynamics, which involves separation of the complex system into a series of individual subsystems–reservoirs. The initial aqueous solution subsequently interacts with the rock in the first reservoir at temperature $T_1$ and pressure $P_1$, altered solution passes to the second reservoir and interacts with the rock at $T_2$ and $P_2$ etc. Thus, the alternating aqueous solution passing through a series of reservoirs changes the rock composition in the reservoirs. The introduction of the mode of successive passing of a certain amount of portions of the solution allows us to form an equilibrium-dynamic model which can be used for studying the processes of formation and alteration of ore mineralization. A similar approach was earlier used by us for modeling the formation of magnetite–chlorite–carbonate rocks with copper-bearing gold–silver solid solution at the Karabash ultramafic massif [11].

Modeling was completed for the K–Mg–Mn–Ca–Al–Si–Ti–Fe–Cu–Ag–Au–Cr–Hg–S–Cl–C–H–O system. The thermodynamic properties of minerals, binary, ternary and quaternary solid solutions, aqueous, and gaseous species considered in the models were the same as in [11] (Tables S2 and S3 of the Supplementary Materials from [11]). Thermodynamic constants for chlorites, ilmenites, pyroxenes, carbonates, olivines, and plagioclases, which are natural binary and ternary solid solutions, as well as solid solutions of quaternary system Ag–Au–Cu–Hg, were calculated considering the activity coefficients of end members for the accepted models of solid solutions [33,35,54,56]. Serpentinite was introduced as an ideal solid solution consisting of antigorite and lizardite. Cr–spinel is represented by an ideal model of solid solution $FeCr_2O_4$–$MgAl_2O_4$–$MgFe_2O_4$–$MgCr_2O_4$ [57].

### 3.2. Initial Data for Thermodynamic Modeling

The formation of the gold-ore mineralization begins with the generation of initial metamorphic ore-forming fluid as a result of the high-temperature zonal metamorphism of early serpentinites. The composition of these serpentinites was specified from the average composition of 19 samples of serpophite–lizardite serpentinites of the Vishnevogorsk–Ilmenogorsk metamorphic complex (Table 4).

**Table 4.** The composition of serpophite–lizardite (1) and antigorite (2) serpentinites (in wt %).

| Components | 1 * | 2 ** |
|---|---|---|
| $SiO_2$ | 38.93 | 39.0 |
| $TiO_2$ | 0.01 | 0.025 |
| $Al_2O_3$ | 1.39 | 1.49 |
| $Cr_2O_3$ | 0.38 | 0.197 |
| $Fe_2O_3$ | 5.33 | 5.90 |
| FeO | 2.2 | 2.4 |
| MnO | 0.07 | 0.07 |
| MgO | 36.98 | 38.0 |
| CaO | 0.82 | 0.04 |
| $K_2O$ | - | 0.01 |
| $H_2O$ | 1.15 | - |
| LOI | 13.23 | 12.5 |
| $CO_2$ | 1.15 | - |
| CuO | - | 0.03 |
| S | - | 0.02 |
| Au (ppm) | | 0.010 |
| Ag (ppm) | | 0.888 |

* data from [14]; ** we used the data of X-ray spectral fluorescent (CPM-35 and XRF 1800), chemical (FeO, CuO, S), ICP-MS (Ag) and chemical spectral (Au) analyses conducted in the CKP Geoanalysist in the Institute of Geology and Geochemistry, UrB RAS (analysists N.P. Gorbunova, L.A. Tatarinova, G.S. Neupokoeva, I.I. Neustroeva, and G.A. Avvakumova).

The calculation of the composition of initial metamorphic fluid was performed for T = 700 °C, P = 10 kbar. In these P–T conditions early serpentinites transform into rocks composed of mainly olivine and enstatite. The calculated composition of the metamorphic water fluid separated during dehydration is given in Table 5. This composition was used in further calculations in the 14-reservoir model. All 14 reservoirs of the model are represented by antigorite serpentinite. The chemical composition of a typical sample of antigorite serpentinite, containing gold mineralization, used in the calculations, is reported in (Table 4). Amount of volatiles: HCl = 0.01 mole.

**Table 5.** Calculated composition of initial metamorphic fluid at 700 °C and 10 kbar.

| Components | Composition, mol/kg $H_2O$ |
|---|---|
| C | 1.811 |
| Mg | $3.980 \times 10^{-3}$ |
| Mn | $6.682 \times 10^{-2}$ |
| Al | $6.552 \times 10^{-8}$ |
| Si | $1.947 \times 10^{-1}$ |
| S | $4.323 \times 10^{-2}$ |
| Cl | $6.932 \times 10^{-1}$ |
| Ca | $2.404 \times 10^{-1}$ |
| Fe | $3.326 \times 10^{-2}$ |
| Cu | $7.608 \times 10^{-3}$ |
| Ag | $5.705 \times 10^{-5}$ |
| Au | $3.519 \times 10^{-7}$ |
| Cr | $6.904 \times 10^{-10}$ |
| H | $4.199 \times 10^{-1}$ |
| O | 4.092 |
| pH | 4.357 |
| $\log f_{O2}$ | −15.05 |
| $\log f_{S2}$ | 0.829 |

### 3.3. Scenarios of Thermodynamic Modeling

We have modeled two scenarios of formation of gold-ore mineralization: metasomatic interaction of metamorphic fluid with serpentinites (scenario 1) and mixing of two fluids: ascending metamorphic and descending meteoric and their discharge in the free space of cracks of schistose serpentinites (scenario 2).

In the calculations under scenario 1, we obtained interaction products of 20 portions of deep-seated metamorphic fluid flowing consistently through nine serpentinite reservoirs at total drop of T from 650 to 250 °C and P from 4.5 to 0.5 kbar (Figure 6). From one reservoir to another, temperature drops by 50 °C and pressure by 0.5 kbar. The simulation was carried out at the rock/water ratio (R/W) = 10. The weight of a portion of metamorphic fluid is 100 g, the weight of each reservoir is 1 kg of serpentinites.

The model for scenario 2 involves two flows: ascending deep-seated metamorphic and descending, resulting from the interaction of meteoric waters with antigorite serpentinites at corresponding temperatures and pressures (Figure 6).

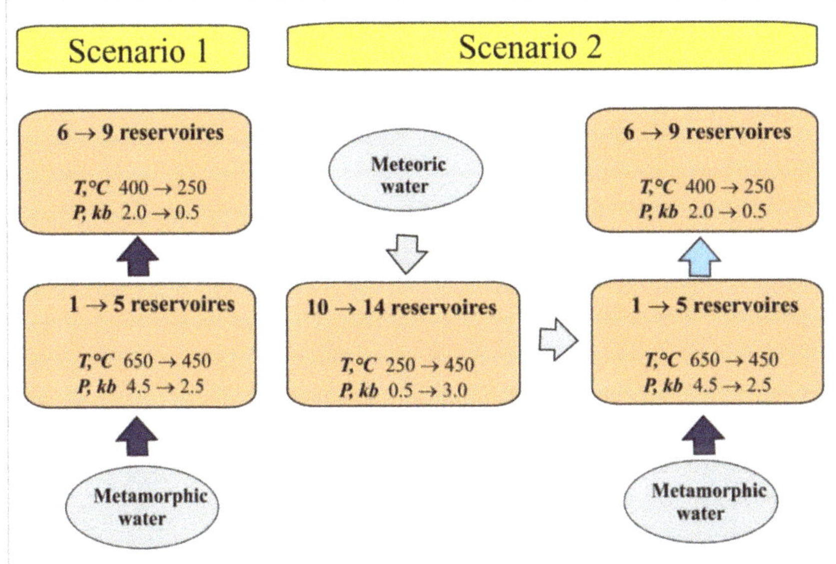


**Figure 6.** Scheme for calculation of 14-reservoir model in scenarios of metasomatic interaction of metamorphic fluid with serpentinite (Scenario 1) and mixing of metamorphic and meteoric fluids (Scenario 2).

The initial data of the model: 100 g of metamorphic fluid, 1 kg of meteoric water, weights of serpentinites are 100 g (reservoirs 1, 2, 8–14) and 5 g (reservoirs 3–7). The increase in the amount of rocks in reservoirs 3–7 is related to the imitation of fluid interaction with the walls of the open crack or weakened zone.

The P, T parameters for two models given in Table 6.

**Table 6.** T, P parameters of reservoirs.

| Reservoir | T, °C | P, kbar |
| --- | --- | --- |
| 1 | 650 | 4.5 |
| 2 | 600 | 4 |
| 3 | 550 | 3.5 |
| 4 | 500 | 3 |
| 5 | 450 | 2.5 |
| 6 | 400 | 2 |
| 7 | 350 | 1.5 |
| 8 | 300 | 1 |
| 9 | 250 | 0.5 |
| 10 | 250 | 0.5 |
| 11 | 300 | 1 |
| 12 | 350 | 1.5 |
| 13 | 400 | 2 |
| 14 | 450 | 3 |

## 4. Results of Thermodynamic Modeling

Scenario of metasomatic interaction of metamorphic fluid with serpentinites simulated the formation of the first type gold-ore mineralization in schistose serpentinites. Figures 7 and 8 show the calculation results after passing of 20 portions of fluid in the flow regime reactor at R/W = 10.

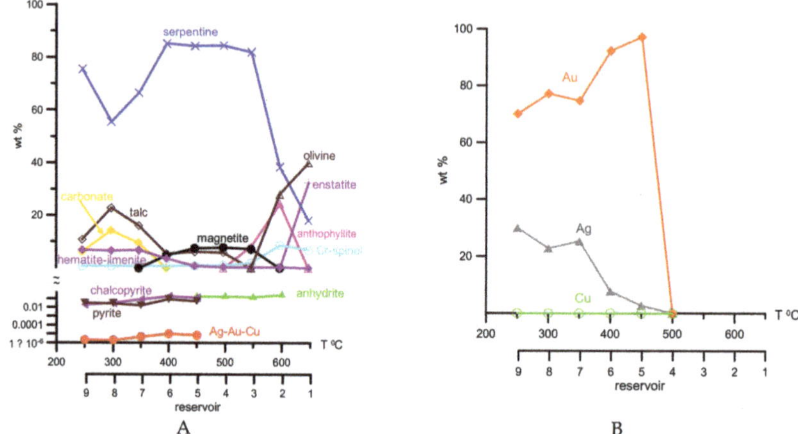

Figure 7. Mineral composition of reservoirs after passing of 20 portions of metamorphic fluid (**A**) and composition of Ag–Au–Cu solid solutions (**B**) in the model under scenario 1.

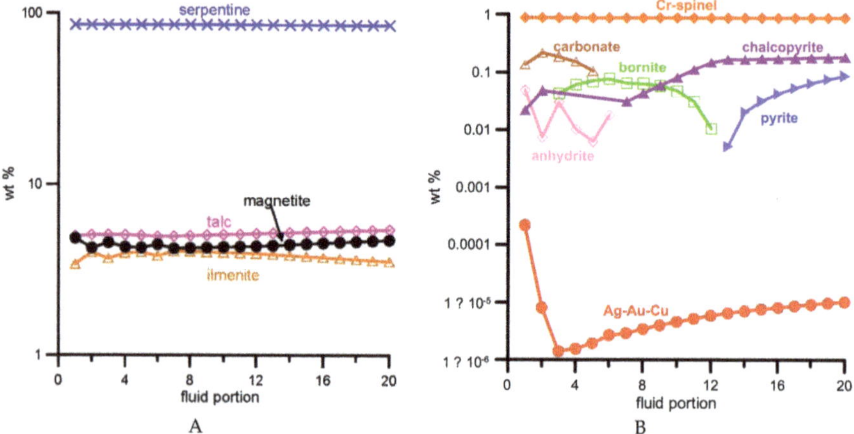

**Figure 8.** Mineral composition of reservoir 6 (T = 400 °C, P = 2 kbar). (**A**) major minerals; (**B**) minor minerals.

At T below 550 °C, olivine–enstatite rock transforms into serpentinites with a small amount of talc and magnetite, and below 400 °C, carbonate appears in serpentinite and the content of talc increases (Figure 7A). The content of accessory minerals in serpentinites is no more than 1 wt %. Among them, at T above 400 °C, anhydrite is formed and at lower temperature, chalcopyrite and pyrite.

In the range of 450–250 °C (reservoirs 5–9) the solid solution Ag–Au–Cu (<0.1 wt % Cu) is deposited. Its fineness varies depending on temperature conditions in the reservoirs and has a tendency to decrease from 950‰ at 450 °C to 700‰ at 250 °C, which corresponds the content of Ag from 0 to 30 wt % (Figure 7B).

A minor amount of gold (to 0.1 ppm) is formed in reservoir 6 at T = 400 °C. The supply of the first portions of fluid does not influence the content of rock-forming minerals (serpentine, talc, magnetite, ilmenite) in this reservoir (Figure 8A). At the same time, secondary carbonate and anhydrite, deposited with the first portions of fluid, with the arrival of additional portions are replaced by bornite and thereafter by chalcopyrite and pyrite (Figure 8B). Further arrival of new portions of fluid increases the content of the Ag–Au–Cu solid solution to 0.1 ppm.

Modeling for the scenario on discharge of metamorphic and meteoric fluids in the open space of cracks reflects the formation of gold-ore mineralization of the second-type in veined magnetite ores. The model involves mixing and further discharge of two types of fluids, metamorphic and surface meteoric, in the open space of cracks. Metamorphic fluid is formed during deserpentinization of lizardite–chrysotile serpentinites, and its initial composition is similar to the fluid from scenario 1. The fluid of meteoric origin is formed during the interaction of submerging meteoric waters [53] with antigorite serpentinites. Oxygen of meteoric fluid is a strong exogenous oxidizer of deep-seated iron in the formation of magnetite ores [58].

Results of the mineral composition modeling for altered serpentinites after passing of 20 portions of fluid through a sequence of reservoirs in the flow reactor regime are given in Figures 9 and 10.

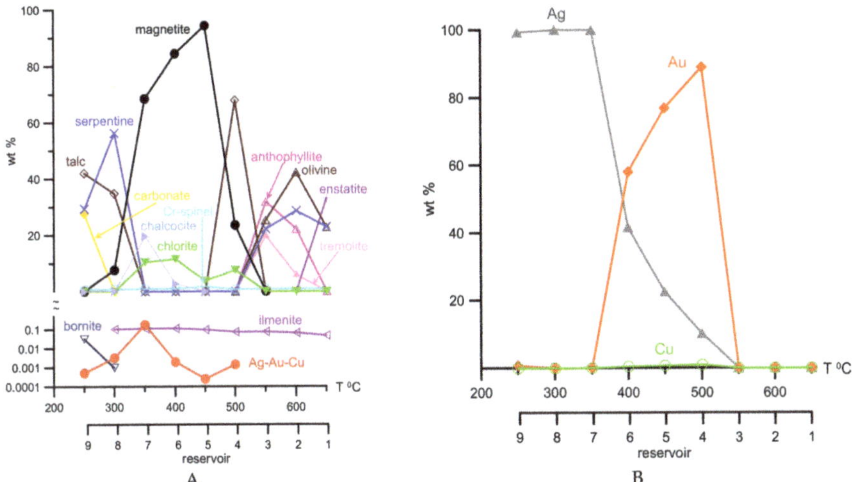

**Figure 9.** Results of the ascending branch of the model in scenario 2. (**A**) mineral composition of reservoirs; (**B**) composition of Ag–Au–Cu solid solutions.

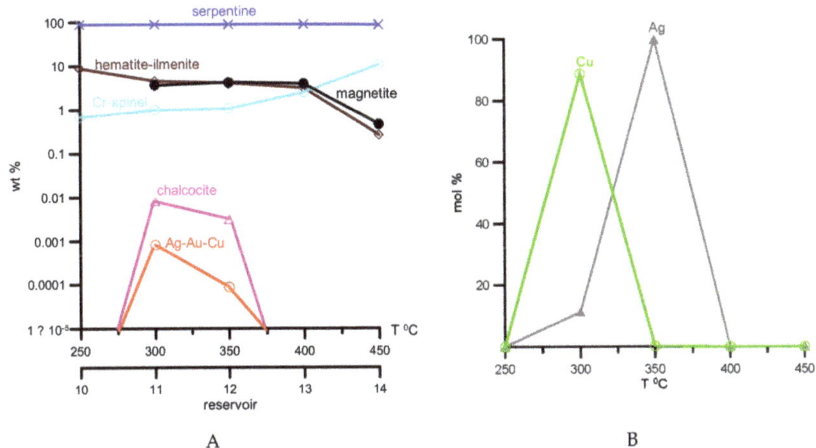

**Figure 10.** Results of descending branch of the model in scenario 2. (**A**) mineral composition of reservoirs; (**B**) composition of Ag–Au–Cu solid solutions.

The high temperature reservoirs (1–3) of the ascending branch of the model are represented by olivine–enstatite rocks containing serpentine (Figure 9A). When the temperature decreases from 700 to 550 °C, olivine and enstatite are replaced by anthophyllite and tremolite. Further decrease in temperature leads to the formation of magnetite together with talc and chlorite (500 °C), and at 350–450 °C magnetite becomes the main mineral with minor chlorite, chalcocite, and Cr–spinel. In the range of 500–250 °C, Ag–Au solid solution also forms and its maximum content in the rock amounts to 0.1 wt % at 350 °C. In this solid solution gold dominates over silver at 400–500 °C, and with decreasing temperature, gold content reduces and becomes mainly silver (Figure 9B). At temperatures below 350 °C, one can observe a mineral association of serpentine with talc, to which at 250 °C carbonate is added.

Mineral composition of rocks of the descending branch of the model (reservoirs 10–14) is represented by serpentinites and small amounts (less than 10 wt %) of Cr–spinel, magnetite and hematite–ilmenite (Figure 10A). At 350–300 °C, accessory chalcocite is present in the rock. In the descending branch of the model Au is absent in the composition of solid solutions. Here they contain only Ag (reservoir 12) and Cu (reservoir 11) (Figure 10B).

In scenario 1, imitating the influence of metamorphic deep-seated fluid on serpentinites, the solution equilibrated with rocks becomes more acidic (pH decreases) with the arrival of the new portions of fluid. After the arrival of 10 portions, pH of the solution attains the steady state. In this state the minimum value of pH = 4 is attained at 450 °C. When the temperature increases to 650 °C or decreases to 250 °C, pH of the solution equals 4.5 (Figure 11A).

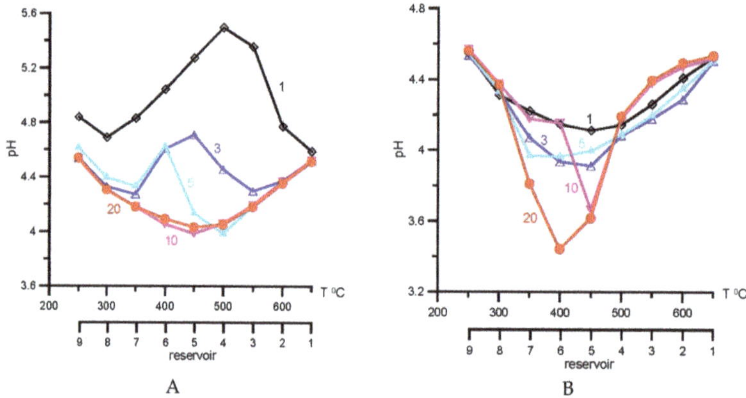

**Figure 11.** Change in pH of reservoirs after the arrival of various amounts of portions of fluid. (**A**) scenario 1; (**B**) scenario 2.

In scenario 2 the behavior of pH in general shows the same tendency as in scenario 1, but here the contribution of the oxidative meteoric fluid supplied to reservoir 5 plays a significant part. The system attains a stationary state after the arrival of more than 20 portions of the solution and minimum value of pH = 3.4 at 400 °C (Figure 11B).

Fugacity values of oxygen and sulfur in both scenarios of modeling have a tendency to reduce with decreasing temperature in the reservoirs (from reservoir 1 to 9) (Figure 12).

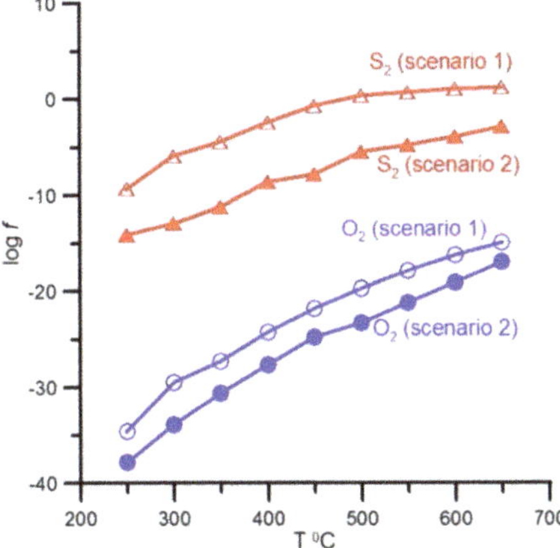



**Figure 12.** Changes in fugacities of oxygen and sulfur in scenarios 1 and 2 after the arrival of 20 portions of fluid.

In reservoirs with T > 450 °C, the fluid is dominated by the chloride complex of gold $AuCl_2^-$ [59] (Figure 13). As the temperature decreases, the dominant role of chloride complex is replaced by hydrosulfide $AuHS^0$ (scenario 1) or hydroxide $AuOH^0$ complexes (scenario 2) [60]. The predominance of gold hydroxide in scenario 2 is related to a more oxidized state of fluid owing to the arrival of meteoric waters.

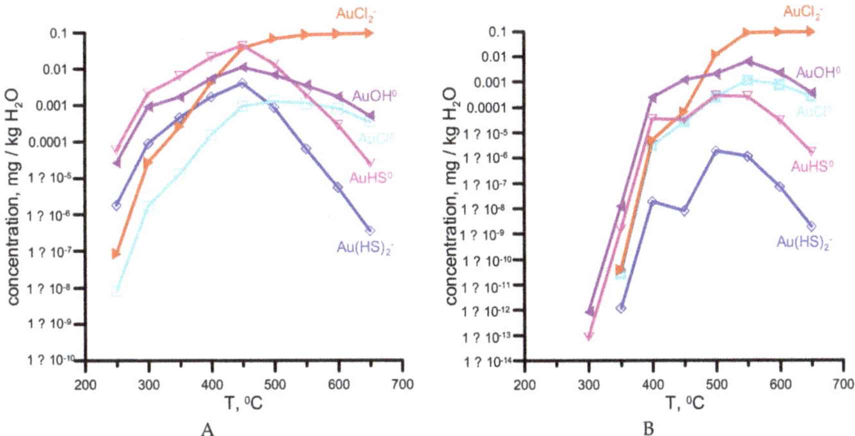



**Figure 13.** Forms of Au transfer in scenarios 1 (**A**) and 2 (**B**).

The content of gold deposited in the rocks under scenarios 1 and 2 differs considerably. In scenario 1 the amount of gold in serpentinites is no more than 0.1 ppm, whereas in scenario 2 its content in magnetite ore amounts to 10–13 ppm (Figure 14A). At the same time, the maximum gold contents in both scenarios are achieved at T = 400 °C (reservoir 6). Consistent accumulation of gold in reservoir 6 with the arrival of new portions of fluid is shown in Figure 14B.

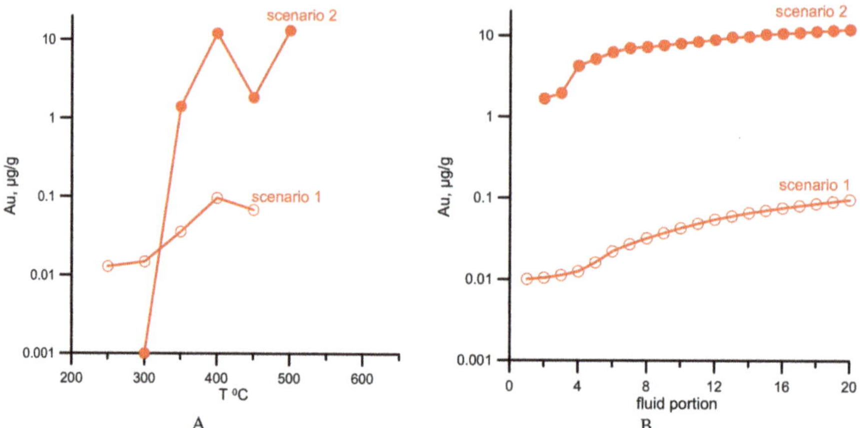


**Figure 14.** The amount of gold in scenarios 1 and 2. (**A**) after the arrival of 20 portions of fluid at different temperatures in reservoirs; (**B**) with the arrival of new portions of fluid in reservoir 6 (400 °C and 2 kbar).

## 5. Discussion

### 5.1. Comparative Analysis of Agreement of Model and Natural Mineral Associations

Our calculations of the formation of gold-ore mineralization can explain some questions that arise when studying serpentinite massifs of ultramafic rocks, including the object of the present study. In particular, calculations for scenario 1 contribute to the solution of the earlier stated problem [61] about the discrepancy of the scales of primary and placer gold content in the areas where ultramafic rocks are spread. Although the placers of streams draining ultramafic massif are of commercial value, findings of native gold in serpentinites are scarce [62–64]. The above-given calculations for scenario 1 show that, in the model of formation of gold mineralization from cooling metamorphic fluid, gold is concentrated from relatively low concentrations of no more than 0.1 ppm (Figure 14). Nevertheless, the washout of extended zones of schistose antigorite serpentinites by streams can lead to the formation of placers with considerable reserves.

Comparative analysis of mineral associations in the first type gold mineralization at the Kagan massif and those obtained in scenario 1 (Table 7) indicates similarities in the low gold content in rocks (to 0.1 ppm).

Moreover, the rocks contain small amounts of magnetite and talc which replaces serpentine. The fineness of Au–Ag solid solutions in calculated and natural associations is no more than 750‰ and they have similar deposition temperatures (300–500 °C). At the same time, accessory chalcopyrite and pyrite occur in paragenesis with Au–Ag solid solutions in scenario 1, but these sulfides were absent in natural samples. Most likely, this difference is due to the very low content of sulfides in antigorite serpentinites of the Kagan massif.

Rocks with a high magnetite content within the model of cooling metamorphic fluid (scenario 1) in the processes of its metasomatic interaction with host serpentinites or discharge in the open space of cracks in the whole range of temperatures and pressures have not been obtained. The mineral association of magnetite, the content of which drastically prevails over serpentine, talc, amphibole and chlorite, was obtained only by strong dilution of metamorphic fluid with weakly mineralized waters in scenario 2. The diluting fluid in this scenario is meteoric water seeping through the bed of serpentinites to the depth.

**Table 7.** Comparative characteristics of minerals and mineral associations of gold mineralization from Kagan massif and model calculations for scenarios 1 and 2.

| Characteristics | Kagan Massif | | Calculated Data | |
| --- | --- | --- | --- | --- |
| | 1 Type of Gold Mineralization | 2 Type of Gold Mineralization | Scenario 1 | Scenario 2 |
| Quantitative ratios of main minerals * | Atg > Ctl, Tc > Chl > Tr, Mag | Mag >> Atg > Tc > Tr, Chl, Ctl, Dol | Srp >> Tc, Mag > Ilm–Hem | Mag >> Srp > Tc > Tr, Chl |
| Quantitative ratios of accessory minerals | Pn > Au(ss) | Ccp > Po, Pn > Cub, Bn, Tal, Sp > Au(ss) | Ccp, Py >> Au(ss) | Ilm–Hem > Cct > Bn, Au(ss) |
| Sequence of mineral deposition ** | Atg, Tr, Mag, Pn, Au(ss) → Tc, Chl, Ctl | Mag, Atg, Ccp, Po, Cub, Tal, Pn, Au(ss) → Tr, Tc, Ctl → Ccp, Bn, Po, Sp, Au(ss) → Chl, Srp, Dol | Srp, Mag, Tc, Ccp, Py, Au(ss) → Srp, Tc, Cb, Ilm–Hem, Au(ss) | Ol, Srp, Ath, Tr → Mag, Chl, Tc, Ilm–Hem, Au(ss) → Mag, Chl, Cct, Au(ss) → Srp, Tc, Cb, Bn, Au(ss) |
| T °C of gold deposition | 500–300 | | 450–250 | 500–350 |
| Composition of solid solutions gold/fineness ‰ | ss(Au,Ag)/772–996 | ss(Au,Ag)/580–960, ss (Au,Ag,Cu)/648–744 ss (Au,Ag,Cu)/280–853 | ss(Au,Ag)/900–950 ss(Au,Ag)/700–800 | ss(Au,Ag)/600–900 ss(Au,Ag)/600 ss(Ag,Au)/0–600 |
| Gold content in rock/ore, ppm | To 0.1 | 0.2–1.2 and more | To 0.1 | To 10 |

* For calculated data at P, T of gold deposition; ** for calculated data from high- to low-temperature reservoir. Ath—anthophyllite; At—gantigorite; Au(ss)—Au–Ag–Cu solid solution; Bn—bornite; Cb—carbonate; Ccp—chalcopyrite; Cct—chalcocite; Chl—chlorite; Ctl—chrysotile; Cub—cubanite; Dol—dolomite; Ilm-Hem—ilmenite–hematite; Mag—magnetite; Ol—olivine; Pn—pentlandite; Po—pyrrhotite; Py—pyrite; Sp—sphalerite; Srp—serpentine; Tal—talnakhite; Tc—talc; Tr—tremolite.

Comparative analysis of mineral associations in the second type of gold mineralization and those obtained in scenario 2 showed both a good similarity of natural and calculated ratios of main and accessory minerals of rocks and identical sequence of their deposition in the open space of cracks (see Table 5). The contents of gold in the model calculations in scenario 2 attain 10–13 ppm and correspond to their level in magnetite ores from the Kagan massif. Calculations for this scenario support the probability of formation of several generations of Au–Ag solutions, covering the whole range of compositions from pure gold to pure silver. However, in the calculations we did not obtain Au–Ag solid solutions with elevated copper content typical of native gold of generation II in magnetite ores.

*5.2. The Possible Sources of Gold, Silver, and Other Metals*

Previous studies of the massifs of ultramafic rocks revealed spatial overlapping of many placers of the Urals with antigorite serpentinites as well as considerable redistribution and concentration of gold during antigoritization [6,7,9]. On the basis of these facts a conclusion was made about the origin of gold-ore mineralization in the crustal period of the history of the formation of ultramafic massifs during regional and local metamorphism, and the gold-productive serpentinite (antigorite) metasomatic formation was also distinguished [7].

Magnetite ores in ultramafic rocks in world literature were described in [65–69]. They are suggested to be of different genesis—magmatic [67], metamorphic [65,69], and hydrothermal [68].

Rare cases of elevated gold contents in magnetite ores are known [70,71]. In the Bou–Azzer mantle ophiolite complex (Anti-Atlas, Morocco), strongly serpentinized periodite is known to contain magnetite veins formed by filling the open space of cracks [71]. Magnetite in the veins is associated with serpentine, magnesite, clinochlore, talc, and andradite. The relatively minor enrichment of magnetite in gold (to 8–14 mg/t) is attributed to the hydrothermal process, and the high mobility of iron is explained by the high water/rock ratio during serpentinization [71].

Magnetite ores in ultramafic rocks enriched with Au and Ag (a few tens of ppb) are widely spread within the massifs of West Mongolia [70]. Veins of massive magnetite ores up to 10 cm thick syngenetic to the processes of serpentinization occur there. Gold and silver are considered to be brought by serpentinized solutions borrowing these elements from chromitites or volcano–sedimentary rocks.

In this work we propose hydrothermal–metasomatic models of the formation of gold mineralization in which deserpentinizing early lizardite-chrysotile serpentinites are the source of metal and water. Metals and water are separated from early serpentinites during high-temperature metamorphism, and the metamorphic fluid moves to the lower temperature discharge region along the schistosity zones in the massif of ultramafic rocks. Metasomatic interaction of the metamorphic fluid with serpentinites in the discharge zones can result in poor gold-ore mineralization in the form of native gold disseminated in schistose serpentinites. A necessary condition for the formation of hydrothermal rich gold-ore mineralization in magnetite veins is mixing of metamorphic and meteoric fluid at T = 500–400 °C and P = 2–3 kbar and discharge of deep-seated and meteoric fluids in the open space of cracks. The mechanism of mixing of the deep-seated and meteoric fluids was used earlier in modeling of the formation of Au–REE mineralization in magnetite–chlorite–carbonate rocks from the Karabash ultramafic massif [11].

### 5.3. Sulfur Fugacity Evolution During Gold-Ore Mineralization Formation in Ultramafic Rocks and Magnetite Ores

The composition of Au–Ag solid solutions deposited under both scenarios at 500–300 °C depends on sulfur fugacity ($fS_2$) in the specified range of temperatures. In the model under scenario 1, Au–Ag solid solutions form in the field of high and very high values of $fS_2$ and are arranged along the covellite (Cv)–digenite (Dg) equilibrium line (Figure 15). The molar fraction of silver in Au–Ag alloys ($X_{Ag}$) in this model, according to experimental data [72], is less than 0.2 (12 wt % Ag). This composition agrees well with that of native gold in schistose serpentinites from the Kagan massif (see Table 2 and Figure 3A).

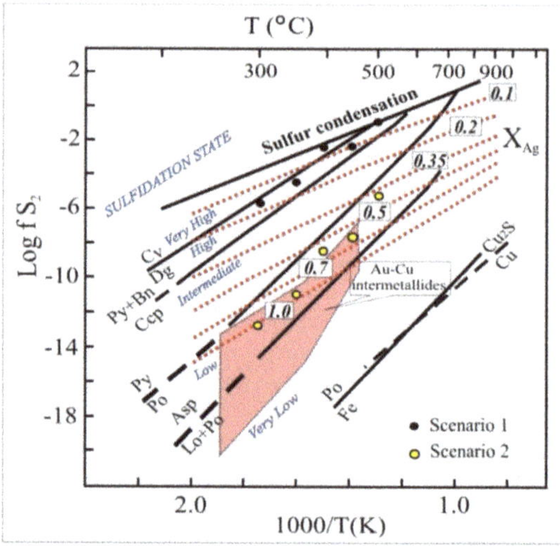


**Figure 15.** Sulfur fugacity versus temperature diagram (modified from [72]) showing the formation conditions of Au–Ag solid solutions at 300–500 °C in Scenarios 1 and 2. Dotted lines show sulfidation of silver in Au–Ag melts with a varying molar fraction of Ag ($X_{Ag}$) from data [73]. Field of Au–Cu intermetallids in rodingites from Zolotaya Gora deposit of Southern Urals is shown [12].

In the model for scenario 2, Au–Ag solid solutions are formed in the field of low values of $fS_2$ below the pyrrhotite–pyrite on the T–log $fS_2$ diagram (see Figure 15). The formation conditions of these solid solutions partially coincide with those of Au–Cu intermetallides from rodingites at the Karabash massif in the Southern Urals [12]. The composition of Au–Ag solid solutions in scenario 2 is

$X_{Ag} > 0.4$ (26.7 wt % Ag) and is in general similar to that of native gold from magnetite ores of the Kagan massif (see Table 3 and Figure 3B).

## 6. Conclusions

1. We have constructed thermodynamic models of the formation of two types of gold-ore mineralization at the Kagan ultramafic massif in the Southern Urals. Metamorphic water which results from dehydration of early serpentinites during high-temperature zonal regional metamorphism (700 °C, 10 kbar) is considered the source of ore-bearing fluid. This water extracts metals from rocks in the high-temperature zones and moves along the faults to the lower temperature region. Metasomatic interaction of the metamorphic fluid with serpentinites in the transit zone in scenario 1 is responsible for the formation at T = 450–250 °C and P = 2.5–0.5 kbar of gold-poor mineralization (to 0.1 ppm) of type 1 disseminated in schistose rocks. A necessary condition for the formation of hydrothermal gold-rich mineralization (to 10–13 ppm and more) in magnetite veins of type 2 in scenario 2 is mixing of metamorphic and meteoric fluid at T = 500–400 °C and P = 2–3 kbar and discharge of mixed fluid in the open space of cracks in schistose serpentinites.

2. The mineral associations, quantitative ratios, and sequence of deposition of the main minerals and composition of Au–Ag solid solutions obtained in scenarios 1 and 2 are, on the whole, similar to the natural types of gold mineralization at the Kagan massif. In mineralization of type 1, serpentinite contains small amounts of magnetite and talc which replaces antigorite that prevails over chrysotile and lizardite. Calculated and natural associations are characterized by Au–Ag solid solutions with the fineness higher than 750‰ and close deposition temperatures (300–500 °C). In type 2, association of magnetite occurs, which drastically dominates over serpentine, talc, amphibole, and chlorite. In this association several generations Au–Ag solid solutions are formed, which cover the whole range of compositions from virtually pure gold to pure silver.

3. The model calculations show that at temperatures higher than 450 °C the fluid with pH = 3.5–4.5 is dominated by the gold chloride complex $AuCl_2^-$, and with decreasing temperature the dominant role of chloride complex is replaced by hydrosulfide $AuHS^0$ (scenario 1) or hydroxide $AuOH^0$ (scenario 2). The predominance of gold hydroxide in scenario 2 is related to a more oxidized state of fluid owing to the arrival of meteoric waters.

4. The composition of Au–Ag solid solutions deposited in both scenarios at 500–300 °C is determined by the value of sulfur fugacity ($fS_2$). In the model for scenario 1, Au–Ag solid solutions are formed at high $fS_2$ values ($10^{-0.4}$–$10^{-6}$) that correspond to the lines of buffer reaction covellite–digenite, and in scenario 2, at low values of $fS_2$ ($10^{-5.5}$–$10^{-14}$) below the buffer line pyrrhotite–pyrite on the T–$fS_2$ diagram.

**Author Contributions:** Conceptualization, V.M., K.C., and G.P.; Methodology, V.M., K.C., and G.P.; Software, K.C.; Investigation, D.V.; Data Curation, V.M., K.C., and G.P.; Writing-Review & Editing, G.P.; Visualization, V.M. and K.C.; Supervision, G.P.

**Funding:** This work was supported by the Russian Foundation for Basic Research (grant 16-05-00407a) and state assignments of IGM SB RAS, IG SB RAS, IGG UB RAS (No AAAA-A18-118052590028-9), IEM RAS (No AAAA-A18-118020590151-3) financed by Ministry of Science and Higher Education of the Russian Federation.

**Acknowledgments:** Our thanks go to Velivetskaya, T.A., and Shanina, S.N., Gorbunova N.P., Tatarinova L.A., and Neupokoeva G.S. for the analytical data. We thank the Reviewers for the important comments and constructive suggestions, which helped us to improve the quality of the manuscript.

**Conflicts of Interest:** The authors declare no conflict of interest.

## References

1. Benevolsky, B.I. *Gold of Russia. Problems of the Use and Reproduction of the Mineral Resource Base*; Geoinformmark: Moscow, Russia, 1995; 87p. (In Russian)
2. Sazonov, V.N.; Ogorodnikov, V.N.; Koroteev, V.A.; Polenov, I.A. *Gold Deposits of the Urals*, 2nd ed.; Ural Mining and Geological Academy: Yekaterinburg, Russia, 2001; 622p. (In Russian)

3.  Ovchinnikov, L.N. *Minerals and Metallogeny of the Urals*; Geoinformmark: Moscow, Russia, 1998; 412p. (In Russian)

4.  Puchkov, V.N. General features relating to the occurrence of mineral deposits in the Urals: What, where, when and why. *Ore Geol. Rev.* **2017**, *85*, 4–29. [CrossRef]

5.  Borodaevsky, N.I. Types of gold deposits related to ultramafic rocks in the Miass and Uchaly districts in the southern Urals. In *200 Years of Gold Industry of the Urals*; Ivanov, A.A., Rozhkov, I.S., Eds.; UFAN SSSR: Sverdlovsk, Russia, 1948; pp. 316–330. (In Russian)

6.  Berzon, R.O. *Gold Resource Potential of Ultramafics*; VIEMS: Moscow, Russia, 1983; p. 47. (In Russian)

7.  Sazonov, V.N. *Gold-Productive Metasomatic Rocks in Mobile Belts: Geodynamic Settings and PTX Parameters of Their Formation and Implications for Forecasting*; Ural Mining and Geological Academy: Yekaterinburg, Russia, 1998; 181p. (In Russian)

8.  Spiridonov, E.M.; Pletnev, P.A. *Zolotaya Gora Deposit of Cupriferous Gold*; Nauchnyi Mir: Moscow, Russia, 2002; 220p. (In Russian)

9.  Murzin, V.V.; Varlamov, D.A.; Shanina, S.N. New Data on the Gold–Antigorite Association of the Urals. *Dokl. Earth Sci.* **2007**, *417*, 1436–1439. [CrossRef]

10. Murzin, V.V.; Varlamov, D.A.; Ronkin, Y.L.; Shanina, S.N. Origin of Au-bearing Rodingite in the Karabash Massif of Alpine-Type Ultramafic Rocks in the Southern Urals. *Geol. Ore Depos.* **2013**, *55*, 278–297. [CrossRef]

11. Murzin, V.; Chudnenko, K.; Palyanova, G.; Kissin, A.; Varlamov, D. Physicochemical model of formation of gold-bearing magnetite–chlorite–carbonate rocks at the Karabash ultramafic massif (Southern Urals, Russia). *Minerals* **2018**, *8*, 306. [CrossRef]

12. Murzin, V.V.; Chudnenko, K.V.; Palyanova, G.A.; Varlamov, D.A.; Naumov, E.A.; Pirajno, F. Physicochemical model of formation of Cu–Ag–Au–Hg solid solutions and intermetallic alloys in the rodingites of the Zolotaya Gora gold deposit (Urals, Russia). *Ore Geol. Rev.* **2018**, *93*, 81–97. [CrossRef]

13. Bazhin, E.A. *Gold Content of the Talovskii Gabbro-Ultramafic Massif (Southern Urals). Problems of Mineralogy, Petrography and Metallogeny; Scientific Readings in Memory of P.N.Chirvinskii: Collection of Scientific Articles*; Issue 15; Perm State Research University: Perm, Russia, 2012; pp. 294–298. (In Russian)

14. Varlakov, A.S.; Kuznetsov, G.P.; Korablev, G.G.; Murkin, V.P. *The Ultrabasites of Vishnyevogorsk-Ilmenogorsk Metamorphic Complex (South Urals)*; Edition of Institute of Mineralogy Ural Division of Russian Academy of Sciences: Miass, Russia, 1998; 195p. (In Russian)

15. Murzin, V.V. Oxygen and Hydrogen Isotopic Composition of the Fluid during Formation of Anthophyllite Metaultramafic Rocks in the Sysert Metamorphic Complex, Central Urals. *Dokl. Earth Sci.* **2014**, *459*, 1548–1552. [CrossRef]

16. Murzin, V.V.; Hiller, V.V.; Varlamov, D.A. The age position of gold–sulfide mineralization in the metahyperbasites of the Sysert' metamorphic complex in the Middle Urals. *Litosfera* **2015**, *4*, 87–92. (In Russian)

17. Reith, F.; Brugger, J.; Zammit, C.M.; Nies, D.H.; Southam, G. Geobiological cycling of gold: From fundamental process understanding to exploration solutions. *Minerals* **2013**, *3*, 367–394. [CrossRef]

18. Kerr, G.; Craw, D. Mineralogy and geochemistry of biologically-mediated gold mobilisation and redeposition in a semiarid climate, Southern New Zealand. *Minerals* **2017**, *7*, 147. [CrossRef]

19. Kalinin, Y.A.; Palyanova, G.A.; Naumov, E.A.; Kovalev, K.R.; Pirajno, F. Supergene remobilization of Au in Au-bearing regolith related to orogenic deposits: A case study from Kazakhstan. *Ore Geol. Rev.* **2019**, *109*, 358–369. [CrossRef]

20. Kalinin, Y.A.; Palyanova, G.A.; Bortnikov, N.S.; Naumov, E.A.; Kovalev, K.R. Aggregation and Differentiation of Gold and Silver during the Formation of the Gold-Bearing Weathering Crusts (on the Example of Kazakhstan Deposits). *Dokl. Earth Sci.* **2018**, *482*, 1193–1198. [CrossRef]

21. Saunders, J.A.; Burke, M. Formation and aggregation of gold (electrum) nanoparticles in epithermal ores. *Minerals* **2017**, *7*, 163. [CrossRef]

22. Murzin, V.V.; Varlamov, D.A. Noble metals in magnetite ores of the Kagan ultramafic massif in the Southern Urals. In *Yearbook-2005/Institute of Geology and Geochemistry*; Collection of Scientific Papers; IGG UrB RAS: Yekaterinburg, Russia, 2006; pp. 379–381. (In Russian)

23. Brown, A.V.; Page, N.J.; Love, A.H. Geology and platinum-group element geochemistry of the Serpentine Hill complex, Dundas Trough, Western Tasmania. *Can. Mineral.* **1988**, *26*, 161–175.

24. Arai, S.; Akizawa, N. Precipitation and dissolution of chromite by hydrothermal solutions in the Oman ophiolite: New behavior of Cr and chromite. *Am. Mineral.* **2014**, *99*, 28–34. [CrossRef]

25. Kapsiotis, A.; Rassios, A.E.; Antonelou, A.; Tzamos, E. Genesis and Multi-Episodic Alteration of Zircon-Bearing Chromitites from the Ayios Stefanos Mine, Othris Massif, Greece: Assessment of an Unconventional Hypothesis on the Origin of Zircon in Ophiolitic Chromitites. *Minerals* **2016**, *6*, 124. [CrossRef]

26. Colás, V.; Padrón-Navarta, J.A.; González-Jiménez, J.M.; Fanlo, I.; Sánchez-Vizcaíno, V.L.; Gervilla, F.; Castroviejo, R. The role of silica in the hydrous metamorphism of chromite. *Ore Geol. Rev.* **2017**, *90*, 274–286. [CrossRef]

27. Dritz, M.; Bochvar, N.R.; Guzei, L.S. *Binary and Multicomponent Copper-Based Systems: Handbook*; Nauka: Moscow, Russia, 1979; 248p. (In Russian)

28. Murzin, V.V.; Varlamov, D.A. Chemical composition of native gold in magnetite ore from the Kagan ultramafic massif (S. Urals). In *Yearbook-Proceedings of the IGG UrB RAS*; IGG UrB RAS: Yekaterinburg, Russia, 2018; Volume 165, pp. 194–199. (In Russian)

29. Wenner, D.; Taylor, H.P., Jr. O/H and $O^{16}/O^{18}$ studies of serpentinization of ultramafic rocks. *Geochim. Cosmochim. Acta* **1974**, *38*, 1255–1286. [CrossRef]

30. Fruh-Green, G.L.; Scambelluri, M.; Vallis, F. O–H isotop ratios of high pressure ultramafic rocks: Implications for fluid sources and mobility in the subducted mantle. *Contrib. Mineral. Petrol.* **2001**, *141*, 145–159. [CrossRef]

31. Murzin, V.V.; Shanina, S.N. Fluid regime and origin of gold-bearing rodingites from the Karabash alpine-type ultrabasic massif, Southern Ural. *Geochem. Int.* **2007**, *45*, 998–1011. [CrossRef]

32. Karpov, I.K.; Chudnenko, K.V.; Kulik, D.A. Modeling chemical mass transfer in geochemical processes: Thermodynamic relations, conditions of equilibria, and numerical algorithms. *Am. J. Sci.* **1997**, *297*, 767–806. [CrossRef]

33. Chudnenko, K.V. *Thermodynamic Modeling in Geochemistry: Theory, Algorithms, Software, Applications*; Academic Publishing House Geo: Novosibirsk, Russia, 2010; 287p, ISBN 978-5-904682-18-7. (In Russian)

34. Kulik, D.A.; Wagner, T.; Dmytrieva, S.V.; Kosakowski, G.; Hingerl, F.F.; Chudnenko, K.V.; Berner, U.R. GEM-Selektor geochemical modeling package: Revised algorithm and GEMS3K numerical kernel for coupled simulation codes. *Comput. Geosci.* **2013**, *17*, 1–24. [CrossRef]

35. Chudnenko, K.; Palyanova, G. Thermodynamic modeling of native formation Cu–Ag–Au–Hg solid solutions. *Appl. Geochem.* **2016**, *66*, 88–100. [CrossRef]

36. Holland, T.J.B.; Powell, R. An internally consistent thermodynamic data set for phases of petrological interest. *J. Metamorph. Geol.* **1998**, *16*, 309–343. [CrossRef]

37. Helgeson, H.C.; Delany, J.M.; Nesbitt, H.W.; Bird, D.K. Summary and critique of the thermodynamic properties of rock-forming minerals. *Am. J. Sci.* **1978**, *278*, 1–229.

38. Robie, R.A.; Hemingway, B.S. *Thermodynamic Properties of Minerals and Related Substances at 298.15 K and 1 Bar ($10^5$ Pascals) Pressure and at Higher Temperatures*; United States Geological Survey: Washington, DC, USA, 1995; 458p.

39. Tagirov, B.R.; Baranova, N.N.; Zotov, A.V.; Schott, J.; Bannykh, L.N. Experimental determination of the stabilities of $Au_2S(cr)$ at 25 °C and $Au(HS)_2^-$ at 25–250 °C. *Geochim. Cosmochim. Acta* **2006**, *70*, 3689–3701. [CrossRef]

40. Shock, E.L.; Sassani, D.C.; Willis, M.; Sverjensky, D.A. Inorganic species in geologic fluids: Correlation among standard molal thermodynamic properties of aqueous ions and hydroxide complexes. *Geochim. Cosmochim. Acta* **1997**, *61*, 907–950. [CrossRef]

41. Stefansson, A. Dissolution of primary minerals of basalt in natural waters. I. Calculation of mineral solubilities from 0 °C to 350 °C. *Chem. Geol.* **2001**, *172*, 225–250. [CrossRef]

42. Dorogokupets, P.I.; Karpov, I.K. *Thermodynamics of Minerals and Mineral Equilibria*; Nauka: Novosibirsk, Russia, 1984; 186p. (In Russian)

43. Reid, R.C.; Prausnitz, J.M.; Sherwood, T.K. *The Properties of Gases and Liquids*; McGraw-Hill Book Company: New York, NY, USA, 1977; 688p.

44. Sverjensky, D.A.; Shock, E.L.; Helgeson, H.C. Prediction of the thermodynamic properties of aqueous metal complexes to 1000 °C and 5 kb. *Geochim. Cosmochim. Acta* **1997**, *61*, 1359–1412. [CrossRef]

45. Shock, E.L.; Helgeson, H.C.; Sverjensky, D.A. Calculation of the thermodynamic and transport properties of aqueous species at high pressures and temperatures: Standard partial molal properties of inorganic neutral species. *Geochim. Cosmochim. Acta* **1989**, *53*, 2157–2183. [CrossRef]

46. Akinfiev, N.N.; Zotov, A.V. Thermodynamic description of chloride, hydrosulfide, and hydroxo complexes of Ag(I), Cu(I), and Au(I) at temperatures of 25–500 °C and pressures of 1–2000 bars. *Geochem. Int.* **2001**, *39*, 990–1006.

47. Akinfiev, N.N.; Zotov, A.V. Thermodynamic description of aqueous species in the system Cu–Ag–Au–S–O–H at temperatures of 0–600 °C and pressures of 1–3000 bar. *Geochem. Int.* **2010**, *48*, 714–720. [CrossRef]

48. Johnson, J.W.; Oelkers, E.H.; Helgeson, H.C. SUPCRT92: Software package for calculating the standard molal thermodynamic properties of mineral, gases, aqueous species, and reactions from 1 to 5000 bars and 0 to 1000 °C. *Comput. Geosci.* **1992**, *18*, 899–947. [CrossRef]

49. Pokrovskii, V.A.; Helgeson, H.C. Thermodynamic properties of aqueous species and the solubilities of minerals at high pressures and temperatures: The system $Al_2O_3$–$H_2O$–NaCl. *Am. J. Sci.* **1995**, *295*, 1255–1342. [CrossRef]

50. Diakonov, I.; Pokrovski, G.; Schott, J.; Castet, S.; Gout, R. An experimental and computational study of sodium–aluminum complexing in crustal fluids. *Geochim. Cosmochim. Acta* **1996**, *60*, 197–211. [CrossRef]

51. Bessinger, B.; Apps, J.A. The Hydrothermal Chemistry of Gold, Arsenic, Antimony, Mercury and Silver. 2003. Available online: http://repositories.cdlib.org/lbnl/LBNL-57395 (accessed on 4 October 2019).

52. Shock, E.L.; Helgeson, H.C. Calculation of the thermodynamic and transport properties of aqueous species at high pressures and temperatures: Correlation algorithms for ionic species and equation of state predictions to 5 kb and 1000 °C. *Geochim. Cosmochim. Acta* **1988**, *52*, 2009–2036. [CrossRef]

53. Chudnenko, K.V.; Pal'yanova, G.A.; Anisimova, G.S.; Moskvitin, S.G. Physicochemical modeling of formation of Ag–Au–Hg solid solutions: Kyuchyus deposit (Yakutia, Russia) as an example. *Appl. Geochem.* **2015**, *55*, 138–151. [CrossRef]

54. Palyanova, G. Physicochemical modeling of the coupled behavior of gold and silver in hydrothermal processes: Gold fineness, Au/Ag ratios and their possible implications. *Chem. Geol.* **2008**, *255*, 399–413. [CrossRef]

55. Zhuravkova, T.V.; Palyanova, G.A.; Chudnenko, K.V.; Kravtsova, R.G.; Prokopyev, I.R.; Makshakov, A.S.; Borisenko, A.S. Physicochemical models of formation of gold–silver ore mineralization at the Rogovik deposit (Northeastern Russia). *Ore Geol. Rev.* **2017**, *91*, 1–20. [CrossRef]

56. Chudnenko, K.V.; Pal'yanova, G.A. Thermodynamic properties of solid solutions in the Ag–Au–Cu system. *Russ. Geol. Geophys.* **2014**, *55*, 349–360. [CrossRef]

57. Yokokawa, H. Tables of thermodynamic properties of inorganic compounds. *J. Natl. Chem. Lab. Ind.* **1988**, *83*, 27–118.

58. Likhachev, A.P. The conditions of magnetite and its ore clusters formation. *Natl. Geol.* **2017**, *4*, 44–53. (In Russian)

59. Zotov, A.; Kuzmin, N.; Reukov, V.; Tagirov, B. Stability of $AuCl_2^-$ from 25 to 1000 °C at pressures to 5000 bar, and consequences for hydrothermal gold mobilization. *Minerals* **2018**, *8*, 286. [CrossRef]

60. Seward, T.M.; Williams-Jones, A.E.; Migdisov, A.A. The chemistry of metal transport and deposition by ore-forming hydrothermal fluids. In *Treatise on Geochemistry*, 2nd ed.; Turekian, K., Holland, H., Eds.; Elsevier: New York, NY, USA, 2014; pp. 29–57.

61. Murzin, V.V. *Gold Mineralization in the Ultramafites of the Urals. Mafic-Ultramafic Complexes of Folded Regions and Related Deposits*; Abstract of the 3D International Conference; Institute of Geology and Geochemistry UB of RAS: Ekaterinburg, Russia, 2009; Volume 2, pp. 61–64.

62. Murzin, V.V.; Sustavov, S.G.; Mamin, N.A. *Gold and Platinum-Group Element Mineralization of Placer Deposits of the Verkh-Neivinsky Massif of Alpine-Type Ultrabasites (the Middle Urals)*; Ural Mining and Geological Academy: Yekaterinburg, Russia, 1999; 93p.

63. Sazonov, V.N.; Murzin, V.V.; Ogorodnikov, V.N.; Volchenko, Y.A. Gold mineralization associated with alpinotype ultrabasites (on the example of the Urals). *Litosfera* **2002**, *4*, 63–77. (In Russian)

64. Zhmodik, S.M.; Mironov, A.G.; Zhmodik, A.S. *Gold-Concentrating Systems of Ophiolite Belts (by the Example of the Sayan-Baikal-Muya Belt)*; Academic Publishing House Geo: Novosibirsk, Russia, 2008; 304p.

65. Paraskevopoulos, G.M.; Economou, M.I. Genesis of magnetite ore occurrences by metasomatism of chromite ores in Greece. *Neues Jahrb. Mineral. Abh.* **1980**, *140*, 29–53.

66. Bakirov, A.G. The relationship between pyrite mineralization and magnetite and sulfide occurrences in ultramafic rocks of the Southern Urals. In *Minerals of Ore Deposits and Pegmatites of the Urals*; UBAS USSR: Sverdlovsk, Russia, 1965; Volume 6, pp. 185–192. (In Russian)

67. Singh, R.N.; Srivastava, S.N.P. Ophiolites and associated mineralization in the Naga Hills, north-eastern India. In *Ophiolites, Proceedings of the International Ophiolite Symposium, Cyprus, Nicosia, 1979*; Panayiotou, A., Ed.; Cyprus Geological Survey: Nicosia, Cyprus, 1980; pp. 758–764.

68. Zucchetti, S.; Mastrangelo, F.; Rossetti, P.; Sandrone, R. Serpentinization and metamorphism: Their relationships with metallogeny in some ophiolitic ultramafics from the Alps. In *Zuffar' Days Symposium in Honor of Piero Zuffardi*; University of Cagliari: Cagliari, Italy, 1988; pp. 137–159.

69. Diella, V.; Ferrario, A.; Rossetti, P. The magnetite ore deposits of the southern Aosta Valley: Chromitite transformed during an alpine metamorphic event. *Ofioliti* **1994**, *19*, 247–256.

70. Agafonov, L.V.; Lkhamsuren, J.; Kuzhuget, K.S.; Oidup, C.K. *Platinum-Group Element Mineralization of Ultramafic-Mafic Rocks in Mongolia and Tuva*; Mongolian State University of Science and Technology: Ulaanbaatar, Mongolia, 2015; 224p. (In Russian)

71. Gahlan, H.A.; Arai, S.; Ahmed, A.H.; Ishida, Y.; Abdel-Aziz, Y.M.; Rahimi, A. Origin of magnetite veins in serpentinite from the Late Proterozoic Bou-Azzer ophiolite, Anti-Atlas, Morocco: An implication for mobility of iron during serpentinization. *J. Afr. Earth Sci.* **2006**, *46*, 318–330. [CrossRef]

72. Einaudi, M.T.; Hedenquist, J.W.; Inan, E.E. Sulfidation state of fluids in active and extinct hydrothermal systems: Transitions from porphyry to epithermal environments. In *Volcanic, Geothermal and Ore-Forming Fluids: Rulers and Witnesses of Processes within the Earth*; Simmons, S.F., Graham, I., Eds.; Society of Economic Geologists Special Publication: Littleton, CO, USA, 2003; pp. 285–314.

73. Barton, P.; Toulmin, J.P. The electrum-tarnish method for the determination of the fugacity of sulfur in laboratory sulfide systems. *Geochim. Cosmochim. Acta* **1984**, *28*, 619–640. [CrossRef]

*Article*

# Experimental Modeling of Noble and Chalcophile Elements Fractionation during Solidification of Cu-Fe-Ni-S Melt

**Elena Sinyakova [1,*], Victor Kosyakov [2], Galina Palyanova [1,3] and Nikolay Karmanov [1]**

[1]    VS Sobolev Institute of Geology and Mineralogy, Siberian Branch of the Russian Academy of Sciences, pr. Akademika Koptyuga 3, 630090 Novosibirsk, Russia

[2]    Nikolaev Institute of Inorganic Chemistry, Siberian Branch of the Russian Academy of Sciences, Lavrent'ev ave. 3, 630090 Novosibirsk, Russia

[3]    Department of geology and geophysics, Novosibirsk State University, Pirogova str., 2, 630090 Novosibirsk, Russia

*    Correspondence: efsin@igm.nsc.ru; Tel.: +7-383-373-05-26

Received: 18 July 2019; Accepted: 29 August 2019; Published: 31 August 2019

**Abstract:** We carried out a directed crystallization of a melt of the following composition (in mol. %): Fe 31.79, Cu 15.94, Ni 1.70, S 50.20, Sn 0.05, As 0.04, Pt, Pd, Rh, Ru, Ag, Au, Se, Te, Bi, and Sb by 0.03. The obtained cylindrical sample consisted of monosulfide solid solution (*mss*), nonstoichometric isocubanite (*icb**), and three modifications of intermediate solid solution ($iss_1$, $iss_2$, $iss_3$) crystallized from the melt. The simultaneous formation of two types of liquids separated during cooling of the parent sulfide melt was revealed. In the first, concentrations of noble metals associated with Bi, Sb, and Te were found. The second is related to Cu and was found to contain a large amount of S in addition to Bi and Sb. We established the main types of inclusions formed during fractional crystallization of Pt-bearing sulfide melt. It was shown that noble metals are concentrated in inclusions in the form of $RuS_2$, $PdTe_2$, $(Pt,Pd)Te_2$, PtRhAsS, and $Ag_2Se$, doped with Ag, Cu, and Pd, in *mss* and in the form of $PtAs_2$; Au-doped with Ag, Cu, and Pd; $Ag_2Te$; and $Pd(Bi,Sb)_xTe_{1-x}$ in *icb** and *iss*. As solid solutions in the base metal sulfides, Rh is present in *mss*, Sn in *iss*.

**Keywords:** Cu-Fe-Ni-S system; platinum-group elements; Au; Ag; metalloid elements; fractional crystallization; drop-shaped inclusions

---

## 1. Introduction

Massive ore bodies of copper-nickel deposits are formed from zones with various phases and chemical compositions. When passing from one zone to another, their phase and chemical composition changes abruptly [1–4]. Such a structure suggests that they resulted from fractional crystallization of magmatic sulfide melt [5–9]. For the Noril'sk deposit, no less than two main types of zonality are observed, in one of which the fugacity of sulfur increases, and in the second, it decreases [1,2,10]. The possibility of several types of zonality is shown in the experiments on directed crystallization of Cu-Fe-Ni-sulfide melts [11–19] (The terms "directed crystallization" and "fractional crystallization" describe the process of gradual solidification of the melt in the absence of mixing in the solid ingot and complete or partial mixing in the melt. From the physicochemical point of view, when modeling the solidification of a cylindrical ingot in laboratory or industrial equipment, the term "directed crystallization" is more acceptable).

The Noril'sk ores contain a wide variety of geochemically important trace elements. The most significant are noble metals (Pt, Pd, Rh, Ru, Ir, Au, Ag), heavy metals (Zn, Sn, Pb), and anion-forming elements (As, Sb, Bi, Se, Te). Noble metals can enter the lattice of the main ore-forming minerals or form independent phases both with the main elements and with other trace elements. Since the Noril'sk ores are an important industrial source of the platinum group elements (PGE), intense research on the regularities of Ag-Au-PGE mineralization has been carried out [1–4,20–24].

Most geologists assume that magmatic sulfide liquids are the source of PGE [2,4,10,25]. In [24], a hypothesis is presented about the pneumatolytic genesis of compounds of noble metals. Both hypotheses are based on the results of geochemical observations and on experimental data [1,2,4,10,17–19,26–35]. Fundamental knowledge about the behavior of noble metals is contained primarily in the phase diagram of the systems related to this problem (see the review in [36–41]). Useful information is contained in experimental data on the composition and structure of samples obtained by prolonged isothermal annealing of partially crystallized samples and further quenching to room temperature [30–32,36–41]. For example, in [36], diagrams of the following systems were described: Pd-Fe-S, Pd-Ni-S, Pd-Cu-S, Pd-Pt-S, Fe-Pt-S, Pt-Fe-As-S, Pt-Pd-As-S, Pd-Ni-As, Pd-Sb-Te, Pt-Pd-Sb, Pd-Bi-Te, Pd-Pt-Sb, and so on. In this case, the composition of the samples is chosen by the experimenter. By contrast, in our works, we studied samples obtained by directed crystallization of melts of a given composition [11–19]. Usually, this method is applied in the case of gradual solidification of a cylindrical sample from one end to another. In this process, the melt is separated from the crystallized mass by a single interface—the crystallization front. The difference in the compositions of the melt and coexisting solid phases leads to fractionation of the components during crystallization, i.e., to a gradual change in the compositions of the melt and solid phases. With a particular melt composition, a phase reaction can run at the crystallization front, in the course of which some of the phases produced from the melt can disappear and new phases can appear. Therefore, crystallization of a multicomponent melt usually produces a sample consisting of several zones with different phase compositions. A similar zoning is also observed in sulfide orebodies.

Below, we describe the main advantage of the directional crystallization method in a quasi-equilibrium regime to study the behavior of elements during the fractional crystallization of sulfide melts. First, an experiment with this process permits a large set of data on the equilibrium phase diagram of a multicomponent (the main and minor elements) system to be obtained. Second, the experiment helps to study the behavior of the main and minor elements both during crystallization and during a subsequent cooling of the crystallized sample. Third, with this method, it is possible to determine the sequence of phase formation during fractional crystallization, which is rather difficult to do using the results of the study of annealed samples.

During the directed crystallization of a sample of a given composition, the "chemical system itself chooses" the crystallization path. This path in the quasi-equilibrium process uniquely depends on the composition of the initial sample and the liquid–solid diagram. This method can be used for systems with an arbitrary number of components, and its results, for constructing both simple and complex diagrams. In experiments on isothermal annealing, the set of quenched samples is studied, the composition of which is determined by the experimenter.

The crystallized zone of the ingot imitates the structure of the ore body. The ingot is used for preparing genetically related samples, the composition of which is determined by the rules of directed crystallization. Thus, we simulated the formation of pyrrhotite-cubanite, pentlandite-bornite, and other mineral varieties of massive ore bodies from the Noril'sk deposit and also studied the behavior of ensembles of minor elements [11–14,16–18,33,42]. In particular, Cu-Fe-Ni sulfide melts with the following ensembles of impurities were crystallized by this method: Pt, Pd, Rh, Ru, Ir, Au, Ag [42], Pt, Pd, Rh, Ru, Ir, Au, Ag, Co, As [17], Pt, Pd, Au, As, Te, Bi, and Sn [18,19,33]. The distribution curves of components were constructed, and phase and chemical compositions of inclusions containing PGE (drop-shaped included) were analyzed.

Results of the geochemical and experimental studies described in the literature are insufficient for the verification of two hypotheses (magmatic and pneumatolytic) about the impurity mineralization of the Noril'sk deposits. To understand the role of these processes, additional results are required. In this paper, we carried out experimental modeling of fractionation crystallization of melt formed from Cu, Ni, S, Pt, Pd, Rh, Ru, Ag, Au, As, Se, Sn, Te, Bi, and Sb, in a closed system, which corresponds to the magmatogenic mechanism of formation of solid ore bodies. New data on the evolution of the studied physicochemical system in this process were obtained.

## 2. Experimental

### 2.1. Sample Preparation

We carried out a directed crystallization of a melt of the following composition (in mol. %): Fe 31.79, Cu 15.94, Ni 1.70, S 50.20, Sn 0.05, As 0.04, Pt, Pd, Rh, Ru, Ag, Au, Se, Te, Bi, and Sb by 0.03. This composition imitates the composition of representative ore samples from the Octyabr'sky mine [26]. The initial sample, 20 g in weight, was prepared from sulfur (99.9999%), additionally purified by vacuum distillation, and other elements with a purity grade of 99.99%. The mixture was heated at a rate of about 100 deg/day up to 700 °C and then for a day up to 1050 °C. The melted sample was kept for 2 days and then quenched in a switched-off furnace. After dry grinding, the powdered sample was replaced into the ampoule, with an 8 mm diameter and a conical end, which was then evacuated and sealed. A scheme showing the experimental apparatus and the procedure of preparing the sample were described in detail in [18]. Crystallization was performed using the Bridgman method in a vertical two-zoned furnace, with a diaphragm, by lowering the ampoule with a homogeneous melt from the hot zone to the cold zone at a rate of $2.3 \times 10^{-8}$ m/s. This regime provided quasi-equilibrium conditions for directed crystallization. The temperature in the lower end of the quartz container was 1025 °C at the beginning of crystallization, and 825 °C at the end. After crystallization, the ampoule was cooled in air at an average rate of ~100 deg/min.

### 2.2. Investigation of Crystallized Samples

The obtained ingot, about 120 mm in length and 8 mm in diameter, was cut perpendicular to the longitudinal axis into 19 parts. These were weighed and the fraction of crystallized melt, *g*, was determined. Seventeen fragments were used to prepare polished sections, which were studied by microscopic and chemical analysis.

Results of the sample study showed that the crystallization front was flat, perpendicular to the axis of the ingot, and homogenous in the averaged chemical composition of components. The average chemical composition of the ingot and inclusions, as well as the local composition phases, were measured using energy dispersion spectrometry (SEM-EDS) on a high-resolution microscope MIRA 3 LMU (Tescan Orsay Holding, Brno–Kohoutovice, Czech Republic), combined with X-ray microanalysis systems INCA Energy 450+ X-Max 80 and INCA Wave 500 (Oxford Instruments Nanoanalysis Ltd, Abingdon, UK) in the Analytical Center for multi-elemental and isotope research SB RAS (analyst N.S. Karmanov, Novosibirsk, Russia). For the analysis, K-series (S, Fe, Cu, Ni) and L-series (Pt, Pd, Rh, Ru, Au, Ag, As, Te, Se, Bi, Sb, Sn) of X-ray radiation were used. As the standards, we used $FeS_2$ (on S), PbTe (on Te), $PtAs_2$ (on As), and the pure elements of Fe, Ni, Cu, Se, Ru, Rh, Pd, Ag, Sn, Sb, Pt, Au, and Bi. Phases smaller than 5 μm were analyzed using a point probe, and larger phases were analyzed in a small raster mode with the size of the scanned area up to 100 $\mu m^2$. The measurements were conducted at an accelerating voltage of 20 kV, electron beam current of 1.5 nA, and live acquisition time of spectra of 30 s. Under these conditions of analysis, the limit of detection (LOD) was 0.4 to 0.5 wt. % for Pt, Au, and Bi, and 0.1 to 0.2 wt. % for the others. The error in determining was no more than 1 to 1.5 relation % for the major components and 2 to 5 relation % for minor. To estimate the average composition of multi-phase areas, we used the total spectrum obtained by scanning the areas of up to 1.5 $mm^2$. To reduce the limit of detection to about 2 times, the accumulation time of spectra was increased to 120 s. The average composition of phase mixtures was calculated by 3 to 5 analyses from various areas of each section along the ingot. The error in determining the major components was 1 to 2 relation % [18].

Some specific features of determining the composition of analyzed phases by the SEM-EDS method are noteworthy. As a result of the low-resolution ability of EDS, the peaks of Au M-series lines significantly overlap the peaks of M-series lines of heavy platinoids and K-series of sulfur.

Unfortunately, the software of the spectrometer in this spectral region (1.9–2.5 keV) does not properly perform the deconvolution of the spectrum, which is probably due to the discrepancy between the model and the real shape of the lines. This leads to a distortion of the analysis results. Therefore, we used the L-series of radiation for these elements, though the lower LOD is 2-fold higher compared to the M-series. In this region of spectra (9–11 keV), we also observed a significant overlapping of lines of the L-series of PGE and Au, but the effect of the deconvolution error of the spectra was much smaller. It is noteworthy that the lack of spectrometer software can lead to a false detection of sulfur in the amount of 0.8 wt. % in gold-bearing phases, as the peak of the S $K_\alpha$ line is located on the "tail" of the Au $M_\alpha$ peak. Additional difficulties arise in the analysis of microphases, the typical size of which is smaller than the X-ray generation region. Errors in the determination of gold-bearing alloys can be related to the possible contamination of low hardness phases during preparation of the samples. These features were taken into account when processing and interpreting data.

The change in melt composition during crystallization was calculated by the formula:

$$c_i^L = \frac{c_{i0} - \int_0^g c_i^S dg}{1 - g} \tag{1}$$

Here, $g$ is the fraction of crystallized melt, $c_{i0}$ is the concentration of the $i$-th component in the initial ingot, $c_i^S$ is the average concentration of the $i$-th component on the surface of the polished section, and the $g$ coordinate is the concentration of the $i$-th component in the melt. The obtained results were used to determine the average distribution coefficients of components between solid phases and the sulfide melt:

$$k_i^j = c_i^j / c_i^L \tag{2}$$

where $c_i^j$ is the average concentration of the $i$-th component in the $j$ phase.

## 3. Results

### 3.1. Behavior of Base Components

Visually, the ingot consists of five zones with different chemical and mineral compositions. Primary zones appeared during the successive crystallization of base metal sulfides (BMS) in accordance with the solid–liquid diagram of the system Cu-Fe-Ni-S [6,43,44]. On cooling of the ingot, the primary minerals completely or partially decayed to form secondary low-temperature minerals. Their sequence formed a second zonality of the crystallized ingot. The examples of microstructures in the cooled sample are shown in Figure 1.

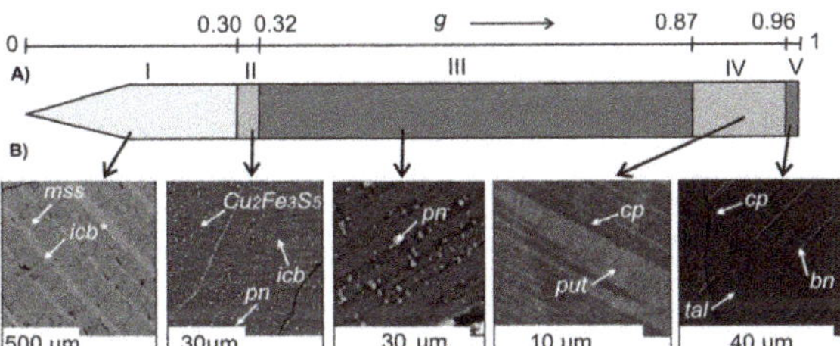


**Figure 1.** Structure of the directly crystallized sample. (**A**) Scheme of zonality. The coordinate of the process is a fraction of the crystallized melt $g$. (**B**) Back-scattered electron images of the polished sections belonging to different zones at $g$ 0.30 (I), 0.32 (II), 0.81 (III), 0.96 (IV), and ~1 (V). Designations of phases: *mss* is monosulfide solid solution $Fe_{43.0}Ni_{2.3}Cu_{1.1}S_{53.5}$, *icb* is isocubanite $CuFe_2S_3$, *icb** is non-stoichiometric isocubanite $Cu_{1.1}Fe_{1.9}S_3$, *pn* is pentlandite $(Fe,Ni)_9S_8$, *cp* is chalcopyrite $CuFeS_2$, *put* is putoranite $(Cu,Ni)_{1.1}Fe_1S_2$, *tal* is talnakhite $(Cu,Ni)_{18}Fe_{16}S_{32}$, bn is bornite $Cu_{4.3}Fe_{1.5}S_{4.2}$.

The change in the composition of the ingot along its length is presented in Table S1. To determine the primary zonality of the sample, we constructed the distribution curves of the components along the ingot and calculated their solid/liquid distribution coefficients. Data for Ni, Cu, and S are shown in Figure 2. Using these data, similar curves for Fe can easily be constructed. The distribution of components in the solid ingot along the zone was described by curved segments, and at the boundary between the zones there is a gap. The dependencies of component concentrations in the melt were described by piecewise continuous curves.

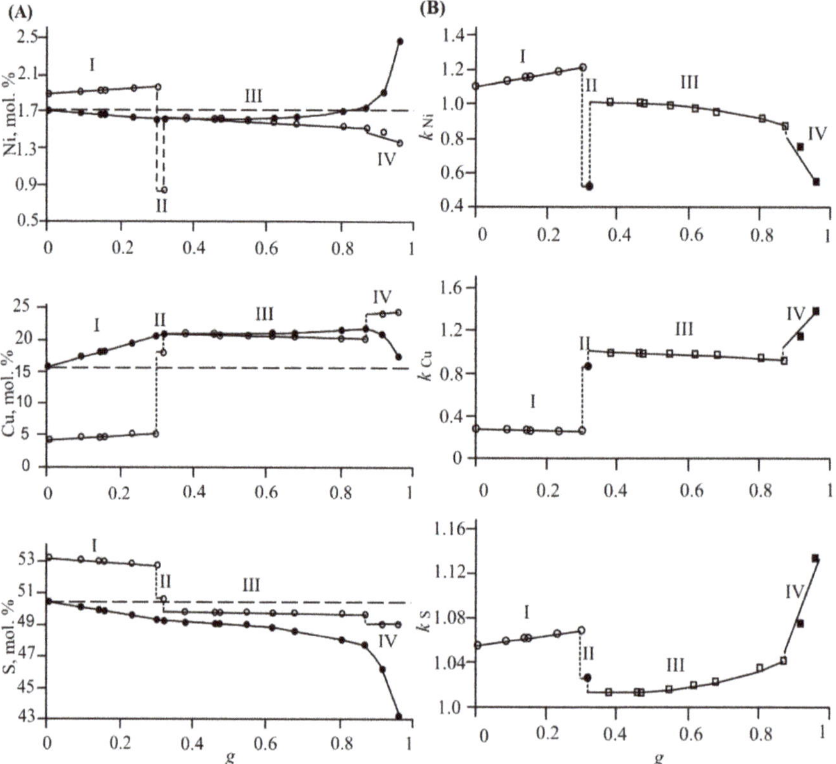

**Figure 2.** Primary chemical zoning of the sample. (**A**) Dependence of the average concentration of Ni, Cu, and S in the solid sample (open circles) and in the melt (closed circles). The composition of the initial ingot is shown by a horizontal dashed line. (**B**) Dependence of the distribution coefficients of Ni, Cu, and S on $g$. Open circles are $\kappa$ (*mss*/L), closed circles are $\kappa$ (*icb**/L), open squares are $\kappa$ (*iss$_1$*/L), and closed squares are $\kappa$ (*iss$_2$*/L). Dashed vertical lines divide the zones.

The average chemical composition of the substance in zone I ($0 \leq g \leq 0.3$) varies from $Fe_{40.54}Ni_{1.89}Cu_{4.38}S_{53.19}$ to $Fe_{39.86}Ni_{2.05}Cu_{5.22}S_{52.87}$. These data demonstrate that a monosulfide solid solution (*mss*) crystallizes from the melt. During crystallization, Fe and Ni mostly pass into *mss* ($k_{Fe}$ = 1.27). Sulfur has a tendency to be concentrated in the solid ingot ($k_S$ = 1.05–1.07), and Cu intensely enriches the melt. ($k_{Cu}$ = 0.25–0.28).

Zone II ($0.3 \leq g \leq 0.32$) occupies 2 vol. % of the sample, but it is clearly reflected in the distribution curves (Figure 2) The average chemical composition of substance in this zone is $Cu_{17.96}Fe_{30.59}Ni_{0.86}S_{50.58}$. It corresponds to nonstoichiometric isocubanite (*icb**) with an idealized formula $Cu_{1.1}Fe_{1.9}S_3$ [13]. The distribution coefficients are $k_{Ni}$ = 0.53 and $k_{Cu}$ = 0.86.

Zone III ($0.32 \leq g \leq 0.87$) occupies most of the ingot. The average composition varies from $Fe_{27.62}Ni_{1.65}Cu_{20.90}S_{49.83}$ to $Fe_{28.57}Ni_{1.52}Cu_{20.26}S_{49.65}$. We attributed this solution to the Ni-containing intermediate solid solution $iss_1$, described by Fleet and Pan (1994). It is worth noting that different forms of the existence of $iss$ were found in the Noril'sk ores [24]. The distribution coefficients of Ni and Cu are close to 1.

The average composition of substance in zone IV ($0.87 \leq g \leq 0.96$) $Fe_{25.44-25.25}Ni_{1.51-1.38}Cu_{24.00-24.32}S_{49.04-49.05}$ corresponds to $iss_2$. An intermediate solid solution of a similar composition is reported in [43]. During crystallization, the $iss_2$ melt becomes enriched with Fe ($k_{Fe}$ = 0.82–0.68) and Ni ($k_{Ni}$ = 0.79–0.56), and the solid phase, in Cu ($k_{Cu}$ = 1.15–1.40) and S ($k_S$ = 1.06–1.13).

The average composition of the substance in zone V ($0.96 \leq g < {\sim}1$) corresponds to $Fe_{23.65}Ni_{2.00}Cu_{26.19}S_{48.16}$. One can see that the composition of this substance is similar to $iss_1$ and $iss_2$ but is highly enriched with copper. We designated it as $iss_3$. This zone is small and crystallized at the very end of the ingot. This complicates the exact construction of the distribution curves in zone V under quasi-equilibrium conditions. This is the reason why zone V is not shown on the distribution curves.

The micrographs shown in Figure 1 characterize the secondary zonality of the sample. The microstructure of the sample in zone I consists of the matrix of *mss* $Fe_{43.0 \pm 0.2}Ni_{2.3 \pm 0.1}Cu_{1.1 \pm 0.1}S_{53.5 \pm 0.1}$ and lamellar inclusions of nonstoichiometric isocubanite $Cu_{1.1}Fe_{1.9}S_3$ (*icb**). On cooling, *icb** decomposed into a mixture of two phases. The main phase is isocubanite of stoichiometric composition (*icb*) with 0.4 mol. % Ni. Thin oriented lamellas of the second phase $Cu_3Fe_4S_{7.1}$ are contained in the matrix of *icb*.

Zone II contained the exsolution products of primary *icb**: Oriented lamellas from $CuFe_2S_3$, phases of composition similar to $Cu_2Fe_3S_5$, and small grains with a content of Ni of 8 mol. %. Most likely, these are inclusions of pentlandite *pn* $(Fe,Ni)_9S_8$.

The microstructure of the sample in zone III consists of a two-lamellar phase and small light inclusions (see Figure 1). We could not determine the exact composition of the phases. The light inclusions contain 18 mol. % Ni and about 48 mol. % S. Most likely, they are pentlandite.

Zone IV is formed of exsolution products of $iss_2$. It contains Ni-bearing chalcopyrite $Fe_{24.1}Ni_{0.8}Cu_{24.6}S_{49.9}$ and putaronite $Fe_{24.8}Ni_{1.9}Cu_{24.2}S_{49.1}$.

Exsolution products of $iss_3$ in zone V are bornite $Fe_{14.5}Ni_{0.2}Cu_{42.2}S_{43.1}$, talnakhite $Fe_{24.1}Ni_{1.1}Cu_{26.4}S_{48.4}$, and chalcopyrite $Fe_{24.1}Ni_{0.8}Cu_{26.1}S_{49.0}$.

### 3.2. Behavior of Microcomponents

### 3.2.1. Solid Solutions of Impurities in BMS

During crystallization of the melt, the impurities may pass into the solid ingot in the form of solid solutions in primary BMS or form independent minerals. In the former case, the distribution of impurities between the solid and liquid is characterized by the values of the distribution coefficients ($k$). In this work, we determined the $k$ for Rh in *mss* and for Sn in *iss*. The contents of other impurities in *mss* and in *iss* in our experiment were below the LOD of the EDS. It is noteworthy that in [37,39,44–46], laser ablation was used to measure the distribution coefficients of Pt, Pd, Ru, Au, As, Te, Bi, Sb, Sn, and Se between *mss* and melt. All of them, besides Ru, were <1.

Figure 3 shows the distribution curves of Rh in *mss* in zone I. The calculated distribution coefficient, $k_{Rh}$, is $5.2 \pm 0.5$, i.e., this element is concentrated in *mss*. In the studies on crystallization of *mss* of another composition, $k_{Rh}$ (*mss/L*) is also greater than 1 and ranges from 1.5 to 6.6 [37,44,46]. We measured the dependence of $k_{Rh}$ on melt composition in the system of Fe-Ni-S in the crystallization region of *mss* [47]. It was shown that the region is separated into two sites, one of them enriched with Ni, $k_{Rh} > 1$, and in the other site, $k_{Rh} < 1$.

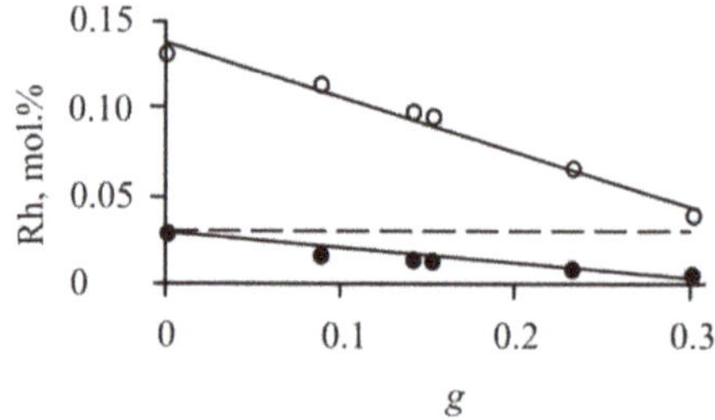

**Figure 3.** Distribution curves of Rh in zone I of the directly crystallized sample. Light circles correspond to the concentration of Rh в *mss*, and dark circles, in the melt. Dashed horizontal line shows the concentration of Rh in the initial melt.

The average concentration (mol. %) of Sn is 0.20 in $iss_1$, 0.25 in $iss_2$, and 0.30 in $iss_3$. Dependence of the distribution coefficient of Sn between $iss_1$ and melt on the fraction of crystallized melt, $g$, is shown in Figure 4.

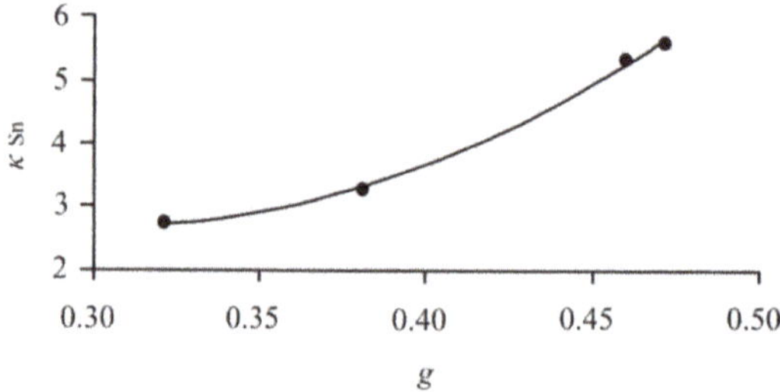

**Figure 4.** Dependence of the distribution coefficient of Sn between $iss_1$ and the melt on the fraction of crystallized melt, $g$.

### 3.2.2. Inclusions of Minor Phases in BMS

The main part of the impurities is present in inclusions smaller than ~100 μm. Most of them are formed from several phases.

Inclusions in *mss* (zone I): It was shown above that in zone I, a primary *mss* partly decomposes on cooling to form lamellas of *icb**. Microphases are present in the matrices of *icb**, *mss*, and at the grain boundaries of *icb** and *mss*, as well as on the surface of pores. We analyzed 147 single-phase inclusions. It was found that these were laurite $RuS_2$, tsumoite BiTe, nevskite Bi(Se,S), naumannite $Ag_2Se$, merenskyites $PdTe_2$, moncheites $(Pt,Pd)Te_2$, minerals of the Pt-Rh-As-S system, most likely platarsite [27], as well as two-phase intergrowths of Au** and $(Pt,Pd)Te_2$ (Table S2). Typical samples of inclusions are shown in Figure 5.

Laurite $RuS_2$ forms numerous faceted crystals <40 μm in size (Figure 5A,B). These contain about 1 mol. % Rh and Fe. The crystals are concentrated near the sample surface. Inclusions of merenskyites, moncheites, and platarsite are present in a large number and are evenly distributed in the cross section of the sample. They have typical sizes of <1–3 μm, which may lead to an increase in the inaccuracy of determining their composition. The inclusions of platinum and palladium tellurides are irregular-shaped and are associated with lamellas of *icb** (Figure 5C,D). Platarsite forms two types of inclusions: Weakly faceted crystals and rosettes (Figure 5G,H). It is worth noting that rosettes have the fifth-order symmetry axis, i.e., can form quasi-crystals. The number of inclusions of naumannite, nevskite, and tsumoite are small. They have a weakly faceted form and are localized in the intergranular cracks in the edge of the sample section (Figure 5B,E,F). Intergrowths of Au and (Pt, Pd)$Te_2$ are <1 μm in size (Figure 5I). They are localized on the surface of the pores.

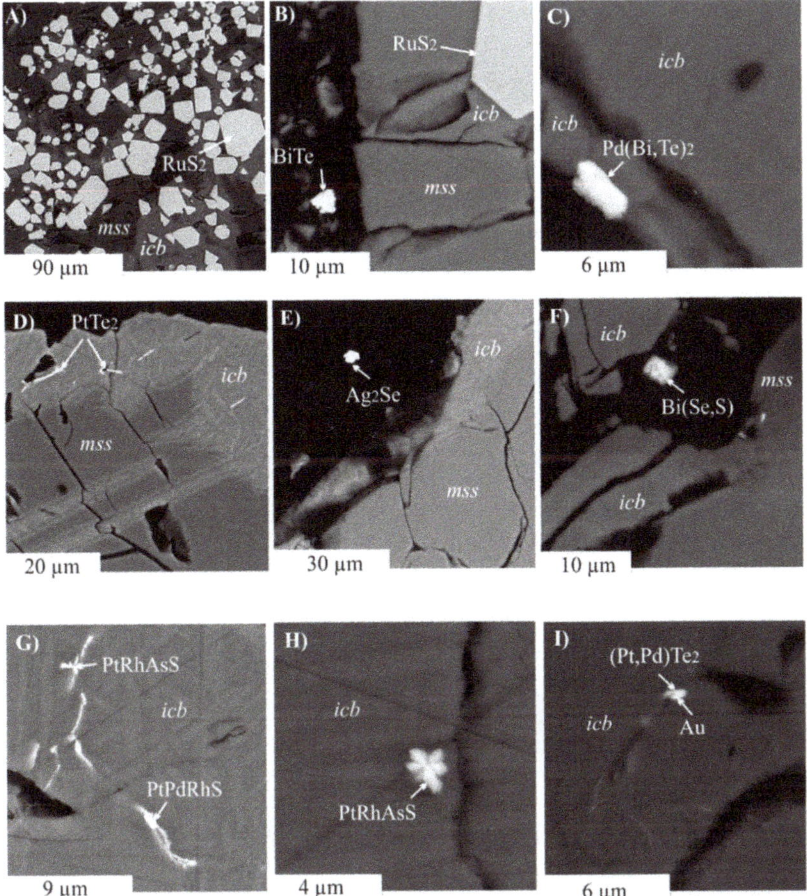

**Figure 5.** Back-scattered electron images of microphases in the *mss* zone. A primary *mss* partly decomposes on cooling to form lamellas of *icb*. Microphases (bright) are present in the matrices of *icb*, *mss*, and at the grain boundaries of *icb* and *mss*, as well as on the surface of pores. Coordinate *g* equals 0.002 (**A,B,F**), 0.09 (**H**), 0.14 (**D,G**), 0.15 (**C,E**), and 0.23 (**I**). Black is pores.

Inclusions in Fe-Cu sulfides (zones II–V): This part of the ingot is formed from Fe- and Cu-rich sulfides (isocubanite, chalcopyrite, talnakhite, putoranite, bornite) and pentlandite. These zones contain numerous small (<20 μm) and large (to 1 mm) pores. As an example, Figure 6 shows the cross section of the sample at $g = 0.38$ (zone III). The small pores are seen to be arranged in parallel rows, which is due to the specific trapping mechanism of gas bubbles, present in the sulfide melt, by a single crystal of $iss_1$. These rows, probably, resulted from the entrapment of bubbles by nanosteps during the layer-by-layer growth of the single crystal from the melt. Large pores are formed by another mechanism. The average size and number of pores increases to the end of the ingot.

Minor minerals form multiphase inclusions. The exception is sperrylite $PtAs_2$ crystals, which may be present in the form of both single-phase inclusions and polyphase intergrowths. The main quantity of minor elements is concentrated in the irregular-shaped inclusions of up to 100 μm in size at the interfaces (Figure 7). These inclusions were found to contain the following minerals: Sobolevskite-kotulskite solid solution $Pd(Bi,Sb)_xTe_{1-x}$, Au with minor Ag, Cu, Pd, hessite $Ag_2Te$, S-rich sperrylite $PtAs_2$, wittichenite $Cu_3BiS_3$, stibiowittichenite $Cu_3SbS_3$, parkerite $Ni_3Bi_2S_2$, tetradymite

$Bi_2Te_2S$, nevskite $Bi(Se,S)$, gersdorffite $NiAsS$, emplectite $CuBiS_2$, and tsumoite $BiTe$. In addition, the matrix of BMS contains small ($\leq$20 μm) single multiphase drop-shaped inclusions, and faceted and non-faceted inclusions of $PtAs_2$. The inclusions of minor minerals are also present in some pores.

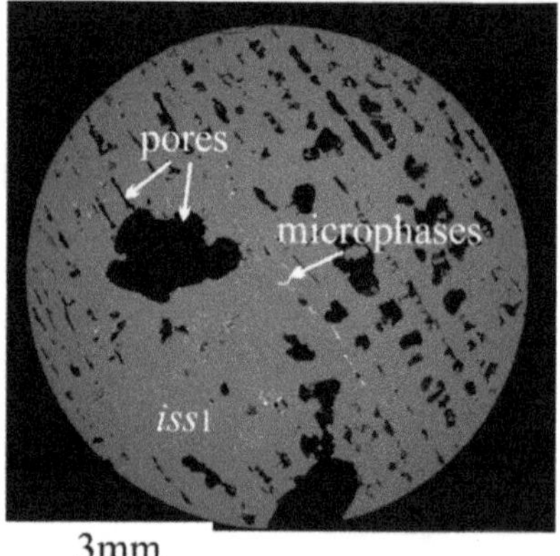


**Figure 6.** Microphotograph of the cross section of the ingot at $g$ = 0.38. Light-gray matrix consists of $iss_1$. Small light inclusions are minor phases. Dark inclusions are pores.

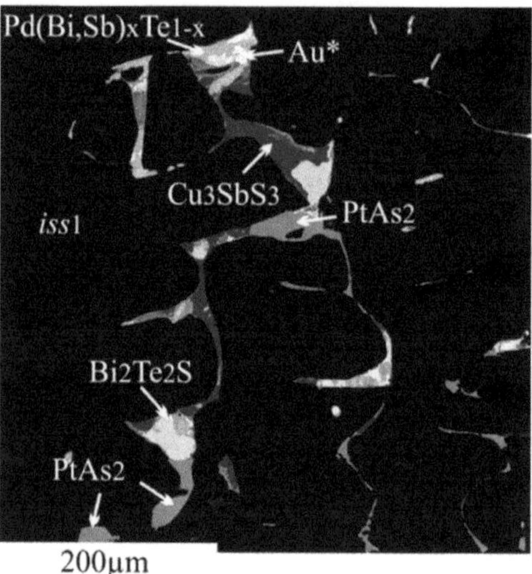


**Figure 7.** Backscattered electron image. The typical microstructure of the sample at $g$ = 0.47. Minor minerals form multiphase inclusions in the $iss_1$ matrix. Black is pores.

Figure 8 shows the pore with multiphase inclusions of minor minerals and sperrylite. It was found that the inclusions in the matrix of BMS and the inclusions associated with pores have similar chemical and mineral compositions, i.e., were formed by the same mechanism.

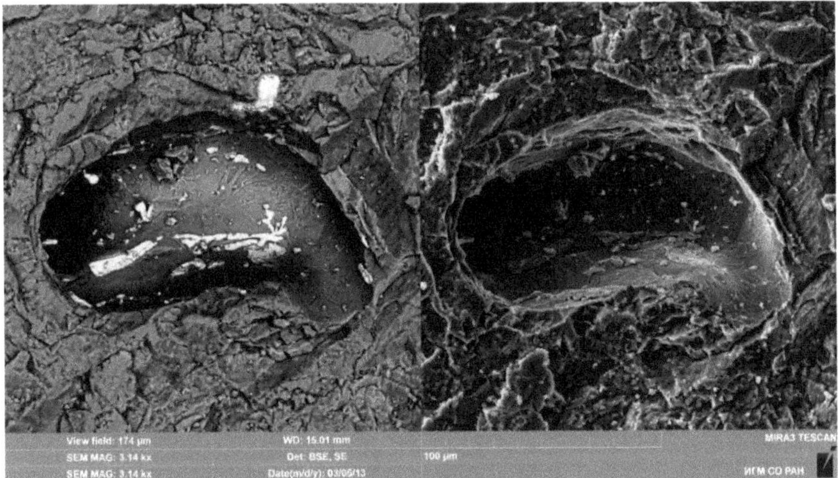

**Figure 8.** Microphotograph of the cleavage surface of the sample from zone II at $g = 0.3$ in reflected electrons on the left and in secondary electrons on the right. There are inclusions of microphases in the gas cavity.

In the matrixes of BMS, there are a great number of drop-shaped multiphase inclusions (of oval and irregular shape with a smooth contour). We analyzed 132 inclusions and divided them into three classes according to their chemical and phase composition: 39 inclusions of class I, 21 of class II, and 72 of class III. Examples of microstructures are shown in Figures 9–11. The phase and chemical compositions are described in Tables S2 and S3.

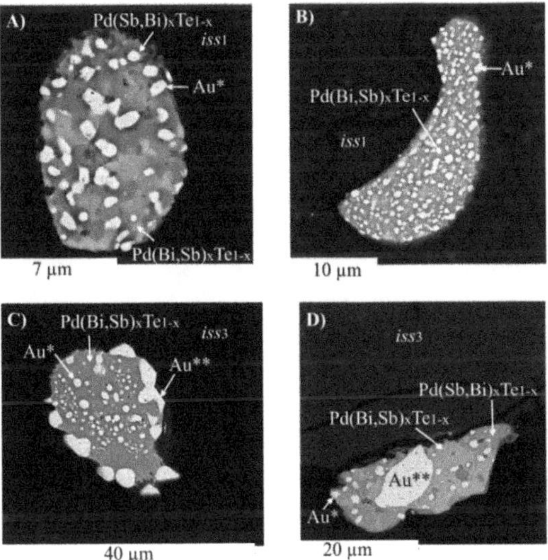


**Figure 9.** Backscattered electron images of the typical microphases of class I inclusions, located in the matrix of $iss_1$ (**A,B**) and $iss3$ (**C,D**). Drop-shaped inclusions consist of sobolevskite-kotulskite solid solution $Pd(Bi,Sb)_xTe_{1-x}$ and Au alloys (Au* and Au**).

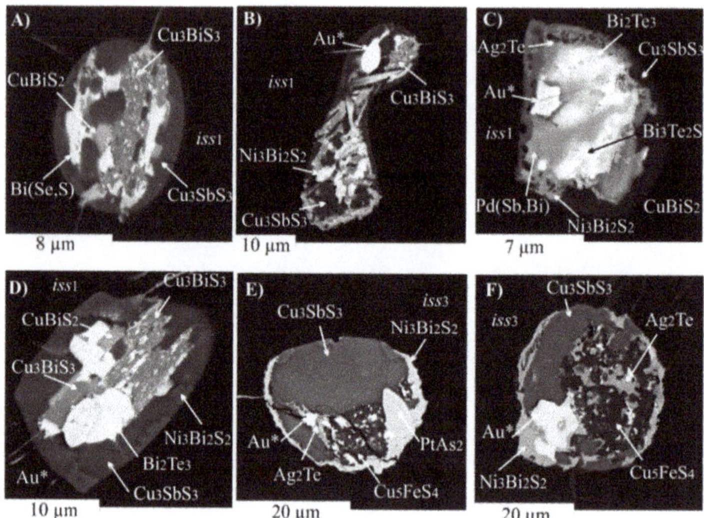

**Figure 10.** Backscattered electron images. Typical microstructure of class II polyphase inclusions, located in the matrix of $iss_1$ (**B–D**) and $iss_3$ (**A,E,F**). Drop shape or more complicated shape inclusions consist of $Cu_3BiS_3$, $Cu_3SbS_3$, $Bi(Se,S)$, $Ni_3Bi_2S_2$, and $CuBiS_2$ sulfosalts. Inside the inclusions Au*, $Ag_2Te$, Bi, and fine-dispersed unknown phases occur.

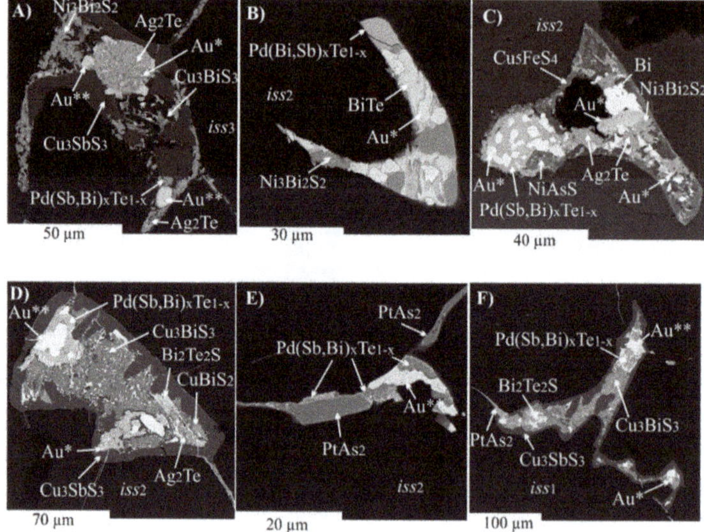

**Figure 11.** Backscattered electron images. Typical structure of composite inclusions of class III in $iss_3$ (**A**), $iss_2$ (**B–E**), and $iss_1$ (**F**). Class III contains composite inclusions from the fragments of I and II classes and refractory microcrystals of $PtAs_2$.

Class I. The matrix of inclusions was formed from sobolevskite-kotulskite solid solution $Pd(Bi,Sb)_xTe_{1-x}$ described in [27,48]. When cooled, the solution separates into phases on the basis of sobolevskite PdTe and kotulskite PdBi (Figure 9). This two-phase matrix contains numerous inclusions $\leq 1$ μm in size from Au alloy (Au*, 84 wt. % Au) (Table S2). The structure in Figure 9A,B was, most likely, formed during eutectic crystallization. The drop-shaped inclusion in Figure 9C has the same composition and structure as in Figure 9A,B. However, on its surface, large (3–5 µm) crystals of high-fineness Au alloy (Au**, 97 wt. % Au) are localized. The melting temperature ($T_m$) of Au** is higher than that of Au*. Figure 9D shows another type of inclusion: Au** crystal of ~10 µm in size has two neighboring drop-shaped inclusions.

Determination of the average chemical composition showed that the inclusions of class I do not contain S and concentrates of Pd, Au, and Ag (Table S3).

Class II. The examples of this class of inclusions of a drop shape or a more complicated shape are shown in Figure 10. They were formed from sulfosalts with idealized formulas of $Cu_3BiS_3$, $Cu_3SbS_3$, $Bi(Se,S)$, $Ni_3Bi_2S_2$, and $CuBiS_2$ (Table S2). Most drop-shaped inclusions are located inside the shell of $Cu_3SbS_3$. Frequently, whiskers were observed on the surface of the shell. Inside the inclusions, fragments of $Cu_3BiS_3$ with a fine-dispersed unknown phase, Au*, $Ag_2Te$, and Bi, occurs.

The inclusions of class II contain a large amount of S and Cu and minor quantities of Sb and Te. Moreover, they contain minor silver and gold and lack Pd. The average composition of inclusions of class II and their phase composition changes along the ingot (Table S3). The constancy of the average chemical composition of inclusions of classes I and II along the ingot is noteworthy.

Class III. The matrix of Fe-Cu sulfides contains compound inclusions of various shapes. Their typical sizes range from ~10 to ~100 µm. The average composition of inclusions varies in a wide range (Table S2). The inclusions in Figure 11A–E consist of fragments that belong to classes I and II. Less frequent are inclusions consisting of combinations of $(I + PtAs_2)$ and $(I + II + PtAs_2)$ (Figure 11E,F).

Let us consider in more detail the behavior of some microminerals of noble metals.

Sobolevskite-kotulskite solid solution $Pd(Bi,Sb)_xTe_{1-x}$ is the main concentrator of Pd. This phase together with Au* and Au** forms, most likely, the eutectic-like structure. It is typical of drop-shaped inclusions of class I. The composition of solid solution varies in a wide range: Pd from 30 to 44 mol. %, Bi from 3 to 20 mol. %, Sb from 11 to 29 mol. %, and Te from 8 to 38 mol. %. The systems of Pd-Te-Bi and Pd-Te-Sb contain continuous regions of solid solutions between PdTe-PdBi and PdTe-PdSb [27,36,48].

The Au alloys are presented in drop-shaped inclusions as numerous non-faceted or weakly faceted crystallites ranging from <1 to ~10 µm in size in the matrix of $Pd(Bi,Sb)_xTe_{1-x}$. Analysis showed that the content of Ag in Au alloys ranges from 3.2 to 18.0 mol. %. Additionally, Au can dissolve copper (4.3–7.6 mol. %) and palladium (0.8 mol. %) (Table S2). These results are consistent with the state diagrams of binary systems, Au-Ag, Au-Cu, and Au-Pd, and the ternary system, Au-Ag-Cu, whose phase diagrams show the presence of wide regions of solid solutions [49,50].

Hessite $Ag_2Te$ is associated mainly with sulfosalts in the inclusions of classes II and III. It is worth noting that in the Ag-Te system, tellurium-rich phases with low melting temperatures are also present [49].

S-rich sperrylite is the main Pt carrier in the crystallized sample. It is the only compound that forms single-phase inclusions. Besides, it is a fragment of compound inclusions. Owing to the different genesis, these varieties differ in the content of sulfur (about 17 and 25 mol. % S, respectively). In addition, all inclusions contain minor contents of Cu, Pd, Sb, and Bi (Table S2). High solubility of S in $PtAs_2$ was found in [36].

There are more than 100 mineral types of sulfosalts (Godovikov, 1992). In our samples, the major sulfosalts are $Cu_3BiS_3$ and $Cu_3SbS_3$. Moreover, they contain $Ni_3Bi_2S_2$, $Bi_2Te_2S$, $Bi(Se,S)$, NiAsS, and $CuBiS_2$. The presence of other finely dispersed sulfosalts is also probable. It is noteworthy that the grains of elementary Bi with minor Cu, Fe, Sb, and Te were observed.

## 4. Discussion

As described above, a crystallized ingot consists of five primary zones with different chemical and phase compositions: *mss* (zone I), *icb** (zone II), *iss*$_1$ (zone III), and *iss*$_2$ (zone IV). Zone V is not described in detail. These zones are sequentially formed with a gradual decrease in the temperature of the melt. In zone I, a phase reaction proceeds with the formation of a monosulfide solid solution from the melt: $L \rightarrow mss$. At the boundary of zones I and II, non-stoichiometric isocubanite is formed by reaction: $L + mss \rightarrow icb^*$. When zone III occurs, the peritectic reaction of the formation of the intermediate solid solution $iss_1$ proceeds: $L + icb^* \rightarrow iss_1$. At the boundary between zones III and IV, $iss_2$ is formed by reaction: $L + iss_1 \rightarrow iss_2$. At the end of the ingot, $iss_3$ is formed, presumably by the reaction: $L + iss_2 \rightarrow iss_3$. We carried out differential thermal analysis studies of samples from different zones of the sample. According to these data, the onset crystallization temperature is 1025 °C for *mss*, 953 °C for *icb**, 947 °C for $iss_1$, and 910 °C for $iss_2$.

In the experimental samples, minor amounts of noble metals were dissolved in BMS (Rh в *mss*, Sn in *iss*) and in other phases (Ag in Au*, Au**; Au, Pt, Pd in Bi). Such forms were observed both in Noril'sk ores and in synthetic samples [2,4,10,18,23,26,33,34,37,51]. However, most parts of the noble metals are present in the inclusions in the form of their own minerals—alloys and compounds with metalloid admixtures (Au*, Au**, $PtAs_2$, $Pd(Bi,Sb)_xTe_{1-x}$, $Ag_2Te$). In addition, inclusions of sulfosalts also exist that do not contain noble metals of $Cu_3BiS_3$, $Cu_3SbS_3$, $Ni_3Bi_2S_2$, $Bi_2Te_2S$, $Bi(Se,S)$, $NiAsS$, and $CuBiS_2$. Using the obtained experimental data, the classification scheme of inclusions in the sample was constructed (Figure 12).

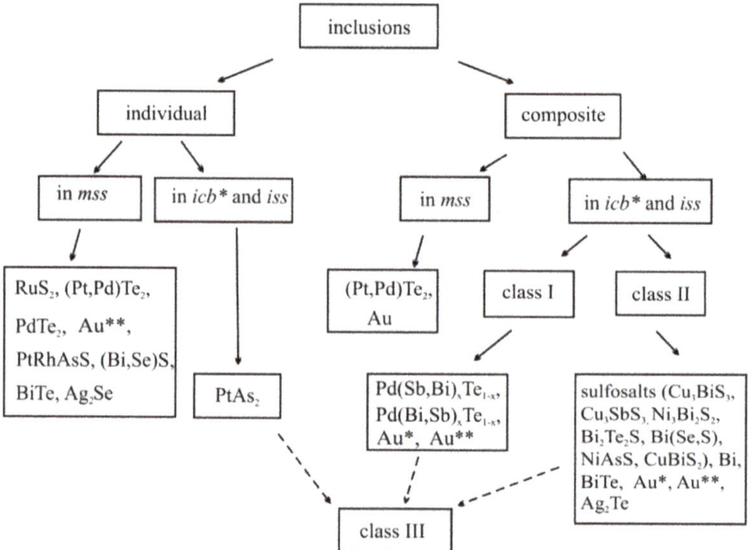


**Figure 12.** Classification scheme of inclusions of minor minerals.

The refractory compounds $RuS_2$, $(Pt,Pd)Te_2$, $PtRhAsS$, $Ag_2Se$, and $PtAs_2$ have melting temperatures higher than the liquidus temperature in the Cu-Fe-Ni-S system. Therefore, they crystallize in sulfide melt and then are trapped during the crystallization of BMS. The existence of two types of drop-shaped inclusions suggests that they resulted from the solidification of liquid drops of different compositions. Most likely, these drops separated as a result of the immiscibility of the parent sulfide melt with impurities.

The main amount of minor minerals is present in composite inclusions in the matrix of Fe-Cu sulfides. The inclusions of class I are formed from a refractory Au* that melts at ~1000 °C and a low-melting $Pd(Bi,Sb)_xTe_{1-x}$ with an estimated melting point of <750 °C (for PdTe $T_m = 746$ °C [52]. It is likely that at liquidus temperatures, drops of this compound with minor Au are present in the sulfide melt. On cooling, these drops are trapped during the crystallization of the BMS matrix. On further cooling, gold crystals are separated from the drops, and solid inclusions are formed during the solidification of $Pd(Bi,Sb)_xTe_{1-x}$.

Sulfosalts of copper form low-melting crystals. For example, the temperature of incongruent decomposition of stibiowittichenite $Cu_3SbS_3$ is 613 °C [53], wittichenite $Cu_3BiS_3$ is 535 °C [54], and tetrahedrite is 574 °C [55]. Au, Ag, and Te are dissolved in sulfosalt drops. On cooling, at first Au* and $Ag_2Te$ ($T_m \sim 960$ °C [56]) crystallize, then sulfosalts solidify.

Class III contains composite inclusions from the fragments of I and II classes and refractory microcrystals of $PtAs_2$. It can be assumed that associations of microcrystals $PtAs_2$ and drops I and II are in the sulfide melt. They solidify to form inclusions of class III.

The system under investigation contains Au* and Au** phases, $T_m$ of which is above the liquidus temperature. Nevertheless, these do not form inclusions. We assume that the gold content in sulfide melt is low at liquidus temperature, and the main amount of Au is concentrated in liquid drops and during solidification is separated as independent phases of Au* and Au**.

During solidification of BMS, small droplets can be trapped on flat areas of the crystallization front. Large liquid drops can fall on the region of the crystallization front with the boundary between the crystallites. In the process of crystallite growth, the drops are trapped, and inclusions of an intricate shape are formed at interfaces (Figure 11B).

Similar inclusions of minor phases were observed in the ores from the Noril'sk and Sudbury deposits [4,23,24,28,57–60] and in synthetic samples [17,18,34,38,39,45]. However, we did not find any results of modeling of fractional crystallization of melts with the formation of two types of PGE–metalloid liquids. Theoretical substantiation of the described phenomena is the data on the phase diagrams of binary, ternary, and more complex systems (e.g., review in [36]).

In this study, we once again demonstrated that in the directed crystallization an ingot is obtained in which phases and phase associations are spatially separated. This makes it possible to reliably and unambiguously determine the sequence of separation of primary phases during fractionation crystallization of a multicomponent sulfide melt and the change in melt composition and phase reactions when passing from one zone to another. As a result of crystallization, one can obtain a number of genetically related samples, allowing determination of the behavior of impurities at different stages of the process. It is difficult to obtain this information of phase processes by other methods. This is especially true of multiphase systems. It is worth noting that laboratory experiments imitate the continuous process of fractionation crystallization of natural sulfide liquids. Nevertheless, the methods of directional crystallization and isothermal annealing and quenching supplement each other.

Many researchers believe that the formation of massive ore bodies during fractionation crystallization takes place in the intrusive cavity in a closed system (for example, [1,2]). We modeled this process. Nowadays, there is also a hypothesis about a significant effect of hydrothermal sulfur-bearing liquids on this process (e.g., [24]). To determine the role of this or that process, additional experimental and theoretical research is necessary.

## 5. Conclusions

On the basis of our experimental data on the directed crystallization of the multicomponent Cu-Fe-Ni sulfide liquid with minor noble metals and As, Te, Se, Bi, Sb, and Sn, the following conclusions can be made:

1.  It was shown that, in directed crystallization of melt, inclusions formed, which are similar to those observed in isothermal experiments and in sulfide ores. This is additional evidence that the minor contents of noble minerals and metalloids were present in the initial sulfide melt after its separation from the silicate melt. There is a probability of low-temperature platinum mineralization of sulfide ores as a result of hydrothermal processes but, most likely, it is realized in the aureoles of disseminated ores surrounding the massive ore bodies.

2.  For the first time, simultaneous formation of two types of liquids separated during cooling of the parent sulfide melt was revealed. In the first, noble metals associated with Bi, Sb, and Te are concentrated. The second is related to Cu and contains a large amount of S in addition to Bi and Sb.

3. We established the main types of inclusions formed during fractional crystallization of Pt-bearing sulfide melt. It was shown that noble metals are concentrated in inclusions in the form of $RuS_2$, $PdTe_2$, $(Pt,Pd)Te_2$, $PtRhAsS$, and $Ag_2Se$, Au** in *mss* and in the form of $PtAs_2$, Au*, Au**, and $Ag_2Te$, $Pd(Bi,Sb)_xTe_{1-x}$ in *icb** and *iss*. As solid solutions in the BMS sulfides, Rh is present in *mss* and Sn in *iss*.

Thus, our experiment showed a more complex behavior of noble metals and metalloid elements during the crystallization of multicomponent sulfide-metalloid melts compared to the earlier reported data of isothermal experiments.

**Supplementary Materials:** The following are available online at http://www.mdpi.com/2075-163X/9/9/531/s1. Table S1: Average concentrations of components in solid phases and in the melt and distribution coefficients of components between the phases and melt, Table S2: Representative SEM/EDS analyses of microminerals, Table S3: EDS results for average composition of the inclusions.

**Author Contributions:** E.S. designed, performed the experiments and wrote the paper, V.K. interpreted the results and wrote the paper, G.P. performed data treatment, N.K. analyzed the compositions of phases, all the authors participated in the manuscript preparation.

**Funding:** This work was supported by state assignment projects of VS Institute of Geology and Mineralogy SB RAS and II.1.64.4 of Institute of Inorganic Chemistry SB RAS, financed by Ministry of Science and Higher Education of the Russian Federation. The research was partially funded by the Siberian Branch of the Russian Academy of sciences, Program of Interdisciplinary Studies, grant number 64 (project 303).

**Acknowledgments:** The authors thank K.A. Kokh for assistance in the experiment. We are grateful to anonymous Reviewers for their valuable comments and suggestions.

**Conflicts of Interest:** The authors declare no conflict of interest.

# References

1. Genkin, A.D.; Distler, V.V.; Gladyshev, G.D.; Filimonova, A.A.; Evstigneeva, T.L.; Kovalenker, V.A.; Laputina, I.P.; Smirnov, A.V.; Grokhovskaya, T.L. *Copper-Nickel Sulfide Ores of the Noril'sk Deposits*; Nauka: Moscow, Russia, 1981; p. 234. (In Russian)

2. Distler, V.V.; Grokhovskaia, T.L.; Evstigneyeva, T.L.; Sluzhenikin, S.F.; Filimonova, A.A.; Dyuzhikov, O.A.; Laputina, I.P. *Petrology of the Sulphide Magmatic Ore-Formation*; Nauka: Moscow, Russia, 1988; p. 230. (In Russian)

3. Naldrett, A.J. *Magmatic Sulfide Deposits: Geology, Geochemistry and Exploration*; Springer: Berlin, Germany, 2004; p. 727.

4. Duran, C.J.; Barnes, S.-J.; Pleše, P.; Prašek, M.K.; Zientek, M.L.; Pagé, P. Fractional crystallization-induced variations in sulfides from the Noril'sk-Talnakh mining district (polar Siberia, Russia). *Ore Geol. Rev.* **2017**, *90*, 326–351. [CrossRef]

5. Vogt, I.H.L. Nickel in igneous rocks. *Econ. Geol.* **1923**, *18*, 307–353. [CrossRef]

6. Craig, J.R.; Kullerud, G. Phase relations in the Cu-Fe-Ni-S system and their application to magmatic ore deposits. *Econ. Geol. Monogr.* **1969**, *4*, 344–358.

7. Naldrett, A.J. Nickel sulfide deposits: Their classification, composition and genesis. *Econ. Geol.* **1981**, *75th Anniversary*, 628–685.

8. Barnes, S.-J.; Makovicky, E.; Makovicky, M.; Rose-Hansen, J.; Karup-Moller, S. Partition coefficients for Ni, Cu, Pd, Pt, Rh, and Ir between monosulfide solid solution and sulfide liquid and the formation of compositionally zoned Ni-Cu sulfide bodies by fractional crystallization of sulfide liquid. *Can. J. Earth Sci.* **1997**, *34*, 366–374. [CrossRef]

9. Mungall, J.E.; Andrews, D.R.A.; Cabri, L.J.; Sylvester, P.J.; Tubrett, M. Partitioning of Cu, Ni, Au, and platinum-group elements between monosulfide solid solution and sulfide melt under controlled oxygen and sulfur fugacities. *Geochim. Cosmochim. Acta* **2005**, *69*, 4349–4360. [CrossRef]

10. Distler, V.V. Platinum mineralization of the Noril'sk deposits. In *Geology and Genesis of Platinoid Deposits*; Nauka: Moscow, Russia, 1994; pp. 7–35. (In Russian)

11. Kosyakov, V.I.; Sinyakova, E.F.; Distler, V.V. Experimental simulation of phase relationships and zoning of magmatic nickel-copper sulfide ores, Russia. *Geol. Ore Depos.* **2012**, *54*, 179–208. [CrossRef]

12. Kosyakov, V.I.; Sinyakova, E.F. Melt crystallization of $CuFe_2S_3$ in the Cu-Fe-S system. *J. Therm. Anal. Calorim.* **2014**, *115*, 511–516. [CrossRef]

13. Kosyakov, V.I.; Sinyakova, E.F. Study of crystallization of nonstoichiometric isocubanite $Cu_{1.1}Fe_{2.0}S_{3.0}$ from melt in the system Cu-Fe-S. *J. Therm. Anal. Calorim.* **2017**, *129*, 623–628. [CrossRef]

14. Kosyakov, V.I.; Sinyakova, E.F. Experimental modeling of pentlandite-bornite ore formation. *Russ. Geol. Geophys.* **2017**, *58*, 1211–1221. [CrossRef]

15. Sinyakova, E.F.; Kosyakov, V.I. Experimental modeling of zoning in copper-nickel sulfide ores. *Dokl. Earth Sci.* **2007**, *417A*, 1380–1385. [CrossRef]

16. Sinyakova, E.F.; Kosyakov, V.I. Experimental modeling of zonality of copper-rich sulfide ores in copper-nickel deposits. *Dokl. Earth Sci.* **2009**, *427*, 787–792. [CrossRef]

17. Sinyakova, E.F.; Kosyakov, V.I. The behavior of noble-metal admixtures during fractional crystallization of As- and Co-containing Cu-Fe-Ni sulfide melts. *Russ. Geol. Geophys.* **2012**, *53*, 1055–1076. [CrossRef]

18. Sinyakova, E.; Kosyakov, V.; Distler, V.; Karmanov, N. Behavior of Pt, Pd, and Au during crystallization of Cu-rich magmatic sulfides. *Can. Mineral.* **2016**, *54*, 491–509. [CrossRef]

19. Sinyakova, E.F.; Kosyakov, V.I.; Borisenko, A.S.; Karmanov, N.S. Behavior of noble metals during fractional crystallization of Cu-Fe-Ni-(Pt, Pd, Rh, Ir, Ru, Ag, Au, Te) sulfide melts. *Russ. Geol. Geophys.* **2019**, *60*, 642–651.

20. Vaulin, L.L.; Sukhanova, E.N. Oktyabr'skoe copper-nickel deposit. *Razved. i Okhrana Nedr* **1970**, *4*, 48–51. (In Russian)

21. Genkin, A.D.; Evstigneyeva, T.L. Associations of platinum-group minerals of the Noril'sk cooper-nickel sulfide ores. *Econ. Geol.* **1986**, *81*, 1203–1212. [CrossRef]

22. Distler, V.V.; Sluzhenikin, S.F.; Cabri, L.J.; Krivolutskaya, N.A.; Turovtsev, D.M.; Golovanova, T.A.; Mokhov, A.V.; Knauf, V.V.; Oleshkevich, O.I. Platinum ores of the Noril'sk layered intrusions: Magmatic and fluid concentration of noble metals. *Geol. Ore Depos.* **1999**, *41*, 241–265.

23. Barnes, S.-J.; Cox, R.A.; Zientek, M.L. Platinum-group element, gold, silver and base metal distribution in compositionally zoned sulfide droplets from the Medvezky Creek mine, Noril'sk, Russia. *Contrib. Mineral. Petrol.* **2006**, *152*, 187–200. [CrossRef]

24. Spiridonov, E.M.; Gritsenko, Y.D. *Epigenetic Low-Grade Metamorphism and Co-Ni-Sb-As Mineralization in the Noril'sk Ore Field*; Nauchnyi Mir: Moscow, Russia, 2009; p. 218. (In Russian)

25. Holwell, D.A.; McDonald, I. A review of the behavior of platinum group elements within natural magmatic sulfide ore systems. *Platin. Met. Rev.* **2010**, *54*, 26–36. [CrossRef]

26. Czamanske, G.K.; Kunilov, V.E.; Zientek, M.L.; Cabri, L.J.; Likhachev, A.P.; Calk, L.C.; Oscarson, R. A proton-microprobe study of magmatic sulfide ores from the Noril'sk-Talnakh district, Siberia. *Can. Mineral.* **1992**, *30*, 249–287.

27. Cabri, L.J. *The Geology, Geochemistry, Mineralogy and Mineral Beneficiation of Platinum-Group Elements*; Canadian Institute of Mining, Metallurgy and Petroleum: Montreal, QC, Canada, 2002; Volume 54, pp. 13–129.

28. Kozyrev, S.M.; Komarova, M.Z.; Emelina, L.N.; Oleshkevich, O.I.; Yakovleva, O.A.; Lyalinov, D.V.; Maximov, V.I. The mineralogy and behavoiur of PGM during processing of the Noril'sk-Talnakh PGE-Cu-Ni-ores. In *The Geology, Geochemistry, Mineralogy and Mineral Beneficiation of Platinum-Group Elements*; Canadian Institute of Mining, Metallurgy and Petroleum: Montreal, QC, Canada, 2002; Volume 54, pp. 757–775.

29. Sluzhenikin, S.; Mokhov, A. Gold and silver in PGE-Cu-Ni and PGE ores of the Noril'sk deposits, Russia. *Miner. Depos.* **2015**, *50*, 465–492. [CrossRef]

30. Peregoedova, A.; Ohnenstetter, M. Collectors of Pt, Pd and Rh in a S-poor Fe-Ni-Cu sulfide system at 760 °C: Experimental data and application to ore deposits. *Can. Mineral.* **2002**, *40*, 527–561. [CrossRef]

31. Peregoedova, A.; Barnes, S.-J.; Baker, D.R. The formation of Pt-Ir alloys and Cu-Pd-rich sulfide melts by partial desulfurization of Fe-Ni-Cu sulfides: Results of experiments and implications for natural systems. *Chem. Geol.* **2004**, *208*, 247–264. [CrossRef]

32. Peregoedova, A.; Barnes, S.-J.; Baker, D.R. An experimental study of mass transfer of platinum-group elements, gold, nickel and copper in sulfur-dominated vapor at magmatic temperatures. *Chem. Geol.* **2006**, *235*, 59–75. [CrossRef]

33. Distler, V.V.; Sinyakova, E.F.; Kosyakov, V.I. Behavior of noble metals upon fractional crystallization of copper-rich sulfide melts. *Dokl. Earth Sci.* **2016**, *469*, 811–814. [CrossRef]

34. Cafagna, F.; Jugo, P.J. An experimental study on the geochemical behavior of highly siderophile elements (HSE) and metalloids (As, Se, Sb, Te, Bi) in a mss-iss-pyrite system at 650 °C: A possible magmatic origin for Co-HSE-bearing pyrite and the role of metalloid-rich phases in the fractionation of HSE. *Geochim. Cosmochim. Acta* **2016**, *178*, 233–258.

35. Bai, L.; Barnes, S.-J.; Baker, D.R. Sperrylite saturation in magmatic sulfide melts: Implications for formation of PGE-bearing arsenides and sulfarsenides. *Am. Mineral.* **2017**, *102*, 966–974. [CrossRef]

36. Makovicky, E. Ternary and quaternary phase systems with PGE. In *The Geology, Geochemistry, Mineralogy and Mineral Beneficiation of Platinum-Group Elements*; Canadian Institute of Mining, Metallurgy and Petroleum: Montreal, QC, Canada, 2002; Volume 54, pp. 131–175.

37. Ballhaus, C.; Tredoux, M.; Spath, A. Phase relations in the Fe-Ni-Cu-PGE-S system at magmatic temperature and application to massive sulphide ores of the sudbury igneous complex. *J. Petrol.* **2001**, *42*, 1911–1926. [CrossRef]

38. Helmy, H.M.; Ballhaus, C.; Berndt, J.; Bockrath, C.; Wohlgemuth-Ueberwasser, C. Formation of Pt, Pd and Ni tellurides: Experiments in sulfide-telluride systems. *Contrib. Mineral. Petrol.* **2007**, *153*, 557–591. [CrossRef]

39. Helmy, H.M.; Ballhaus, C.; Fonseca, R.O.C.; Wirth, R.; Nagel, T.J.; Tredoux, M. Noble metal nanoclusters and nanoparticles precede mineral formation in magmatic sulfide melts. *Nat. Commun.* **2013**, *4*. [CrossRef] [PubMed]

40. Karup-Moller, S.; Makovicky, E.; Barnes, S.-J. The metal-rich portions of the phase system Cu-Fe-Pd-S at 1000 °C, 900 °C and 725 °C: Implications for mineralization in the Skaergaard intrusion. *Mineral. Mag.* **2008**, *72*, 941–951. [CrossRef]

41. Vymazalova, A.; Laufek, F.; Kristavchuk, A.V.; Drabek, M. The system Ag-Pd-Te phase relation and mineral assemblages. *Mineral. Mag.* **2015**, *79*, 1813–1832. [CrossRef]

42. Sinyakova, E.F.; Kosyakov, V.I.; Borisenko, A.S. Effect of the presence of As, Bi, and Te on the behavior of Pt metals during fractionation crystallization of sulfide magma. *Dokl. Earth Sci.* **2017**, *477*, 1422–1425. [CrossRef]

43. Fleet, M.E.; Pan, Y. Fractional crystallization of anhydrous sulfide liquid in the system Fe-Ni-Cu-S, with application to magmatic sulfide deposits. *Geochim. Cosmochim. Acta* **1994**, *58*, 3369–3377. [CrossRef]

44. Ebel, D.S.; Naldrett, A.J. Crystallization of sulfide liquids and interpretation of ore composition. *Can. J. Earth Sci.* **1997**, *34*, 352–365. [CrossRef]

45. Helmy, H.M.; Ballhaus, C.; Wohlgemuth-Ueberwasser, C.; Fonseca, R.O.C.; Laurenz, V. Partitioning of Se, As, Sb, Te and Bi between monosulfide solid solution and sulfide melt—Application to magmatic sulfide deposits. *Geochim. Cosmochim. Acta* **2010**, *74*, 6174–6179. [CrossRef]

46. Liu, Y.; Brenan, J. Partitioning of platinum-group elements (PGE) and chalcogens (Se, Te, As, Sb, Bi) between monosulfide-solid solution (MSS), intermediate solid solution (ISS) and sulfide liquid at controlled $fO2$–$fS2$ conditions. *Geochim. Cosmochim. Acta* **2015**, *159*, 139–161. [CrossRef]

47. Sinyakova, E.F.; Kosyakov, V.I.; Nenashev, B.G. Coefficients of rhodium partition between melt and monosulfide solid solution during directional crystallization of melt in the Fe-FeS-NiS-Ni system. *Dokl. Earth Sci.* **2004**, *397*, 649–653.

48. Cook, N.J.; Ciobanu, L.; Merkle, R.K.W.; Bernhardt, H.-J. New data on sobolevskite, taimyrite and $Pt_2CuFe$ (tulameenite?) in complex massive talnakhite ore, Noril'sk orefield, Russia. *Can. Mineral.* **2002**, *40*, 329–340. [CrossRef]

49. Massalski, T.B.; Okamoto, H.; Subramanian, P.R.; Kacprzak, L. *Binary Alloy Phase Diagrams*, 2nd ed.; ASM International, Materials Park: Novelty, OH, USA, 1990; p. 3242.

50. Effenberg, G.; Ilyenko, S. Ternary alloy systems. Noble metal systems. Selected systems from Ag-Al-Zn to Rh-Ru-Sc, Landolf-Börnstein—Group IV. *Phys. Chem.* **2006**, *11B*. [CrossRef]

51. Barnes, S.-J.; Ripley, E.M. Highly siderophile and strongly chalcophile elements in magmatic ore deposits. *Rev. Mineral. Geochem.* **2016**, *81*, 725–774. [CrossRef]

52. Okamoto, H. Pd-Te (Palladium-Tellurium). *J. Phase Equilibria Diffus.* **2013**, *34*, 72–73. [CrossRef]

53. Ilyasheva, N.A. Study of the $Cu_2S$-$Sb_2S_3$ system at 320–400 °C. *Neorg. Mater.* **1973**, *9*, 677–679.

54. Malevskii, A.Y. Study of the system Cu-Bi-S. In *Proceedings of the Meeting on Experimental and Technical Mineralogy and Petrography*; Nauka: Moscow, Russia, 1971; pp. 302–308. (In Russian)

55. Ilyasheva, N.A. Specific features of crystallization of tetrahedrite in the Cu-Sb-S system. *Neorg. Mater.* **1984**, *20*, 563–568. (In Russian)

56. Krachek, F.C.; Ksanda, C.J.; Cabri, L.J. Phase relations in the silver-tellurium system. *Am. Mineral.* **1966**, *51*, 14–28.
57. Evstigneeva, T.L.; Genkin, A.D.; Kovalenker, V.A. A new bismuthide of palladium, sobolevskite, and the nomenclature of minerals of the system PdBi-PdTe-PdSb. *Zapiski Vses. Mineralog. Obshch.* **1975**, *104*, 568–579. (In Russian) [CrossRef]
58. Dare, S.A.S.; Barnes, S.-J.; Prichard, H.M.; Fisher, P.C. The timing and formation of platinum-group minerals from the Creigton Ni-Cu-Platinum-group elements sulfide deposit, Sudbury, Canada: Early crystallization of PGE-rich sulfarsenides. *Econ. Geol.* **2010**, *105*, 1071–1096. [CrossRef]
59. Dare, S.A.S.; Barnes, S.-J.; Prichard, H.M.; Fisher, P.C. Chalcophile and platinum-group element (PGE) concentrations in the sulfide minerals from the McCreedy East deposit, Sudbury, Canada, and the origin of PGE in pyrite. *Miner. Depos.* **2011**, *46*, 381–407. [CrossRef]
60. Dare, S.A.S.; Barnes, S.-J.; Prichard, H.M.; Fisher, P.C. Mineralogy and geochemistry of Cu-Rich ores from the McCreedy East Ni-Cu-PGE deposit (Sudbury, Canada): Implications for the behavior of platinum group and chalcophile elements at the end of crystallization of a sulfide liquid. *Econ. Geol.* **2014**, *109*, 343–366. [CrossRef]

 *minerals*

*Article*

# X-ray Photoelectron Spectroscopy (XPS) Study of the Products Formed on Sulfide Minerals Upon the Interaction with Aqueous Platinum (IV) Chloride Complexes

**Alexander Romanchenko *, Maxim Likhatski and Yuri Mikhlin**

Institute of Chemistry and Chemical Technology, Siberian Branch of the Russian Academy of Sciences, Krasnoyarsk 660036, Russia; lixmax@icct.ru (M.L.); yumikh@icct.ru (Y.M.)
* Correspondence: romaas82@mail.ru; Tel.: +7-391-205-1928

Received: 8 November 2018; Accepted: 6 December 2018; Published: 8 December 2018

**Abstract:** The interaction of aqueous solutions bearing platinum-group elements (PGEs) with sulfides is important for understanding the formation and weathering of PGE ore deposits, mineral processing, and synthesis of nanomaterials. Here, the surface species formed upon the contact of the main sulfide minerals (pyrite, pyrrhotite, galena, chalcopyrite and valleriite) with the solutions of $H_2PtCl_6$ (pH 1.5, 20 °C) have been studied using X-ray photoelectron spectroscopy (XPS). Uptake of Pt increased gradually with increasing interaction time, and depended, as well as the composition of immobilized products, on the mineral nature and the state of its surface, e.g., the chemical pre-treatment. The highest rate of Pt deposition was observed on galena and valleriite and the lowest on pyrite and pyrrhotite. The preliminary moderate oxidation of pyrrhotite promoted Pt deposition, which, however, was hindered under harsh reaction conditions. The pre-oxidation of pyrite in all cases resulted in a decrease of the Pt deposition. Initially, Pt(IV) chloride complexes adsorb onto the mineral surface, and then the reduction of Pt(IV) to Pt(II) and substitution of chloride ions with sulfide groups occur forming sulfides of Pt(II) and then, Pt(IV). The reduction of Pt species to the metallic state was observed at valleriite after 24 h, probably due the negative charge of the sulfide nanolayers of this sulfide-hydroxide composite mineral.

**Keywords:** XPS; sulfide minerals; platinum; deposition; valleriite; chloride complexes

---

## 1. Introduction

Platinum-group elements (PGEs) are widely used in various areas, first of all in catalysis; their chalcogenides are also of interest for electrocatalysis [1–3]. The increasing industrial demand for PGEs and their limited natural sources require the investigation of the genesis of PGE ores and new approaches to the recovery technologies [4], secondary processes and environmental behavior of engineered PGE species [5,6]. Platinum-group element ores are mainly associated with ultramafic-mafic rocks and, little is known about the interaction of Pt-bearing aqueous solutions with sulfide minerals [7–20]. Platinum-group element deposits of hydrothermal origin exist, and mechanisms of hydrothermal transportation of PGEs [9–12], including the biological transport and transformation [21–24] have been studied since the end of the 1980s.

Platinum-group elements are commonly associated with base metal sulfides, particularly pentlandite, pyrrhotite, chalcopyrite, pyrite [15,25,26]. Palladium is hosted in base metal sulfides (mainly pentlandite) and occurs both within the crystal lattice and as nanometer-sized inclusions of discrete PGE minerals [27]. Platinum usually has low concentrations within base metal sulfides and mainly forms discrete platinum-group minerals (PGMs). Most work has been focused on the experimental study and mathematical modeling of Pt dissolution, and stability of its complex compounds with ligands such as $HS^-$, $Cl^-$, $S_2O_3{}^{2-}$, $OH^-$, $NH_3$ [11,12,17,18,22,28,29]. At low temperature conditions, acidic pHs and high redox potential, Pt transport by chloride ions is the most suitable [18,30].

In addition to the understanding of mobilization and deposition of PGEs, the interaction of Pt-bearing solutions with base metal sulfides is important for mineral processing of PGEs by heap leaching [31–37] and concentrator plants [35]. Platinum-group elements transfer into solution by leaching (both under technological and natural conditions) can then be taken up on the surface of sulfide minerals. However, in contrast to the vast literature devoted to the interaction of aqueous solutions of gold and silver with sulfides (see, e.g., [38–40] and references therein), only limited studies describe the deposition of PGEs, on the surface of pyrite, galena and sphalerite by X-ray photoelectron spectroscopy (XPS) [16,41]. By the interaction of the $Na_2PdCl_4$ solution (pH 4) with pyrite, sulfide species of Pd(II) are formed at room temperature along with retaining mixed aqua- and hydrochloride complexes of Pd(II) [16]. Recently [41], the interaction of the acidic solution of $H_2PdCl_4$ solution with pyrrhotite and pyrite before and after their preliminarily leaching under different conditions was studied. Palladium-oxyhydroxides are not formed, though it is not completely proved since the differences in the binding energies of the Pd 3d line of oxides and sulfides are not significant [42,43]. So far, no studies analyzing Pt species immobilized from aqueous solutions onto sulfide minerals, and kinetics of their interaction have been carried out yet.

The aim of this study to investigate the chemical species of Pt formed on the surface of sulfide minerals, including the preliminarily oxidized ones, upon their contact with aqueous Pt(IV)-chloride complexes. This knowledge is important for understanding processes leading to the loss of PGEs in the processing of sulfide ores, affinage, and the behavior of PGEs in tailing ponds where they are leached under near-surface conditions [5,6]. The interaction of metal sulfides with the solutions of Pt-containing complex compounds is of interest for preparation of composite nanomaterials and their potential use in (electro)catalysis, electronics [1–3].

## 2. Materials and Methods

For this study, natural sulfide minerals from Noril'sk are studied, i.e., pyrrhotite, [$Fe_9S_{10}$], containing impurities of pentlandite, [$(Ni,Fe)_9S_8$] (up to 3 wt.%) and chalcopyrite, [$CuFeS_2$] (1–2 wt.%), and chalcopyrite. Pyrite, [$FeS_2$], and galena, [PbS], originated from the Sukhoi Log deposit, and the Sanzheevskoe deposit, respectively. Specimens of valleriite, [$(CuFeS_2)\cdot1.5[(Mg,Al)(OH)_2]$] from the Nadezhda deposit (Noril'sk provenance) contained about 50% of the target mineral and varying minor quantities of pyrrhotite, chalcopyrite, magnetite, serpentine, pyrite; in more detail this material was characterized in refs. [39,44–50]. Plates with the size of approximately $2 \times 4 \times 5$ mm were cut from the minerals, and the surface was polished with sandpaper, washed with distilled water and wiped with wet filter paper to remove fine particles directly before Pt deposition. Further, the sample was conditioned in the solution of 1 mmol/L $H_2PtCl_6$ + 30 mmol/L HCl, pH 1.5, unless otherwise stated, without stirring at 20 °C for a time varying from 1 h to 48 h, then washed with distilled water and transferred into the photoelectron spectrometer chamber. The Pt solutions were prepared by dissolving of crystalline $H_2PtCl_6$ (Dragcvetmet, Moscow, Russia) in a HCl solution. All solutions were prepared using analytical grade reagents and deionized water (Millipore).

For the preliminary oxidation the minerals were prepared in the same manner as for XPS. To oxidize, leaching was made using the solution of 0.25 M $FeCl_3$ + 1 M HCl during 30 min at 50 °C, and then, the sample was rinsed with the solution of 1 M HCl to remove Fe ions and, consequently, to prevent the formation of the products of Fe(III) hydrolysis. Then, the sample was rinsed with distilled water and placed into the $H_2PtCl_6$ solution for a specified period of time, usually, 2–4 h. The electrochemical oxidation was carried out using an Elins P30 SM potentiostat (Elins, Moscow, Russia) in the 1 M HCl solution at a constant potential over 10 minutes. A Pt wire was used as a counter electrode and a saturated Ag/AgCl electrode was used as the reference electrode;

all potentials are given relative to that. After the electrochemical oxidation the mineral sample was rinsed with distilled water and used further for depositing Pt.

The XPS were acquired using a SPECS spectrometer equipped with a PHOIBOS 150 MCD-9 analyzer (SPECS, Berlin, Germany) at an electron take-off angle of 90° employing monochromatic Mg K$\alpha$ radiation (1253.6 eV) of an X-ray tube operated at 180 W. The analyzer pass energy was 10 eV for narrow scans, and 20 eV for survey spectra. The C 1s peak at 284.45 eV from hydrocarbon contaminations was used as a reference. The Pt $4f_{7/2,5/2}$, S $2p_{3/2,1/2}$ doublets were fitted, after the subtraction of the Shirley-type background, with the coupled peaks with the Gaussian–Lorentzian peak profiles, the spin–orbit splitting of 3.30 eV and 1.19 eV, and the branching ratios of 0.75 and 0.5, respectively, using the CasaXPS software package (version 2.3.16, Casa Software, Teignmouth, UK). The element concentrations on the surface were determined from the wide scan using the empirical sensitivity coefficients. Some photoelectron spectra were measured at the Russian–German laboratory at the dipole magnet beamline of the synchrotron radiation facility BESSY II (Berlin, Germany). Tapping-mode atomic force microscopy (AFM) studies were conducted using a Solver P-47 multimode scanning probe microscope (Nanotekhnologiya MDT, Moscow, Russia) in air at room temperature. A silicon cantilever with the resonance frequencies of 150–250 kHz was used as a probe.

## 3. Results

### 3.1. Analysis of the Surface of Pyrite and Pyrrhotite Using X-ray Photoelectron Spectroscopy and Atomic Force Microscopy

The XPS spectra show that both Pt and Cl can be detected on the surface of sulfide minerals after the aqueous media treatment (Figure 1). Typically, Cl$^-$ ions are weakly adsorbed on the metal sulfide surface and/or can easily be removed by washing with water [39]. Thus, Cl observed in the spectra largely belongs to adsorbed Pt–Cl complexes.

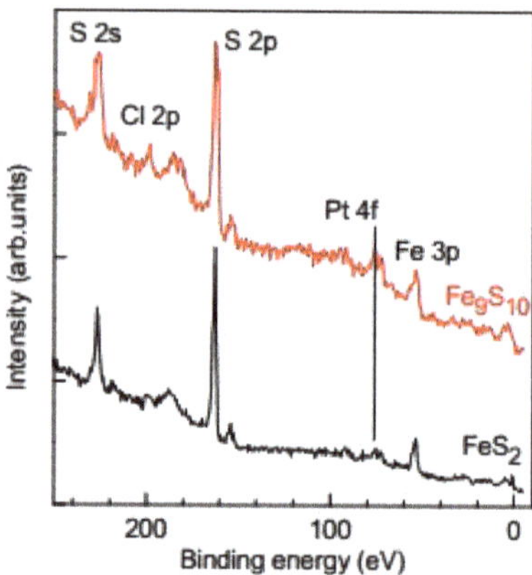


**Figure 1.** Fragments of the X-ray photoelectron survey spectra of pyrite and pyrrhotite after the interaction with 1 mmol/L H$_2$PtCl$_6$ + 0.03 mol/L HCl solution for 4 h.

Atomic force microscopy (AFM) is used to visualize immobilized Pt products and shows strong changes of the mineral substrate relief due to surface etching. Platinum species are hardly distinguished, especially after the short periods of Pt deposition. Pyrite is a more convenient mineral for the AFM study since its surface is to a lower extent subjected to changes by chemical attack. Figure 2 illustrates regions with the particles of Pt-containing products, in which particles with the lateral size of 50–150 nm cover about 50% of the surface area.

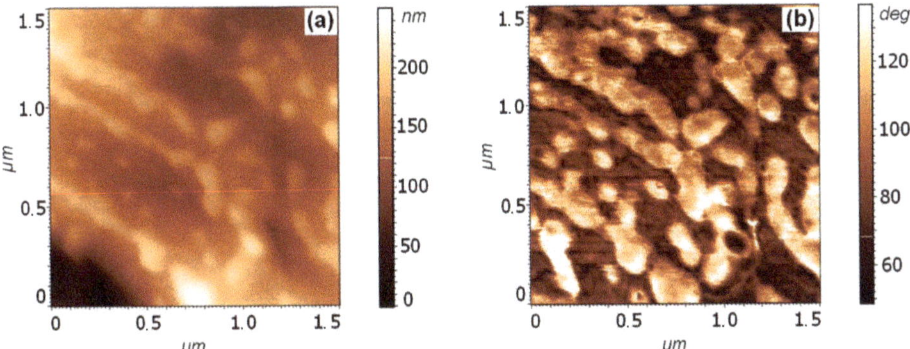

**Figure 2.** Atomic force microscopy images of the pyrite surface after the interaction with the solution of $H_2PtCl_6$ for 14 h; (**a**) relief, (**b**) phase contrast.

## 3.2. Deposition of Platinum Revealed by XPS

### 3.2.1. Pyrrhotite and Pyrite

Figure 3a,b presents the data on atomic concentrations of Pt, S, Cl, Fe and O on the surface of pyrite and pyrrhotite after the interaction with 1 mmol/L $H_2PtCl_6$ solutions. The concentrations of Cl and Pt increase with time. The Fe concentration on pyrrhotite gradually decreases with time but it insignificantly changes on pyrite during 24 h.; this is due to various rates of oxidation of these minerals.

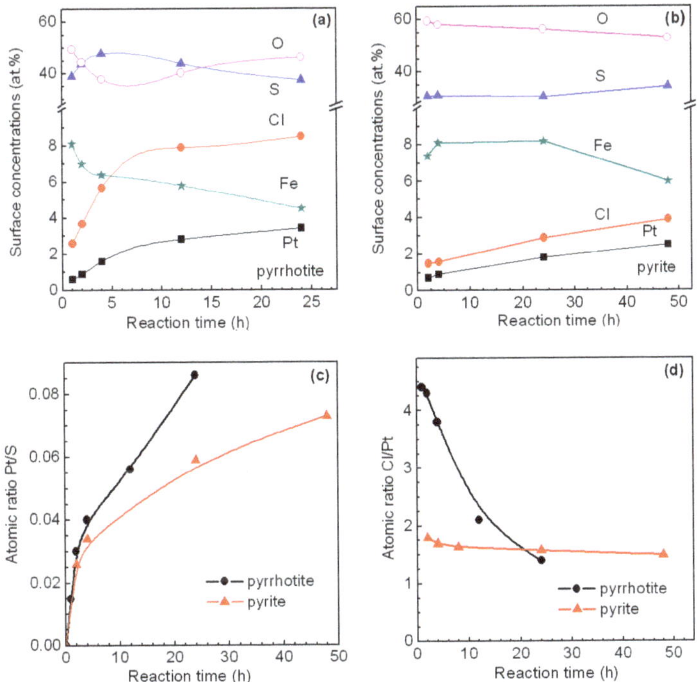

**Figure 3.** Atomic concentrations of Pt, Cl, S, O and Fe on the surface of (**a**) pyrrhotite and (**b**) pyrite and the atomic concentration ratios of (**c**) Pt/S and (**d**) Cl/Pt on pyrrhotite and pyrite after different interaction time with the $10^{-3}$ M $H_2PtCl_6$ solution.

Figure 3c shows that an increase of Pt concentration with time is similar on the surfaces of pyrite and pyrrhotite. The deposition rate is relatively high over 3–5 h, and it gradually decreases but does not drop to zero, so the Pt concentration does not reach saturation. The ratios Cl/Pt show the rate of substitution of $Cl^-$ ions with sulfide ions and on the influence of this process on the Pt accumulation on the mineral surface. The atomic ratio Cl/Pt is about 1.8 in the initial deposition stage and slightly decreases to 1.5 after 48 h at pyrite. At pyrrhotite, the value of Cl/Pt of approximately 4.3 is apparently higher in the initial stage, and it decreases down to 1.5 over 24 h.

### 3.2.2. Galena, Chalcopyrite and Valleriite

The results of the surface analysis for galena, chalcopyrite and valleriite are presented in Figure 4 and also in Table S1 (Supplementary materials). The deposition of Pt on galena and valleriite is considerably higher than at pyrite and pyrrhotite. The amounts of the Pt normalized to the sulfide phase, that is the ratios Pt/S (Figure 4b), suggest that valleriite accumulates Pt faster than galena and other minerals. The deposition of Pt on chalcopyrite is 1.5 times slower than on galena, while it is 2 times slower on pyrite and pyrrhotite. The Cl/Pt ratios presented in Figure 4c implies a negative correlation between the surface amount of Cl and the accumulation of Pt on the minerals.

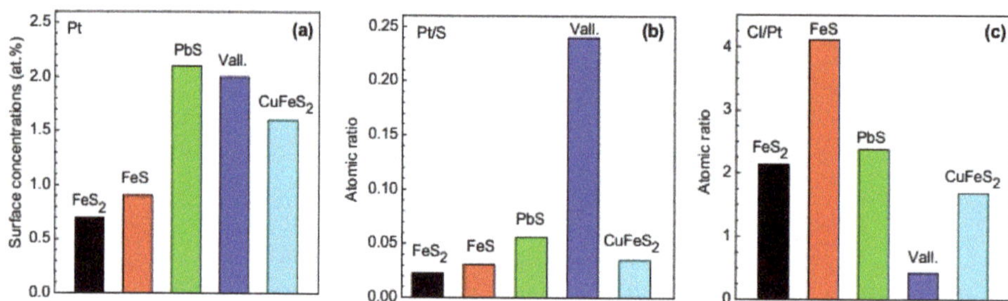

**Figure 4.** Atomic concentrations of (**a**) Pt, and the ratios (**b**) Pt/S, and (**c**) Cl/Pt at the surface of sulfide minerals after 2 h interaction with 1 mmol/L $H_2PtCl_6$ solution.

### 3.3. *Species of the Deposited Platinum*

### 3.3.1. XPS of the Reference Materials

Photoelectron spectra from reference samples Pt (chloride complexes, Pt sulfide and metallic nanoparticles) precipitated onto highly oriented pyrolytic graphite (HOPG) were preliminarily measured for the reliable assignment of components of Pt 4f spectra. Two components can be revealed in the Pt 4f spectra of pure aqueous $H_2PtCl_6$ dried at HOPG, with the component at the binding energy of 72.9 eV being attributable to Pt(II) binding $Cl^-$ anions, and the high-energy component at 75.0 eV corresponding to $PtCl_6^{2-}$ ions (Figure 5a). For the samples of Pt-sulfide hydrosol obtained by mixing the solutions of $Na_2S$ and $H_2PtCl_6$, two signals with the binding energy of ~74 eV and 72.6 eV can be assigned to Pt(IV)–S and Pt(II)–S compounds [51–53], respectively (Figure 5b). The binding energies can vary in a small range, depending on the S species bound to Pt. Along with minor lines of sulfite and sulfate which are likely to be products of $Na_2S$ oxidation, two low-energy doublets with S $2p_{3/2}$ at 162.6 eV and 163.3 eV are found in the S 2p spectra (Figure S1, Supplementary materials). These binding energies are typical for disulfide and polysulfide anions, respectively, or other S species carrying similar local negative charges. In its turn, Pt bound with polysulfide could have the binding energy by several tenth eV lower than Pt-sulfide species; this may result in a shift or broadening in the corresponding Pt 4f spectra. The Pt 4f spectra of the metallic Pt particles obtained via the reduction by formaldehyde show the binding energy of $Pt^0$ metallic nanoparticles to have the value of about 71.5 eV (Figure 5c). The same energy was found for these Pt nanoparticles deposited on pyrite, indicating the absence of the substrate influence on the chemical shift (Figure 5d). The spectra also contain a minor component from Pt(II)–Cl complexes (73 eV) and a signal at ~72 eV, which is probably due to unidentified product of incomplete reduction of $H_2PtCl_6$.

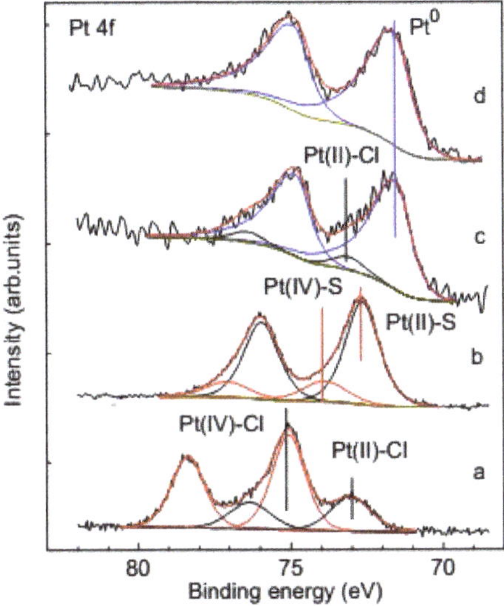


**Figure 5.** X-ray photoelectron spectroscopy (XPS) Pt 4f$_{7/2,5/2}$ spectra from reference Pt compounds deposited on highly oriented pyrolytic graphite (HOPG): (**a**) $H_2PtCl_6$, (**b**) $H_2PtCl_6$ + $H_2S$, (**c**) metallic particles, and on pyrite, (**d**) metallic particles.

### 3.3.2. Change of Platinum Species in the Deposition Process on Pyrite

Figure 6 shows Pt 4f spectra for pyrite interacted with the 1 mmol/L $H_2PtCl_6$ solution. The fitting of the spectra suggests that Pt on the surface is in the form of Pt(II) and Pt(IV) bound both with S and Cl, the proportions of which depending on the deposition duration. The high energy component with the binding energy of 75.2 eV corresponding to the adsorbed complexes of tetravalent Pt is present during the first two hours of the reaction (~25% of total Pt), and then it almost vanishes. Amount of the Pt(IV) species is significantly lower on pyrite than on pyrrhotite (Figures 6 and 7). The share of Pt(II)–Cl species having the binding energy of about 73 eV stays in the range of 30–40%, and thus adsorbed chloride complexes prevail during a 24 h reaction. The Pt 4f$_{7/2}$ component arising afterwards at 74.1 eV can be attributed to Pt(IV) sulfide, possibly PtS$_2$ (about 18% of total Pt). The further course of the reaction results in a noticeable decrease of the Pt-chloride species and increase in the amount of Pt(II) sulfide (~72.5 eV). However, the photoelectron spectra acquired using synchrotron radiation with the photon excitation energy of 250 eV (Figure S2, Supplementary materials) having a higher surface sensitivity show slightly different composition of the surface products. The main difference consisting in the presence of the component at 75.0 eV from Pt(IV)–Cl species with the relative intensity up to 8%. Thus, independent of the reaction time, a small amount of the adsorbed $PtCl_6^{2-}$ ions seems to be always present on the mineral surfaces.

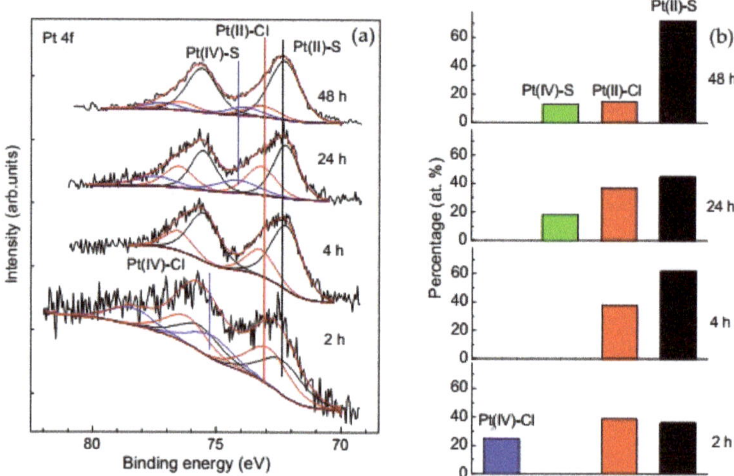

**Figure 6.** (**a**) Normalized Pt 4f spectra of pyrite treated with 1 mmol/L $H_2PtCl_6$ solution for 2 h, 4 h, 24 h and 48 h, and (**b**) distribution of the surface Pt species.

### 3.3.3. Change of Platinum Species in the Deposition Process on Pyrrhotite

About 80% of Pt immobilized on the pyrrhotite surface (Figure 7) is represented by divalent Pt bound to Cl and S, with the former being dominant for short deposition (e.g., 60% vs. 22% after 2 h), and the latter increasing with time and prevailing after 24 h reaction (36% vs. 40%). The amount of Pt(IV) tends to decrease within the range of 30% to 20% with increasing the interaction time. Concurrently, the component at 75.2 eV shifts towards 74.1 eV, suggesting that the Pt(IV)–Cl adsorbates transform to Pt(IV)–S species.

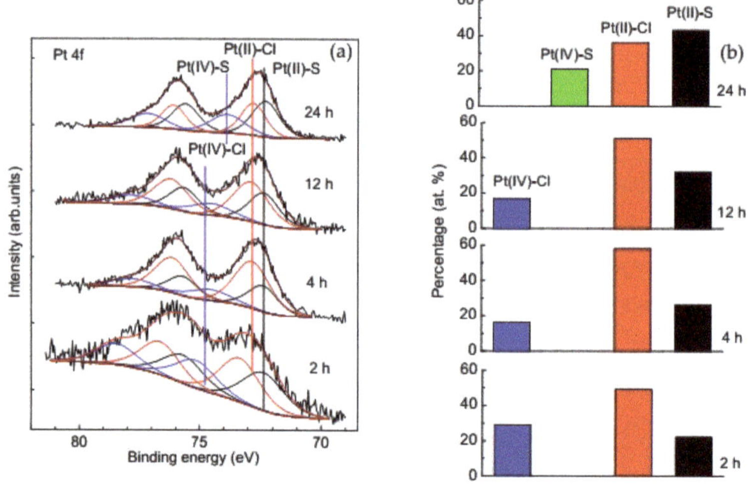

**Figure 7.** (**a**) Normalized Pt 4f spectra for pyrrhotite samples after 2, 4, 12 and 24 h of the interaction with 1 mmol/L $H_2PtCl_6$ solution, and (**b**) corresponding distribution of the Pt species.

Both the reduction and sulfidization of immobilized Pt(IV)–Cl species are faster on pyrite than pyrrhotite, despite pyrite being much more chemically inert. At the same time, it should be taken in mind that the total uptake of Pt is somewhat higher at pyrrhotite.

*3.4. Influence of the Preliminary Modification of the Mineral Surface on the Platinum Deposition*

One can expect that the state of a mineral surface is modified, for example, the weathering and mineral processing could affect the interaction with Pt-bearing solutions. The preliminary oxidative leaching of pyrrhotite in the 0.25 M FeCl$_3$ + HCl solution (30 min at 50 °C) leads to a significant increase in the amount of the deposited Pt (Figure 8a). The pre-treatment in 1 M HCl under the non-oxidative dissolution conditions (−50 mV, 10 min), which produces a massive metal-deficient layer on the pyrrhotite surface [44], also results in an increased amount of Pt on the surface but to a smaller extent than after the oxidative treatment (Figure 8). After the anodic pre-oxidation at 0.5 V (region of potentials of the passive dissolution of pyrrhotite) a significant decrease in the Pt deposition, with Pt/S being two times lower as compared to unreacted pyrrhotite is observed.

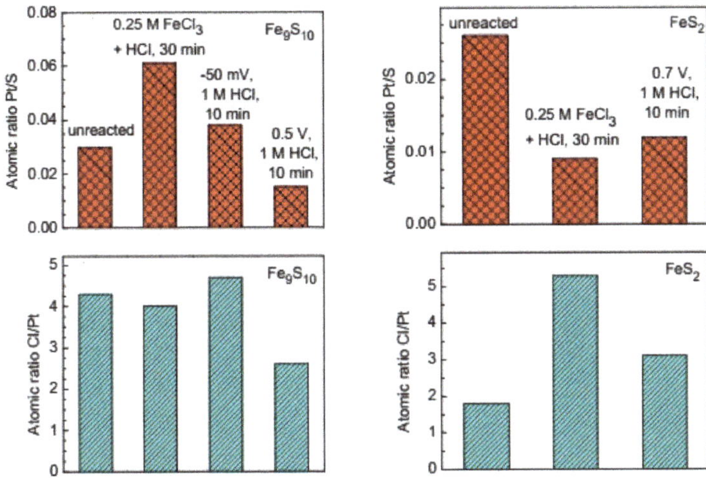

**Figure 8.** Pt/S and Pt/Cl atomic ratios for pyrrhotite (**left panels**) and pyrite (**right panels**) after various chemical pre-treatment followed by 2 h interaction with 1 mmol/L H$_2$PtCl$_6$ solution.

Figure 9 shows the Pt 4f spectra and forms of Pt deposited on the preliminarily oxidized pyrrhotite; the Pt concentration on the oxidized pyrite is low, so the spectra were of poor quality and are not shown here. On the base of these data, we hypothesize that the enhanced uptake of Pt at pyrrhotite mainly as the sulfide species is due to excessive S in the Fe-depleted surface layers produced by the chemical pre-treatment [44,45]. The Fe depletion and S enrichment of the reacted pyrite are much lower [44], and the oxidation promotes mainly increasing surface concentration of adsorbed chloride ions, which probably retard the adsorption of the Pt complex ions.

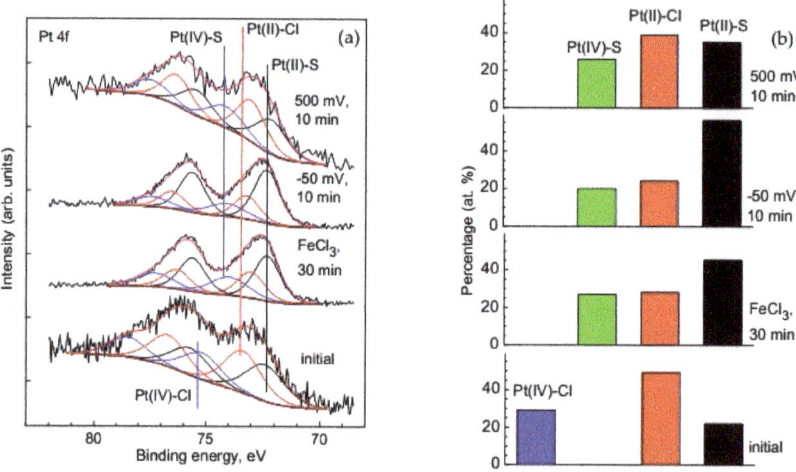

**Figure 9.** (**a**) Normalized XPS Pt 4f spectra of pyrrhotite after the deposition of Pt from 1 mmol/L H$_2$PtCl$_6$ solution onto unreacted surface and onto the surface preliminarily treated in 0.25 mol/L FeCl$_3$ + HCl during 30 min, 1 mol/L HCl at the potential of $-50$ mV, and at the potential of 500 mV (10 min), and (**b**) corresponding diagram of the distribution of the Pt forms.

### 3.5. Platinum Species on Different Sulfide Minerals

In the Pt 4f spectra of galena (Figure 10) two distinct components are shown one with the binding energy of 72.3 eV and contribution of 65%, and another at 73.4 eV amounting of 35%, almost independent of the deposition time. The first component corresponds to Pt(II)–S species, and the second originates from Pt(II) atoms bound to Cl$^-$, apparently in adsorbed PtCl$_4{}^{2-}$ ions. Neither tetravalent Pt nor metallic Pt were observed on galena under the given conditions.

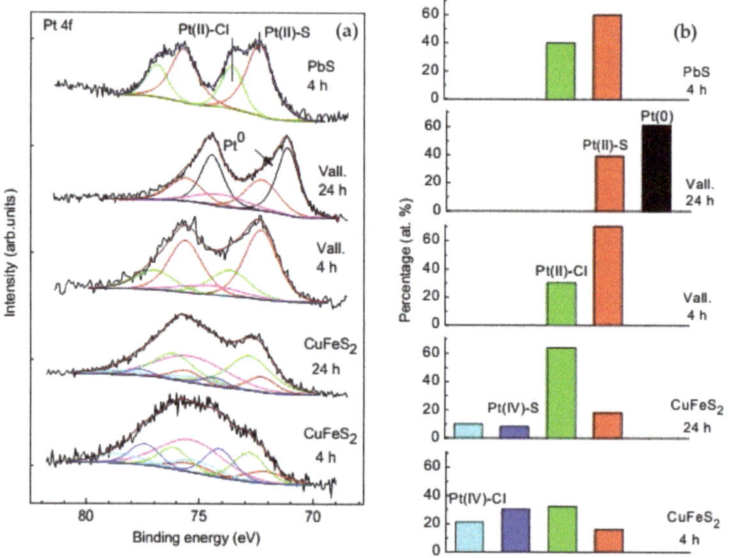

**Figure 10.** (**a**) XPS Pt 4f spectra of the sulfide minerals reacted with 1 mmol/L H$_2$PtCl$_6$ solution and (**b**) corresponding distribution diagrams of the platinum forms.

The spectra of Pt collected from valleriite after 4 h of interaction with the $H_2PtCl_6$ solution are rather similar to galena, with the two doublets being somewhat wider and contribution of the chloride complex being lower. Surprisingly, metallic Pt of ~60% from the entire amount of Pt, is found to form on valleriite as the reaction progresses, which is evidenced by the appearance of the Pt $4f_{7/2}$ peak at 71.5 eV. The remainder is Pt(II) bound with sulfur (the binding energy of 72.3 eV), while contributions of Pt–Cl species is lower. Noteworthy, the Pt 4f bands are overlapped with Cu 3p band at ~75 eV, which is weak for valleriite. Chalcopyrite, has almost the same composition as sulfide nanolayers in valleriite, but the uptake of Pt is much lower, with Pt–Cl species dominating and Pt(IV) species notably contributing especially in the initial stages; rather intense Cu 3p line overlaps the Pt 4f spectra, making difficult their unanimous fitting.

## 4. Discussion

The process of Pt immobilization onto sulfide minerals appears to proceed via a sequence of stages. The first is the adsorption of Pt-chloride complexes. The following stages include the substitution of chloride ions with sulfide (and, maybe, di- and polysulfide) anions and reduction of tetravalent Pt to divalent Pt, and to metallic state in the case of valleriite. It is possible that there exist more Pt products in addition to those described above, including intermediates of the substitution reactions, involving some $OH^-$ or/and $H_2O$ ligands, etc. However, exact compositions of the Pt species cannot be unequivocally derived from the above XPS data because of contributions of numerous surface S-, Cl-, O-bearing entities at the sulfide substrates and errors of the fitting procedures.

As it was demonstrated for pyrrhotite and pyrite, the total and relative contents of Pt products differ as a function not only of the reaction time but also of mineral nature, the state of its surface, in particular, a measure of oxidation and composition of the modified metal sulfide. The key factor seems to be the activity of the surface sulfur anions involved in the reduction of Pt(IV) and the substitution of $Cl^-$ ions in the coordination sphere of Pt. In a number of cases, e.g., for galena and pyrrhotite, Pt complexes may be reduced by aqueous bisulfide ions (hydrogen sulfide) formed due to acidic dissolution of the metal sulfides. Both these effects are related with reactivity of metal sulfide minerals, which is commonly known to increase in the order pyrite < chalcopyrite < valleriite < galena, pyrrhotite, that generally agrees with increasing the uptake of Pt (Figure 5). Furthermore, oxidation of pyrrhotite, including in ambient air, results in its passivation [54–56], while both non-oxidative and oxidative preliminary leaching causes the formation of more active over-stoichiometric sulfide, di- and polysulfide anions, which facilitate the fixation of Pt via the reduction and sulfidization reaction.

The formation of Pt(IV)–S species in the next stages can be explained by oxidation of the Pt(II)–S by di- and polysulfide groups as reported in refs [57,58]. The oxidative capability of the -S–S- species arising at the sulfide mineral surfaces upon their prolonged oxidation seems to be in fact mediated by Fe(III) entities and oxygen.

The largest uptake of Pt was observed at valleriite, the most important mineral of so-called "coppery" ores of the Noril'sk deposits (about 7% of the total resources), which form veins in sedimentary and metamorphic rocks on the margins of the massive magmatic ores and is known for increased contents of PGEs [59,60]. Valleriite is a sulfide-hydroxide composite mineral where the sulfide nanolayers with the composition $CuFeS_n$ close to chalcopyrite alternate with brucite-like hydroxide nanolayers $[Mg,Al(OH)_2]$ [61–63]. The hydroxide layers are believed to have positive charge due to the substitution of $Mg^{2+}$ for $Al^{3+}$, and so the sulfide nanolayers are negative to support electro neutrality [54], but this has not been confirmed experimentally yet. We believe that the localization of electron density at the sulfide layers makes them a strong reducing agent, which is capable of transforming Pt compounds to the metallic state over its immobilization. This mechanism corroborates with the fact that chalcopyrite being the structural analogue of the sulfide nanolayers and a precursor of valleriite in the geochemical processes, exhibits much weaker reducing properties towards Pt(II,IV) species.

## 5. Conclusions

1.  The interaction of the aqueous solutions of $H_2PtCl_6$ with the surface of sulfide minerals results in the deposition of Pt on their surface. The amount of the deposited Pt increases with time.

2.  The highest rate of Pt uptake is observed on galena and valleriite.

3.  The preliminary moderate oxidation and non-oxidative leaching of pyrrhotite creating metal-deficient surface layers usually promotes Pt deposition. In the case of pyrite, the preliminary oxidation decreases the amount of the deposited Pt.

4.  The main Pt phases immobilized on sulfide minerals are sulfides and chloride complexes of Pt(II), and, to a smaller extent, chloride complexes of tetravalent Pt; no Pt(IV) was observed on galena and valleriite. As the reaction progresses, the amount of Pt(IV) usually decreases while the amount of the sulfide forms increases.

5.  Metallic Pt was found to form only on valleriite, probably owing to the negative charge localized at sulfidic nanolayers.

6.  The di- and polysulfide surface species arising on sulfide minerals upon their oxidation are capable of oxidizing immobilized Pt(II)–S to Pt(IV)–S species.

**Supplementary Materials:** The following are available online at http://www.mdpi.com/2075-163X/8/12/578/s1, Figure S1: XPS the S 2p spectra from a droplet of sol formed in the reaction $H_2PtCl_6$ + $H_2S$ dried on pyrolytic graphite, Figure S2: SR photoelectron Pt 4f spectra of the surfaces of pyrrhotite (a) and pyrite (b) after the interaction with solution of 1 mmol/L $H_2PtCl_6$ + 0.03 mol/L HCl for 12 h. Table S1: Atomic concentrations of the elements on the surfaces of different sulfide minerals after 2 h interaction with 1 mmol/L M $H_2PtCl_6$.

**Author Contributions:** A.R. and Y.M. conceived and designed the study; A.R. prepared samples for all experiments; M.L. performed synthetic, AFM experiments; A.R. and Y.M. conducted XPS measurements; A.R. and Y.M. wrote the paper.

**Funding:** This research was funded by Russian Foundation for Basic Research and the Government of Krasnoyarsk Territory, Krasnoyarsk Regional Fund of Science to the research project: The investigation of interaction aqueous solutions of gold, platinum and palladium sulfide nano- and mesoscale particles with sulfide and oxide minerals, project number 17-45-240759.

**Conflicts of Interest:** The authors declare no conflict of interest. The funding sponsors had no role in the design of the study; in the collection, analyses, or interpretation of data; in the writing of the manuscript, and in the decision to publish the results.

## References

1.  Dey, S.; Jain Vimal, K. Platinum Group Metal Chalcogenides: Their syntheses and applications in catalysis and materials science. *Platin. Metals Rev.* **2004**, *48*, 16–29.

2.  Chia, X.; Adriano, A.; Lazar, P.; Sofer, Z.; Luxa, J.; Pumera, M. Layered platinum dichalcogenides ($PtS_2$, $PtSe_2$, and $PtTe_2$) electrocatalysis: Monotonic dependence on the chalcogen size. *Adv. Funct. Mater.* **2016**, *26*, 4306–4318. [CrossRef]

3.  Bhatt, R.; Bhattacharya, S.; Basu, R. Growth of $Pd_4S$, PdS and $PdS_2$ films by controlled sulfurization of sputtered Pd on native oxide of Si. *Thin Solid Films* **2013**, *539*, 41–46. [CrossRef]

4.  Jha, M.K.; Lee, J.; Kim, M.; Jeong, J.; Kim, B.; Kumar, V. Hydrometallurgical recovery/recycling of platinum by the leaching of spent catalysts: A review. *Hydrometallurgy* **2013**, *133*, 23–32. [CrossRef]

5.  Junge, M.; Oberthür, T.; Kraemer, D.; Melcher, F.; Pina, R.; Derrey, I.T.; Manyeruke, T.; Strauss, H. Distribution of platinum-group elements in pristine and near-surface oxidized Platreef ore and the variation along strike, northern Bushveld Complex, South Africa. *Miner. Depos.* **2018**. [CrossRef]

6.  Kraemer, D.; Junge, M.; Oberthür, T.; Bau, M. Improving recoveries of platinum and palladium from oxidized Platinum-Group Element ores of the Great Dyke, Zimbabwe, using the biogenic siderophore Desferrioxamine B. *Hydrometallurgy* **2015**, *152*, 169–177. [CrossRef]

7.  Kraemer, D.; Junge, M.; Oberthür, T.; Bau, M. Oxidized ores as future resource for platinum group metals: Current state of research. *Chem. Ing. Tech.* **2017**, *89*, 53–63. [CrossRef]

8.  Gammons, C.H.; Bloom, M.S.; Yu, Y. Experimental investigation of the hydrothermal geochemistry of platinum and palladium: I. Solubility of platinum and palladium sulfide minerals in $NaCl/H_2SO_4$ solutions at 300 °C. *Geochim. Cosmochim. Acta* **1992**, *56*, 3881–3894. [CrossRef]

9.  McDonald, I.; Ohnenstetter, D.; Rowe, J.P.; Tredoux, M.; Pattrick, R.A.D.; Vaughan, D.J. Platinum precipitation in the Waterberg deposit, Naboomspruit, South Africa. *S. Afr. J. Geol.* **1999**, *102*, 184–191.

10. Oberthür, T.; Melcher, F.; Fusswinkel, T.; van den Kerkhof, A.M.; Sosa, G.M. The hydrothermal Waterberg platinum deposit, Mookgophong (Naboomspruit), South Africa. Part 1: Geochemistry and ore mineralogy. *Mineral. Mag.* **2018**, *82*, 725–749. [CrossRef]

11. Gammons, C.H.; Bloom, M.S. Experimental investigation of the hydrothermal geochemistry of platinum and palladium: II. The solubility of PtS and PdS in aqueous sulfide solutions to 300 °C. *Geochim. Cosmochim. Acta* **1992**, *57*, 2451–2467. [CrossRef]

12. Mountain, B.W.; Wood, S.A. Chemical controls on the solubility, transport, and deposition of platinum and palladium in hydrothermal solutions: A thermodynamic approach. *Econ. Geol.* **1988**, *83*, 492–510. [CrossRef]

13. Gammons, C.H. Experimental investigations of the hydrothermal geochemistry of platinum and palladium: V. Equilibria between platinum metal, Pt(II), and Pt (IV) chloride complexes at 25 to 300 °C. *Geochim. Cosmochim. Acta* **1992**, *59*, 1655–1667. [CrossRef]

14. Gammons, C.H. Experimental investigation of the hydrothermal geochemistry of platinum and palladium: IV. The stoichiometry of Pt (IV) and Pd (II) chloride complexes at 100 to 300 °C. *Geochim. Cosmochim. Acta* **1994**, *59*, 3881–3894. [CrossRef]

15. Godel, B.; Barnes, S.-J. Platinum-group elements in sulfide minerals and the whole rocks of the JM Reef (Stillwater Complex): Implication for the formation of the reef. *Chem. Geol.* **2008**, *248*, 272–294. [CrossRef]

16. Hyland, M.M.; Bancroft, G.M. Palladium sorption and reduction on sulphide mineral surfaces: An XPS and AES study. *Geochim. Cosmochim. Acta* **1989**, *54*, 117–130. [CrossRef]

17. Plyusnina, L.P.; Likhoidov, G.G.; Shcheka, Z.A. Experimental modeling of platinum behavior under hydrothermal conditions (300–500 °C and 1 kbar). *Geochem. Int.* **2007**, *45*, 1124–1130. [CrossRef]

18. Wood, S.A.; Mountain, B.W.; Pan, P. The aqueous geochemistry of platinum, palladium and gold: Recent experimental constraints and a re-evaluation of theoretical prediction. *Can. Mineral.* **1992**, *30*, 955–982.

19. Jaireth, S. The calculated solubility of platinum and gold in oxygen-saturated fluids and the genesis of platinum-palladium and gold mineralization in the unconformity-related uranium deposits. *Miner. Depos.* **1992**, *27*, 42–54. [CrossRef]

20. Watkinson, D.H.; Melling, D.R. Hydrothermal origin of platinum-group mineralization in low-temperature copper sulfide-rich assemblages, Salt Chuck Intrusion, Alaska. *Econ. Geol.* **1992**, *87*, 175–184. [CrossRef]

21. Reith, F.; Zammit, C.M.; Shar, S.S.; Etschmann, B.; Bottrill, R.; Southam, G.; Ta, C.; Kilburn, M.; Oberthür, T.; Ball, A.S.; et al. Biological role in the transformation of platinum-group mineral grains. *Nat. Geosci.* **2016**, *9*, 1–6. [CrossRef]

22. Reith, O.F.; Campbell, S.G.; Ball, A.S.; Pring, A.; Southam, G. Platinum in Earth surface environments. *Earth-Sci. Rev.* **2014**, *131*, 1–21. [CrossRef]

23. Pawlak, J.; Łodyga-Chruścińska, E.; Chrustowicz, J. Fate of platinum metals in the environment. *J. Trace Elem. Med. Biol.* **2014**, *28*, 247–254. [CrossRef] [PubMed]

24. Reith, F.; Shuster, J. *Geomicrobiology and Biogeochemistry of Precious Metals*; MDPI: Basel, Switzerland, 2018; pp. 114–131; ISBN 978-3-03897-347-8.

25. Piña, R.; Gervilla, F.; Barnes, S.-J.; Oberthür, T.; Lunar, R. Platinum-group element concentrations in pyrite from the Main Sulfide Zone of the Great Dyke of Zimbabwe. *Miner. Depos.* **2016**, *51*, 853–872. [CrossRef]

26. Dare, S.A.S.; Barnes, S.-J.; Prichard, H.M.; Fisher, P.C. Chalcophile and platinum-group element (PGE) concentrations in the sulfide minerals from the McCreedy East deposit, Sudbury, Canada, and the origin of PGE in pyrite. *Miner. Depos.* **2011**, *46*, 381–407. [CrossRef]

27. Junge, M.; Wirth, R.; Oberthür, T.; Melcher, F.; Schreiber, A. Mineralogical siting of platinum-group elements in pentlandite from the Bushveld Complex, South Africa. *Miner. Depos.* **2015**, *50*, 41–54. [CrossRef]

28. Kubrakova, I.V.; Tyutyunnik, O.A.; Koshcheeva, I.Y.; Sadagov, A.Y.; Nabiullina, S.N. Migration behavior of platinum group elements in natural and technogeneous systems. *Geochem. Int.* **2017**, *55*, 108–124. [CrossRef]

29. Torres, R.; Lapidus, G.T. Platinum, palladium and gold leaching from magnetite ore, with concentrated chloride solutions and ozone. *Hydrometallurgy* **2016**, *166*, 185–194. [CrossRef]

30. Fuchs, W.A.; Rose, A.W. The geochemical behavior of platinum and palladium in the weathering cycle in the Stillwater complex, Montana. *Econ. Geol.* **1974**, *69*, 332–346. [CrossRef]

31. Adams, M.; Liddell, K.; Holohan, T. Hydrometallurgical processing of Platreef flotation concentrate. *Miner. Eng.* **2011**, *24*, 545–550. [CrossRef]

32. Liddell, K.S.; Adams, M.D. Kell hydrometallurgical process for extraction of platinum group metals and base metals from flotation concentrates. *J. S. Afr. Inst. Min. Metall.* **2012**, *112*, 31–36.

33. Liddell, K.; Newton, T.; Adams, M.; Muller, B. Energy consumption for Kell hydrometallurgical refining versus conventional pyrometallurgical smelting and refining of PGM concentrates. *J. S. Afr. Inst. Min. Metall.* **2011**, *111*, 127–132.

34. Mpinga, C.N.; Eksteen, J.J.; Aldrich, C.; Dyer, L. Direct leach approaches to Platinum Group Metal (PGM) ores and concentrates: A review. *Miner. Eng.* **2015**, *78*, 93–113. [CrossRef]

35. Mwase, J.M.; Petersen, J.; Eksteen, J.J. A conceptual flowsheet for heap leaching of platinum group metals (PGMs) from a low-grade ore concentrate. *Hydrometallurgy* **2012**, *111*, 129–135. [CrossRef]

36. Mwase, J.M.; Petersen, J.; Eksteen, J.J. Assessing a two-stage heap leaching process for Platreef flotation concentrate. *Hydrometallurgy* **2012**, *129*, 74–81. [CrossRef]

37. Mwase, J.M.; Petersen, J.; Eksteen, J.J. A novel sequential heap leach process for treating crushed Platreef ore. *Hydrometallurgy* **2014**, *141*, 97–104. [CrossRef]

38. Mikhlin, Y.L.; Romanchenko, A.S. Gold deposition on pyrite and the common sulfide minerals: An STM/STS and SR-XPS study of surface reactions and Au nanoparticles. *Geochim. Cosmochim. Acta* **2007**, *71*, 5985–6001. [CrossRef]

39. Mikhlin, Y.L.; Romanchenko, A.S.; Likhatski, M.N.; Karacharov, A.; Erenberg, S.B.; Trubina, S.V. Understanding the initial stages of precious metals precipitation: Nanoscale metallic and sulfidic species of gold and silver on pyrite surfaces. *Ore Geol. Rev.* **2011**, *42*, 47–54. [CrossRef]

40. Mycroft, J.R.; Bancroft, G.M.; McIntyre, N.S.; Lorimer, J.W. Spontaneous deposition of gold on pyrite from solutions containing Au (III) and Au (I) chlorides. Part I: A surface study. *J. Electroanal. Chem.* **1990**, *292*, 139–152. [CrossRef]

41. Romanchenko, A.S.; Mikhlin, Y.L. An XPS study of products formed on pyrite and pyrrhotine by reacting with palladium(II) chloride solutions. *J. Struct. Chem.* **2015**, *56*, 531–537. [CrossRef]

42. Gulyaev, R.V.; Stadnichenko, A.I.; Slavinskaya, E.M.; Koscheev, S.V.; Ivanova, A.S.; Boronin, A.I. In situ preparation and investigation of $Pd/CeO_2$ catalysts for the low-temperature oxidation of CO. *Appl. Catal. A Gen.* **2012**, *439*, 41–50. [CrossRef]

43. Kibis, L.S.; Stadnichenko, A.I.; Koscheev, S.V.; Zaikovskii, V.I.; Boronin, A.I. Highly oxidized palladium nanoparticles comprising $Pd^{4+}$ species: Spectroscopic and structural aspects, thermal stability, and reactivity. *J. Phys. Chem. C* **2012**, *116*, 19342–19348. [CrossRef]

44. Mikhlin, Y.L.; Tomashevich, Y.V.; Pashkov, G.L.; Okotrub, A.V.; Asanov, I.P.; Mazalov, L.N. Electronic structure of the non-equilibrium iron-deficient layer of hexagonal pyrrhotite. *Appl. Surf. Sci.* **1998**, *125*, 73–84. [CrossRef]

45. Mikhlin, Y. Reactivity of pyrrhotite surfaces: An electrochemical study. *Phys. Chem. Chem. Phys.* **2000**, *2*, 5672–5677. [CrossRef]

46. Mikhlin, Y.; Romanchenko, A.; Shagaev, A. Scanning probe microscopy studies of PbS surfaces oxidized in air and etched in aqueous acid solutions. *Appl. Surf. Sci.* **2006**, *252*, 5245–5258. [CrossRef]

47. Mikhlin, Y.; Kuklinskiy, A.; Mikhlina, E.; Kargin, V.; Asanov, I. Electrochemical behaviour of galena (PbS) in aqueous nitric acid and perchloric acid solutions. *J. Appl. Electrochem.* **2004**, *34*, 37–46. [CrossRef]

48. Mikhlin, Y.L.; Tomashevich, Y.V.; Asanov, I.P.; Okotrub, A.V.; Varnek, V.A.; Vyalikh, D.V. Spectroscopic and electrochemical characterization of the surface layers of chalcopyrite ($CuFeS_2$) reacted in acidic solutions. *Appl. Surf. Sci.* **2004**, *225*, 395–409. [CrossRef]

49. Mikhlin, Y.L.; Romanchenko, A.S.; Tomashevich, E.V.; Volochaev, M.N.; Laptev, Y.V. XPS and XANES study of layered mineral valleriite. *J. Struct Chem.* **2017**, *58*, 1137–1143. [CrossRef]

50. Laptev, Y.V.; Shevchenko, V.S.; Urakaev, F.K. Sulphidation of valleriite in $SO_2$ solutions. *Hydrometallurgy* **2009**, *98*, 201–205. [CrossRef]

51. Muhler, M.; Paal, Z. Sulfided platinum black by XPS. *Surf. Sci. Spectra* **1996**, *4*, 125–129. [CrossRef]

52. Dembowski, J.; Marosi, L.; Essig, M. Platinum disulfide by XPS. *Surf. Sci. Spectra* **1994**, *2*, 133–137. [CrossRef]

53. Karhu, H.; Kalantar, A.; Väyrynen, I.J.; Salmi, T.; Murzin, D.Y. XPS analysis of chlorine residues in supported Pt and Pd catalysts with low metal loading. *Appl. Catal. A* **2003**, *247*, 283–294. [CrossRef]

54. Mikhlin, Y.; Tomashevich, Y.; Vorobyev, S.; Saikova, S.; Romanchenko, A.; Félix, R. Hard X-ray photoelectron and X-ray absorption spectroscopy characterization of oxidized surfaces of iron sulfides. *Appl. Surf. Sci.* **2016**, *387*, 796–804. [CrossRef]

55. Mikhlin, Y.; Nasluzov, V.; Romanchenko, A.; Tomashevich, Y.; Shor, A.; Félix, R. Layered structure of the near-surface region of oxidized chalcopyrite (CuFeS$_2$): Hard X-ray photoelectron spectroscopy, X-ray absorption spectroscopy and DFT+U studies. *Phys. Chem. Chem. Phys.* **2017**, *19*, 2749–2759. [CrossRef] [PubMed]

56. Mikhlin, Y.; Romanchenko, A.; Tomashevich, Y.; Shurupov, V. Near-surface regions of electrochemically polarized chalcopyrite (CuFeS$_2$) as studied using XPS and XANES. *Phys. Procedia* **2016**, *84*, 390–396. [CrossRef]

57. Bonnington, K.J.; Jennings, M.C.; Puddephatt, R.J. Oxidative addition of S-S bonds to dimethylplatinum(II) complexes: Evidence for a binuclear mechanism. *Organometallics* **2008**, *27*, 6521–6530. [CrossRef]

58. McCready, M.S.; Puddephatt, R.J. Oxidative addition of functional disulfides to platinum(II): Formation of chelating and bridging thiolate–carboxylate complexes of platinum(IV). *Inorg. Chem. Commun.* **2011**, *14*, 210–212. [CrossRef]

59. Yakubchuk, A.; Nikishin, A. Noril'sk-Talnakh Cu-Ni-PGE deposits: A revised tectonic model. *Miner. Depos.* **2004**, *39*, 125–142. [CrossRef]

60. Harris, D.C.; Cabry, L.J.; Stewart, J.M. A "valleriite-type" mineral from Noril'sk, Western Siberia. *Am. Mineral.* **1970**, *55*, 2110–2114.

61. Evans, H.T.; Allman, R. The crystal structure and crystal chemistry of valleriite. *Zeitschrift fur Kristallographie* **1968**, *127*, 73–93. [CrossRef]

62. Hughes, A.E.; Kakos, G.A.; Turney, T.W.; Williams, T.B. Synthesis and structure of valleriite, a layered Metal Hydroxide/Sulfide Composite. *J. Solid State Chem.* **1993**, *104*, 422–436. [CrossRef]

63. Li, R.; Cui, L. Investigations on valleriite from Western China: Crystal chemistry and separation properties. *Int. J. Miner. Process.* **1994**, *41*, 271–283. [CrossRef]

*Article*

# Colloidal and Deposited Products of the Interaction of Tetrachloroauric Acid with Hydrogen Selenide and Hydrogen Sulfide in Aqueous Solutions

Sergey Vorobyev [1], Maxim Likhatski [1], Alexander Romanchenko [1], Nikolai Maksimov [1], Sergey Zharkov [2,3], Alexander Krylov [2] and Yuri Mikhlin [1,*]

[1] Institute of Chemistry and Chemical Technology of the Siberian Branch of the Russian Academy of Sciences, Krasnoyarsk 660036, Russia; yekspatz@yandex.ru (S.V.); lixmax@icct.ru (M.L.); romaas82@mail.ru (A.R.); burmakina@ksc.krasn.ru (N.M.)

[2] Kirensky Institute of Physics of the Siberian Branch of the Russian Academy of Sciences, Krasnoyarsk 660036, Russia; zharkov@iph.krasn.ru (S.Z.); shusy@iph.krasn.ru (A.K.)

[3] Electron Microscopy Laboratory, Siberian Federal University, Krasnoyarsk 660041, Russia

* Correspondence: yumikh@icct.ru; Tel.: +7-391-205-1928

Received: 11 October 2018; Accepted: 29 October 2018; Published: 31 October 2018

**Abstract:** The reactions of aqueous gold complexes with $H_2Se$ and $H_2S$ are important for transportation and deposition of gold in nature and for synthesis of AuSe-based nanomaterials but are scantily understood. Here, we explored species formed at different proportions of $HAuCl_4$, $H_2Se$ and $H_2S$ at room temperature using in situ UV-vis spectroscopy, dynamic light scattering (DLS), zeta-potential measurement and ex situ Transmission electron microscopy (TEM), electron diffraction, X-ray diffraction (XRD), X-ray photoelectron spectroscopy (XPS), and Raman spectroscopy. Metal gold colloids arose at the molar ratios $H_2Se(H_2S)/HAuCl_4$ less than 2. At higher ratios, pre-nucleation "dense liquid" species having the hydrodynamic diameter of 20–40 nm, zeta potential $-40$ mV to $-50$ mV, and the indirect band gap less than 1 eV derived from the UV-vis spectra grow into submicrometer droplets over several hours, followed by fractional nucleation in the interior and coagulation of disordered gold chalcogenide. XPS found only one $Au^+$ site (Au $4f_{7/2}$ at 85.4 eV) in deposited AuSe, surface layers of which partially decomposed yielding $Au^0$ nanoparticles capped with elemental selenium. The liquid species became less dense, the gap approached 2 eV, and gold chalcogenide destabilized towards the decomposition with increasing $H_2S$ content. Therefore, the reactions proceed via the non-classical mechanism involving "dense droplets" of supersaturated solution and produce $AuSe_{1-x}S_x/Au$ nanocomposites.

**Keywords:** gold selenide; gold sulfoselenide; colloids; nanoparticles; nucleation; liquid intermediates; deposition

## 1. Introduction

Mixed silver–gold sulfides and selenides, including nanoparticulate ones, have attracted intense interest as minerals and natural sources of precious metals [1–14] and due to high ionic conductivity, tunable optical characteristics, thermoelectric and other properties promising for materials applications [15–19]. Gold chalcogenides, in particular gold selenide, are much less examined, despite aqueous Au chalcogenide-derived complexes play the crucial role as carriers of gold in hydrothermal fluids and brines (e.g., [20–25] and references therein), and AuSe, Au(Se,S) and Au(Te,Se,S) phases in intergrowths with native gold have been found in the high-sulfidation epithermal deposits formed from acidic fluids[14]. Solid gold sulfide $Au_2S$ is metastable [2,5,26,27]. Gold selenide that is the only thermodynamically stable compound in the binary Au–Se system [3,11,28–37] has two crystallographic modifications α-AuSe and β-AuSe, which crystallize in the monoclinic space group C2/m with different lattice constants. In both structures, a half of Au atoms are in a square-planar coordination to four Se atoms, and the others are linearly bound to two Se neighbors. Rabenau and Schulz [31] and then other authors [34,35] have assumed that the gold in the first sites is $Au^{3+}$, while $Au^+$ is located in the linear environment, and each Se atom is linked with one $Au^+$ and two $Au^{3+}$ cations. On the other hand, X-ray absorption spectroscopy [36] and the theoretical calculations [37] have found fully occupied Au 5d bands and therefore only monovalent $Au^+$ atoms.

It is known from chemistry textbooks (for example [38]) that direct interaction between aqueous gold complexes and selenide ions results in precipitation of unstable gold selenides with unclear compositions. Preparation of nanoscale materials containing gold selenide have been described in a few publications [17,39–45]. Nath and co-workers [39] have synthesized "AuSe nanoalloys" via fusion of elemental Au and Se nanoparticles in a micellar solution, typically with an excess of Au NPs. Prokeš et al. [40] have used mixed dispersions of gold nanoparticles and elemental selenium, or Se powder suspended in $HAuCl_4$ solutions in various proportions as precursors for laser ablation, and detected in gas phase, by using mass-spectrometry, numerous gold-selenide nanoclusters $Au_mSe_n$ leaning to compositions with Au/Se = 2 and Au/Se = 1. Machogo et al. [42] have recently reported the preparation of AuSe nanostructures mixed with metallic nanogold and unreacted Se by the reaction of $HAuCl_3·3H_2O$ with elemental Se in oleylamine. It has been found in several studies aimed at synthesis of hybrid nanostructures that gold commonly deposits onto metal selenide nanoparticles in elemental form [43–45]. Particularly, Cueva et al. [45] have concluded from X-ray photoelectron spectroscopy (XPS) analysis that the shells deposited on CdSe NPs are most probably composed of AuSe or AuSeCl, which are instable and transform into dot-shaped $Au^0$ particles under the electron beam in Transmission electron microscopy (TEM). Several studies on adsorption of selenium [46,47] and alkylselenides on gold and Au NPs [48,49] have revealed no AuSe phases and, for example, demonstrated that alkylselenides replace alkylsulfides due to favored Au–Se bonding.

The products of the reaction of sulfide ions with gold tetrachloride depend on the initial ratio of the reagents [50–55]. Metallic Au NPs emerging at the molar ratios from 0.5 to 1.5 show a second localized surface plasmon resonance (LSPR) in the near-infrared region, which has been suggested to assign to gold-sulfide/gold core/shell particles [50], aggregation of gold nanoparticles [54], or, and this is most likely, to longitudinal plasmon resonance in non-spherical Au NPs [55]. Gold sulfide is the main final product at the $Na_2S/HAuCl_4$ ratio higher than 2 but its formation is retarded, probably by slow nucleation and crystal growth of a metastable $Au_2S$ phase [55–58]. The reaction was concluded to proceed via non-classical pathway involving a series of pre-nucleation intermediates (10–40 nm "clusters" and submicrometer droplets made up of "dense liquid" due to spinodal decomposition of supersaturated solutions), whose existence was confirmed using in situ atomic force microscopy, dynamic light scattering (DLS) and other techniques [57,58]. The question arises as to whether similar mechanisms occur in the selenide solutions, considering the stronger Au–Se bonding and thermodynamic stability of AuSe.

Here, we studied the interaction of aqueous tetrachloroauric acid with hydrogen selenide and their mixtures with hydrogen sulfide with focus on the media with initial molar chalcogenide-to-gold ratio of 3, when the reaction may be expected to proceed via the formation of the liquid intermediates. We employed XPS, TEM, electron and X-ray diffraction and other techniques to shed light onto morphology, composition, and chemical bonding in gold-selenide and sulfoselenides deposited from the solutions. The results of the research are important for understanding the fundamentals of nucleation, transportation, and deposition of noble metals in the nature, and in a future synthesis of gold chalcogenide nanomaterials.

## 2. Materials and Methods

The gold precursor was prepared by diluting the 0.1 mol aqueous stock solution of hydrogen tetrachloroaurate (98% purity, the Krasnoyarsk Plant of Non-Ferrous Metals) with deionized water to obtain 0.3 mmol/L HAuCl$_4$ directly before the reaction. Hydrogen selenide and hydrogen sulfide were obtained via hydrolysis of Al$_2$Se$_3$ and Al$_2$S$_3$ prepared by sintering Al powder with elemental selenium or sulfur in a quartz crucible. The gaseous hydrogen chalcogenides were absorbed by deionized water preliminary de-aerated by purging with Ar (grade 5.5) for 30 min; the saturated solutions were sealed and kept at 5 °C for several days before application. The H$_2$Se (or H$_2$S) solutions were diluted with de-aerated water to a desired concentration, which was verified by titration with elemental iodine and starch as indicator. In a typical procedure, a required volume (0.075–0.75 mL) of about 50 mmol/L H$_2$Se or H$_2$S, or their mixture was quickly added to 0.3 mmol/L HAuCl$_4$ solution (25 mL) under vigorous agitation during 2 min and then left still at room temperature. In several experiments, 1 mmol/L cetyltrimethyl ammonium bromide (CTAB) was added afterwards in an attempt to stabilize the colloids formed. After predetermined time intervals, a portion of the solution was taken for examination.

UV-vis absorption spectra were collected in the range 200–1600 nm from the reaction mixtures loaded in a quartz cell with the light pass length of 2 mm using a Shimadzu UV 3600 spectrometer (Shimadzu, Kyoto, Japan). Hydrodynamic diameter (volume weighted mean size $Z_{av}$) and zeta-potentials of colloidal products were determined using a Zetasizer Nano ZS spectrometer (Malvern, Cambridge, UK) at scattering angle of 173° in a folded polystyrene cell or polycarbonate cell with Pd electrodes.

TEM, energy dispersive X-ray analysis (EDS), and selected area electron diffraction (SAED) studies were performed using a JEOL JEM 2100 instrument (JEOL, Tokyo, Japan) operating at an accelerating voltage of 200 kV. To prepare a sample for examination, a droplet of the reaction solution was placed onto a Cu grid covered with a layer of amorphous carbon and allowed to dry in air at room temperature.

For XPS studies, a reaction solution droplet was dried at highly oriented pyrolytic graphite (HOPG). Generally, the spectra were taken both without washing and after water washing the residue; the results were essentially the same, and only those from washed specimens are presented here. The spectra were acquired using a SPECS spectrometer (SPECS, Berlin, Germany) equipped with a PHOIBOS 150 MCD-9 analyzer at electron take-off angle 90° employing monochromatic Al K$\alpha$ radiation (1486.6 eV) of an X-ray tube operated at 200 W. The analyzer pass energy was 10 eV for narrow scans, and 20 eV for survey spectra. C 1s peak at 284.45 eV from HOPG was used as a reference. The Au 4f$_{7/2,5/2}$, Se 3d$_{5/2,3/2}$, S 2p$_{3/2,1/2}$ doublets were fitted, after subtraction of Shirley-type background, with coupled peaks with Gaussian-Lorentzian peak profiles, the spin-orbit splitting of 3.67 eV, 0.88 eV, and 1.19 eV, and the branching ratios of 0.75, 0.667, and 0.5, respectively, using CasaXPS software (version 2.3.16, Casa Software, Teignmouth, UK).

X-ray diffraction was measured from residues precipitated from the solutions, rinsed with water, and deposited on Si support employing XPERT-PRO diffractometer (PANalytical, Almelo, The Netherlands) using Cu K$\alpha$ radiation (30 mA, 40 kV). The Raman spectra in the backscattering geometry were recorded on a Horiba Jobin-Yvon T64000 spectrometer (Horiba, Kyoto, Japan) equipped with a liquid nitrogen cooled charge-coupled device detection system in subtractive dispersion mode. Ar$^+$ ion laser Spectra-Physics Stabilite 2017 with $\lambda$ = 514.5 nm and power of 1 mW on a sample was used as an excitation light source.

## 3. Results

### 3.1. UV-Vis Absorption Spectroscopy

Figure 1 shows UV-vis spectra of the solutions with various initial molar ratios of H$_2$Se to HAuCl$_4$, and also H$_2$S to HAuCl$_4$ after 1 h reaction. As the quantity of H$_2$Se increased, absorption peaks of tetrachloroaurate at 228 nm and 315 nm disappeared and the surface plasmon resonance of metallic gold nanoparticles arose near 530 nm, but, in contrast to the reduction of HAuCl$_4$ with sodium sulfide solution [50–57], no LSPR was observed in the near-IR spectral region. The LSPR vanished as the H$_2$Se/HAuCl$_4$ ratio approached 2, and the spectra only slightly varied at the ratios of 3 and higher.

The effect of the relative concentration of $H_2S$ was similar but the absorption was notably lower than for the selenide systems in all the wavelength range. Previously, we found that the absorption ($\alpha$) of the solutions with high $Na_2S/HAuCl_4$ ratios plotted as $(\alpha h\nu)^{1/2}$ vs. photon energy ($h\nu$) was linear, such as that of amorphous semiconductors with an indirect band gap. The gap width ($E_g$) of 1.43 eV determined from the UV-vis spectra [55–57] was attributed to $Au_2S$-like species in "dense droplets" of supersaturated solutions, although the exact nature of this soft material is obscure. The effect of the ratios of $H_2Se$ and $H_2S$ to $HAuCl_4$ on the band gap is shown Figure 1c. The $E_g$ values increased up to 1.07 eV and 2.0 eV for $H_2Se$ and $H_2S$, respectively, as the initial chalcogen/gold ratios changed from 2 to 3.5, and then slowly decreased. The gap widths are in reasonable agreement with the values of 0.6 eV and 0.7 eV reported for bulk $\alpha$- and $\beta$-AuSe, respectively [37], and those ranged from 1.3 eV to 2.6 eV for $Au_2S$ [50–57]. Therefore, this parameter has a physical meaning related to Au–Se and Au–S bonding and can be used at least for monitoring the reaction.

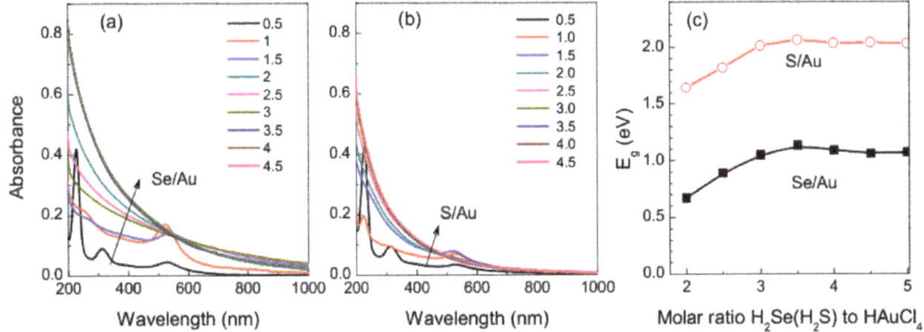

**Figure 1.** UV-vis absorption spectra of the reaction solutions measured after 1 h reaction at various molar ratios (**a**) $H_2Se$ to $HAuCl_4$ and (**b**) $H_2S$ to $HAuCl_4$ shown in the legend. (**c**)The indirect band gaps determined from $(\alpha h\nu)^{1/2}$ vs photon energy ($h\nu$) plots for the UV-vis spectra as a function of the ratio of $H_2Se$ ($H_2S$) to $HAuCl_4$.

The spectra of the reaction mixtures with the molar ratio $(H_2Se + H_2S)/HAuCl_4 = 3$ presented in Figure 2 demonstrate that the optical absorption increased and the $E_g$ magnitude gradually reduced upon increasing the proportion of $H_2Se$, and the $E_g$ vs. $H_2Se/(H_2Se + H_2S)$ is asymmetric due to stronger Au–Se bonding in comparison with Au–S bonding.

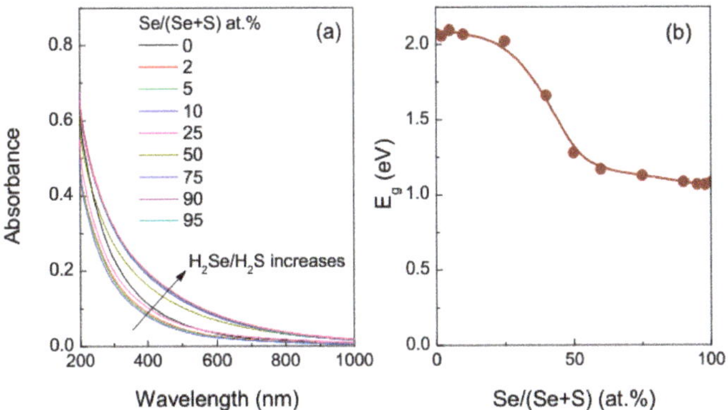

**Figure 2.** (**a**) UV-vis absorption spectra of the mixed $H_2Se/H_2S$ reaction media with the total molar ratio hydrogen chalcogenide to $HAuCl_4$ of 3 and (**b**) the indirect band gaps determined from the UV-vis spectra as a function of the ratio $H_2Se/H_2S$ after 1 h reaction.

The media formed are stable at least for several hours, depending on the composition, with the absorption increasing and the $E_g$ parameter slowly decreasing with the reaction time (Figure 3).

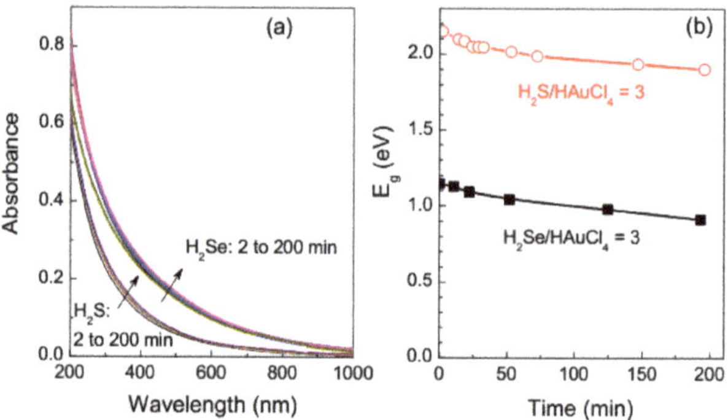



**Figure 3.** (**a**) UV-vis absorption spectra of the reaction solutions with the molar ratios $H_2Se/HAuCl_4 = 3$ and $H_2S/HAuCl_4 = 3$, and (**b**) the in direct band gaps calculated from the UV-vis spectra as a function of reaction time.

### 3.2. Dynamic Light Scattering and Zeta-Potential Measurement

DLS and zeta-potential measurements (Figure 4) reveal nanoscale entities with the average hydrodynamic diameter ($Z_{av}$) of 20–40 nm and negative surface charges after 1 h reaction at the ratios of $H_2Se$ (or $H_2S$) to $HAuCl_4$ below 2. The above UV-vis spectra and previous results [50–57] suggest that these values correspond to gold nanoparticles, although the reaction seems to be not completed and $Au^+$-chalcogenide intermediates are also present. Both $Z_{av}$ and zeta potential (ranged from $-50$ mV to $-40$ mV) moderately varied with the $H_2Se/HAuCl_4$ ratio. On the contrary, the hydrodynamic diameter gradually decreased with increasing concentration of $H_2S$ and sharply enhanced from 20 nm to about 300 nm as the $H_2S/HAuCl_4$ ratio reached 3; correspondingly, zeta-potential magnitude altered from $-40$ mV to $-10$ mV and then to $-55$ mV. Such behavior can be explained in terms of a lower stability of Au–S intermediates and a higher rate of the reaction yielding $Au^0$ with sulfide than selenide (see Discussion below). In both the selenide and sulfide media with the chalcogenide-to-gold proportion of 3 and higher, the values of zeta potential are close to $-50$ mV, which may be considered as a signature of the "dense" liquid intermediates.

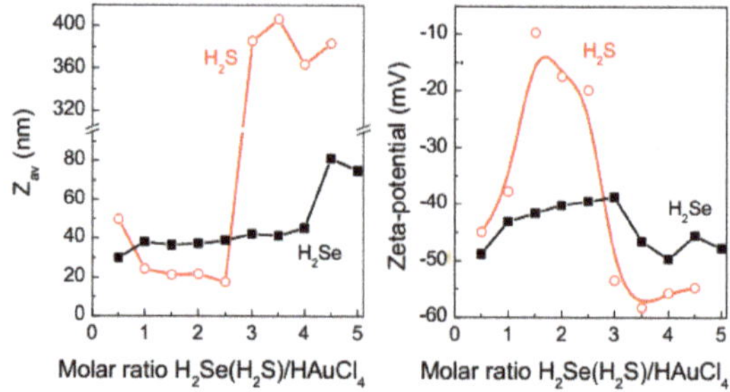



**Figure 4.** Hydrodynamic diameters $Z_{av}$ and zeta-potentials of the colloidal species at various ratios $H_2Se$ and $H_2S$ to $HAuCl_4$ determined in 1 h after mixing the reactants.

Figure 5 shows evolution of $Z_{av}$ with time after addition of $H_2Se$ or $H_2S$ to aqueous $HAuCl_4$ at the molar ratio of 3 to 1. The hydrodynamic diameter rather slowly increased in the gold-selenide medium during 6–8 h, and rapidly afterwards (the big species observed in the first minutes and then dissolved likely originate from local supersaturation immediately after mixing the reactants). Similar behavior was previously found for the interaction of tetrachloroaurate with sodium sulfide solutions and interpreted in terms of slow coalescence of dense liquid nanoscale droplets ("clusters") into submicrometer droplets [57,58]. However, in the reaction involving $H_2S$, the sulfide-containing species rapidly grew up to approximately 500 nm in less than 2 h. Nucleation (within the submicrometer droplets) and precipitation of both gold chalcogenides is not instant but gradual and poorly reproducible, unless very high concentrations of chalcogenide ions or an injection of electrolyte were applied; this is illustrated in Figure 4 for the case of $H_2S$. Some insight into the nucleation and coagulation in $Na_2S$ + $HAuCl_4$ solutions was reported elsewhere [58] but the phenomena require further investigation.

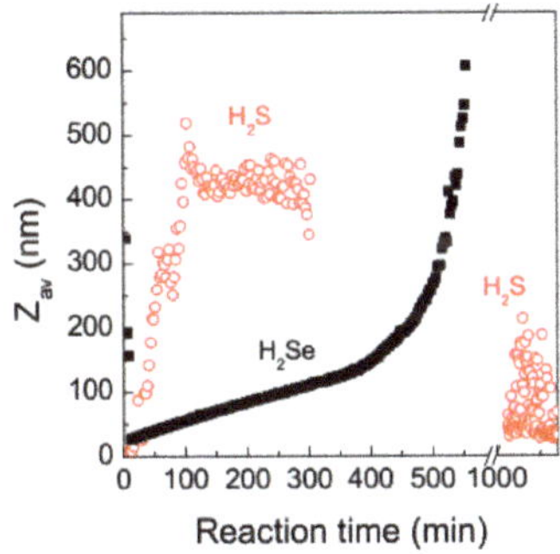


**Figure 5.** Evolution of average hydrodynamic diameter ($Z_{av}$) with time after mixing aqueous solutions of $H_2Se$ or $H_2S$ with $HAuCl_4$ at the molar ratio 3 to 1.

The effect of substitution of $H_2S$ for $H_2Se$ on the properties of colloidal products is difficult to elucidate because of very different reaction rates. In the initial stages, when smaller dense liquid species are believed to arise and submicrometer droplets are absent yet [55–58], a steady increase in $Z_{av}$ from 20 nm to 40 nm and zeta-potential magnitude from about $-35$ mV to $-50$ mV as the relative concentration of $H_2Se$ increased (Figure S1, Supplementary data).

### 3.3. TEM

The reaction products were further examined using ex situ TEM, electron diffraction, XPS, XRD, Raman scattering. These techniques are commonly employed, in particular, in wet chemical synthesis to characterize nanoscale materials, which, however, may be modified by immobilization and drying.

Representative TEM images of the species formed via the interaction of tetrachloroauric acid with hydrogen selenide, hydrogen sulfide and their 1-to-1 mixture at the initial molar ratio hydrogen chalcogenide/$HAuCl_4$ of 3 are given in Figure 6. The products of the reaction with hydrogen selenide are about 40 nm aggregates composed of irregular 5–10 nm particles of two sorts, dense and low-contrast, along with lesser contrast loose material. High-resolution TEM (HR-TEM) (Figure 6b) and the relevant Fourier transforms from the less dense species revealed interplanar distances of 3.2–3.3 Å. At the same time, electron diffraction (Figure 6c) showed diffuse circles from elemental gold (d = 2.33 Å, 2.02 Å, 1.43 Å, etc., PDF 00-004-0784), which is present as the larger dense NPs. The circles corresponding to d = 3.29 Å, 1.72 Å, 1.31 Å likely originate from a gold-selenide compound; meanwhile, the strongest reflections from either $\alpha$-AuSe ($d_{310}$ = 2.70 Å, $d_{003}$ = 2.74 Å, $d_{313}$ = 1.79 Å, PDF 00-020-0457), or $\beta$-AuSe ($d_{111}$ = 2.79 Å, $d_{002}$ = 4.02 Å, etc., PDF 00-020-0458) are absent.

In the case of the mixture of $H_2Se$ and $H_2S$, TEM shows (Figure 6d,e) preferentially separated Au NPs of 5–6 nm in size with the interplanar distances of ~2.2–2.3 Å (observed in HR-TEM), and low-contrast species with the distances of 3.2–3.3 Å and 3.0–3.1 Å. Electron diffraction revealed, in addition to gold, weak reflections at 3.48 Å, 3.10 Å, 1.72 Å. The products of the reaction with $H_2S$

(Figure 6g–i) represented a network-like structure formed by low-contrast wires and ~3 nm Au NPs, the quantity of which visibly grew under the electron beam; very similar images were obtained with $Na_2S$ solution [55]. In addition to metallic gold, electron diffraction shows very weak diffuse reflections at 2.83 Å and possibly some others attributable to gold sulfide (i.e., $d_{111}$ = 2.89 Å, PDF 04-007-4652).

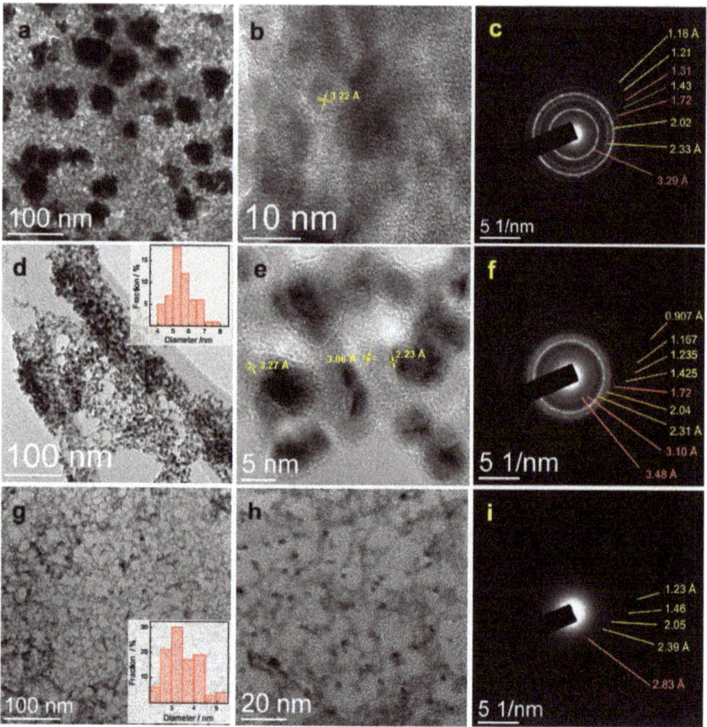

**Figure 6.** TEM images and selected area electron diffraction patterns of the air-dried solutions of (a–c) 0.3 mmol/L $HAuCl_4$ + 0.9 mmol/L $H_2Se$, (d–f) 0.3 mmol/L $HAuCl_4$ + 0.45 mmol/L $H_2Se$ + 0.45 mmol/L $H_2S$; (g–i) 0.3 mmol/L $HAuCl_4$ + 0.9 mmol/L $H_2S$ after about 1 h reaction. The data illustrate the formation of gold nanoparticles via post-synthetic decomposition of disordered gold chalcogenides (less contrast material), see the text for detail.

EDS analysis (Figure S2, Supplementary data) of the Au–Se products found average atomic Se/Au ratios an order of 2, with essentially inhomogeneous spatial distribution of the elements that varied from less than 1 at the dense nanoparticles attributable to $Au^0$ to ~3 for the low-contrast material; a small amount of Cl (atomic proportion Cl/Au of ~0.1) was also detected. For the mixture (1:1) of $H_2Se$ and $H_2S$, the average composition of the products approached (Se + S)/Au~1, and also was spatially inhomogeneous. The products formed with $H_2S$ had the atomic S/Au proportion less than 1 that decreased with time due to rapid decay and volatization of S.

The results suggest that some share of Se and especially S might evaporate in vacuum under electron beam, but gold-selenide species observed with TEM/SAED/EDS still contain excessive selenium relative to AuSe. The interplanar distance of ~3.2 Å is a signature of the Au–Se-based structures, and this parameter decreases to 3.0 Å for Au–(Se,S) and 2.9 Å for Au–S; these values may be close to Au–Au distances [11,12,30–32] but different from *d* values in the gold chalcogenides. The data describe immobilized intermediates somewhat distorted due to drying and evaporation of chalcogens, density of which decreased with substitution of Se for S, as it is seen in TEM images. Furthermore, decomposition of the gold chalcogenide species yielded gold nanoparticles, the dimensions of which decreased from Au–Se to Au–S systems, probably because of more stable Au–Se bonding, slower nucleation, and smaller number of Au nuclei, and then larger Au NPs, or denser liquid species, or both factors, which can be in fact interrelated.

### 3.4. X-ray Diffraction Analysis and Raman Scattering

XRD patterns (Figure 7a) show a very wide maximum and a shoulder corresponding to interplanar distances of roughly 2.8 Å and 1.8 Å, respectively, which are near to $d_{310}$ and $d_{313}$ planes in $\alpha$-AuSe (PDF 00-020-0457), or to $d_{111}$ (and $d_{200}$) plane in $Au_2S$. At the same time, reflections of elemental gold are absent. This means that highly are bulk products of the reactions with high contents of hydrogen chalcogens, whereas $Au^0$ NPs disordered gold chalcogenides observed in TEM, SAED, and XPS (below) is surface species formed ex situ.

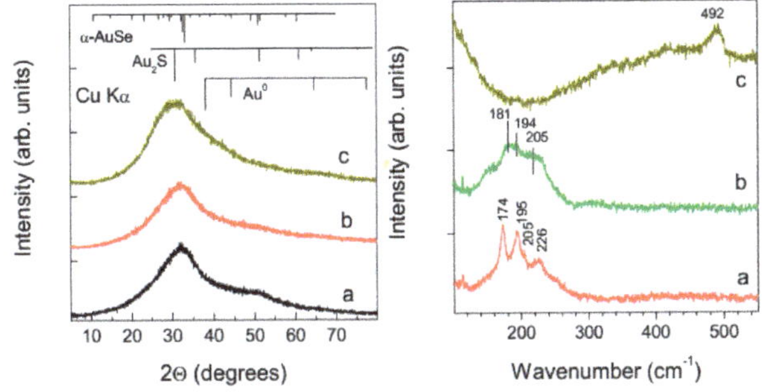

**Figure 7.** X-ray diffraction patterns (left panel; the reflections of $Au^0$, $Au_2S$, and $\alpha$-AuSe are shown for comparison) and Raman scattering (right panel) of the products deposited from aqueous solutions of (a) 0.3 mmol/L $HAuCl_4$ + 0.9 mmol/L $H_2Se$, (b) 0.3 mmol/L $HAuCl_4$ + 0.45 mmol/L $H_2Se$ + 0.45 mmol/L $H_2S$; (c) 0.3 mmol/L $HAuCl_4$ + 0.9 mmol/L $H_2S$ after 1 h reaction and dried in air.

Raman spectra of the residue obtained from the $HAuCl_4$ + $3H_2Se$ medium detected a set of Au–Se stretches (174 cm$^{-1}$, 195 cm$^{-1}$, 205 cm$^{-1}$, 226 cm$^{-1}$) characteristic of AuSe crystals overlapped with a broad band attributable to a disordered material. The lines widened and slightly shifted if the mixture of $H_2Se$ and $H_2S$ was used, suggesting further disordering of gold chalcogenide phase. The Se–Se stretching that is expected to be around 290–330 cm$^{-1}$ [59] was not detected. The spectrum from the products of the reaction $HAuCl_4$+ $3H_2S$ showed a broad band near 490 nm that is due to S–S stretching frequencies, probably from S adsorbed at Au NPs, while Au–S stretches near 270 cm$^{-1}$ [55,59] are absent. This agrees with previous findings of a range of Au–S and S–S vibrations at lower concentrations of sulfide when metallic nanogold formed, and disappearance of the bands at the molar ratio 1-to-3 [55]. As the Raman spectra are in fact surface-enhanced due to an intimate contact of chalcogenides with gold nanoparticles, no enhancement of Raman spectra of $Au_2S$ was observed with $H_2S$ because the contact between $Au_2S$ and Au NPs was insignificant, in contrast to the systems containing $H_2Se$ (see TEM images in Figure 6).

### 3.5. X-ray Photoelectron Spectroscopy

X-ray photoelectron spectra (Figure 8, see also Table S1, Supplementary data) characterize the net composition and chemical state of atoms in surface layers of roughly 2 nm thick, which, though, may be modified in the ultra-high vacuum conditions and under X-ray irradiation. The binding energy (BE) of the main peak Au $4f_{7/2}$ (84.14 $\pm$ 0.05 eV) is slightly higher than that of bulk metal (84.0 eV); this may be due to either the final state effect that is characteristic of nanoparticles less than about 5 nm, or the chemical shift caused by the transfer of electrons from Au to chalcogen atoms [60,61]. A second, less intense peak at 85.4 $\pm$ 0.1 eV decreased several times and slightly shifted to a higher BE upon substitution of $H_2Se$ for $H_2S$ in the reaction media. Also, there is a minor signal at 86.4 eV from $Au^{3+}$–Cl species.

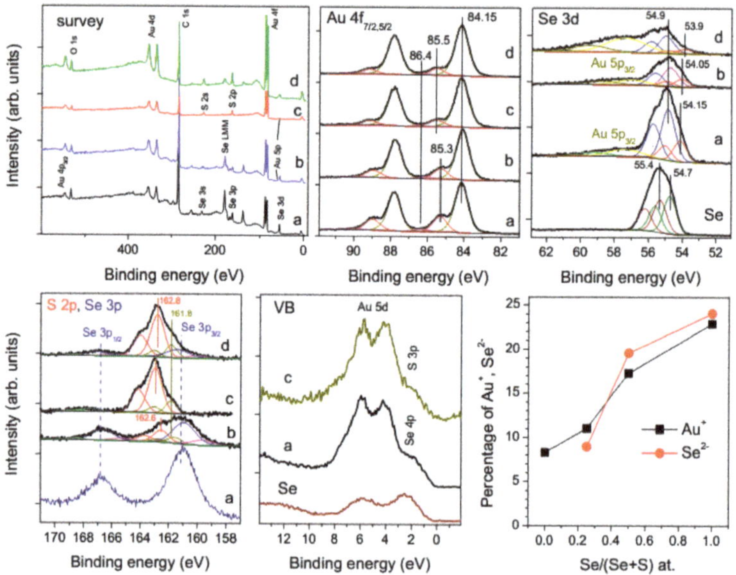

**Figure 8.** X-ray photoelectron spectra from sols produced via the reaction of aqueous $HAuCl_4$ (0.3 mmol/L) with hydrogen selenide and sulfide, dried at HOPG and washed with water (a) $3H_2Se$, (b) $1.5H_2Se + 1.5H_2S$, (c) $3H_2S$, (d) $3H_2S$ and then $H_2Se$, and from elemental Se powder. The lower right panel shows percentage of $Au^+$ (BEs about 85.4 eV) and selenide (54.2 eV) for these products.

The Se 3d spectra, which are partially overlapped with Au $5p_{3/2}$ bands, can be fitted using two doublets at 54.9 eV (more intense) and $54.1 \pm 0.1$ eV, with the relative intensity of the weaker low-energy component decreased and its position slightly shifted to lower BEs with increasing share of $H_2S$ in the reaction media. For comparison, the Se 3d spectrum of a commercial powder of elemental Se was fitted using two doublets with the Se $3d_{5/2}$ peaks at 54.7 eV and 55.3 eV and 55.3 eV, which may be assigned to mono- and polymeric $Se^0$ species [59].

The BE of the main maximum S $2p_{3/2}$ at about 163 eV (note that S 2p spectra are overlain by Se 3p bands in case of $H_2Se/H_2S$ mixture) can be ascribed to polysulfide anions or polymeric S adsorbed at gold. A smaller component at 161.8 eV originates from monosulfide ions in gold chalcogenide phases or/and chemisorbed atomic S [56]; the proportion of monosulfide is higher in a blend with $H_2Se$. Insignificant contributions of oxysulfur compounds are observed around 168 eV.

The atomic chalcogen/Au ratio is close to 1 for the reaction $H_2Se + HAuCl_4$ and somewhat increases with growing content of $H_2S$, while the intensities of the smaller Au 4f component at 85.5 eV and the Se 3d band at 54.2 eV enhance with increasing the relative concentration of $H_2Se$ and retain their atomic ratio of about 1 under various conditions (Figure 8). The findings, which seemingly disagree with an excess of chalcogens found with EDS and also with a higher volatility of sulfur as compared with selenium, can be rationalized in terms of surface decomposition of gold chalcogenides, and different lateral resolution and probing depths of the techniques. The assignment of the Au 4f components is not straightforward, particularly, since AuSe contains two sorts of Au atoms, the oxidation state of which is a matter of dispute [35–45]. The Au $4f_{7/2}$ peaks at 84.0–84.4 eV reported for gold–gold selenide [39–41], gold-CdSe [44,45]), adlayers of Se or alkyl selenides at Au [45–49,59], etc., originate from $Au^0$ surfaces but not AuSe phases. In nanostructured AuSe with admixture of Au [42], the values of 83.7 eV and 84.9 eV have been found, along with Se $3d_{5/2}$ line at 54.5 eV, and ascribed to $Au^0$ and $Au^+$, respectively; nevertheless, the authors assert the mixed valence state of gold ($Au^+Au^{3+}Se_2$). The BEs from 84.2 eV to 84.7 eV attributed to $Au^+$ in a series Ag–Au sulfides, selenides and sulfoselenides increase with decreasing content of Ag and the transfer of electron density from Ag to Au atoms [60,61]. These values should be higher in AuSe and $AuSe_{1-x}S_x$ compounds where silver is absent, so we assign the component at 85.4 eV and the Se 3d doublet at 54.2 eV to $Au^+$ and $Se^{2-}$ centers, respectively. The valence band spectra (Figure 8) are similar for the products of the reactions of $HAuCl_4$ with $H_2Se$ and $H_2S$, with low if any density of Au 5d states near the Fermi level, suggesting that the Au 5d band is completely occupied, and $Au^{3+}$ centers do not occur in the AuSe phases.

The largest contribution to the VB and Au 4f spectra comes from metallic Au NPs (BE of 84.1 eV) capped with elemental selenium (BE of 54.8 eV) or a mixture of Se and S and formed via decomposition of AuSe(S) phases and partial evaporation of chalcogens. The yield of $Au^0$ is lower in the Au–Se system and increases with addition of $H_2S$ because of stronger Au bonding to Se than S. This is mainly surface phenomena as UV-vis spectra show no plasmon peaks from metallic gold NPs in the solutions with the ratio $H_2Se(H_2S)$ to $HAuCl_4$ of 3 or larger, and XRD revealed no gold reflections for bulk precipitates. Capping ligands, in particular, CTAB (see Table S1, Supplementary data) to some extent inhibited decomposition of AuSe, and their effect is worthy of further investigation.

## 4. Discussion

The reaction $H_2Se(H_2S)$ + $HAuCl_4$ proceeds similarly to the $Na_2S$ + $HAuCl_4$ [50–58], and we assume an analogous mechanism involving pre-nucleation dense liquid intermediates at the molar ratios of hydrogen chalcogenide to $HAuCl_4$ of 3 and higher. The Au–Se bonding is more favorable than the Au–S so one could expect a higher rate of nucleation of AuSe and shorter lifetime of liquid intermediates. Instead, we observed much slower growth of the nanoscale clusters, subsequent nucleation of gold chalcogenides in the interior of the droplets, and their coagulation in the reaction with $H_2Se$ than with $H_2S$. This likely relates with a denser gold chalcogenide material as seen in TEM, larger concentrations of chalcogenide (Se), and a higher magnitude of the negative zeta-potentials of the nanoscale fluid clusters in the initial stages; these factors appeared to impede coalescence of the clusters into submicrometer droplets. At the same time, the stronger Au–Se bonding retards ex situ surface decomposition of AuSe yielding $Au^0$NPs, in comparison with gold-sulfide system. Noteworthy, the reaction is faster with $H_2S$ than with $Na_2S$ [55–58] probably due to lower pH and so concentration of $S^{2-}$ anion. The substitution of selenium for sulfur also causes gradual increase of the band gap derived from UV-vis absorption spectra. Such parameters as interplanar distances in HR-TEM and electron diffraction, the binding energies of photoelectron Se and Au lines change steadily too, so it seems likely, but not certain, that S can partially replace Se in the dense liquid intermediates; the effect was observed for any order of mixing the reactants.

The crystal structures of α- and β-AuSe are well understood [30–32], but their electronic structure, particularly, the oxidation state of gold is a matter of dispute. The X-ray photoelectron Au 4f and valence band spectra (Figure 8) imply only one form of $Au^+$ and $Se^{2-}$ both in linear and plain-square coordination of Au with Se (see also [42]). These findings agree with the results of quantum chemical calculations [37] and X-ray absorption spectra [36], but contradict [197]Au Mössbauer spectra [35], in which two types of Au atoms have been found both in α-AuSe and β-AuSe. The discrepancy is likely due to liquid helium temperature used in the Mössbauer experiment. The spectra of bulk crystalline materials [60,61], including those measured from gold selenides and sulfoselenides (will be published elsewhere), support the above interpretation. On the other hand, XPS, as well as TEM and other techniques evidenced the formation of elemental gold via decomposition of surface layers of disordered gold selenides in air and vacuum, so adjacent gold–selenide phase might be distorted.

The non-classical pathway of gradual nucleation within a dense droplet and crystal growth, arrested possibly due to a lack of solutes accumulated within other liquid species, is believed to prevent crystallization of gold chalcogenides and results in their compromised structure inclined to decompose to metallic gold. Such reactions can play a role in the deposition of gold colloids in several types of ores [62,63], even although the hydrothermal processes proceed at higher temperatures and in a different time scale, resulting, in particular, in crystallization and annealing of the solid phases. The formation of crystalline gold selenide, including the nanoparticles which are interesting also for materials science, is possible under ambient conditions but it may require stabilizing agents. This was not a target of this research, and we just demonstrated here that, for example, addition of CTAB stabilizes AuSe, although not very effectively. It is likely that CTAB reacts with surfaces of dense liquid intermediates rather than solid AuSe NPs, and it is necessary to find out conditions, possibly the moment for injection of capping ligands, in order to bypass the formation of long-live liquid intermediates.

## 5. Conclusions

The products of the room-temperature reaction of $HAuCl_4$ with $H_2Se$ and $H_2S$ in aqueous solutions depend on relative concentrations of the reactants. Colloidal gold nanoparticles showing plasmon maxima near 530 nm form at the molar ratios of hydrogen chalcogenide to gold of less than 2. At the ratios higher than 3, pre-nucleation dense liquid entities of 20–40 nm in the average hydrodynamic diameter and negative zeta potential grow into submicrometer droplets with time; gradual nucleation and coagulation of disordered gold chalcogenide within the droplets were faster with $H_2S$. The soft species can be characterized from UV-vis absorption spectra as having an indirect band gap less than 1 eV in $H_2Se$ media and 2 eV in the case of $H_2S$; the gap reduces as the reaction proceededor $H_2Se$ was partially substituted for $H_2S$. Surface layers of amorphous gold–selenide AuSe, which form upon drying the reaction media, tend to decompose yielding metallic gold nanoparticles capped with elemental selenium; meanwhile, no metal gold was found in the bulk precipitate using XRD diffraction. As $H_2Se$ was replaced with $H_2S$, the products became less stable towards the surface decomposition, and the size of $Au^0$ nanoparticles decreased. UV-vis spectra, XPS, TEM and electron diffraction offer hints that sulfur can partially replace selenium both in dense liquid species and precipitated gold chalcogenide phases. XPS Au 4f and valence band spectra suggest that AuSe phase contains only one state of $Au^+$ with the BE of 85.4 eV. Introduction of a capping ligand (CTAB) somewhat retards decomposition of AuSe. The reaction is an example of the non-classical mechanism of nucleation and crystal growth of inorganic materials involving dense liquid intermediates, which can contribute to the transfer and deposition of gold in the nature. The process can be also employed to prepare, in particular, thin films of $AuSe_{1-x}S_x/Au^0$ nanohybrides with varying contents of $Au^0$ and the band gap of gold chalcogenide.

**Supplementary Materials:** The following are available online at http://www.mdpi.com/2075-163X/8/11/492/s1, Figure S1: Hydrodynamic diameters $Z_{av}$ and zeta-potentials of the colloidal species at various ratios $H_2Se$ and $H_2S$ to determined in 10 min after the mixing, Figure S2: STEM images and corresponding EDS analysis data, Table S1: Atomic ratios and fitting parameters derived from XPS.

**Author Contributions:** S.V., M.L. and Y.M. conceived and designed the study; S.V. and M.L. performed synthetic, DLS, zeta-potential experiments; S.Z. performed TEM, SAED, EDS experiments, A.K. collected Raman spectra; A.R. and Y.M. conducted XPS measurements; N.M. performed UV-vis experiments; S.V. and Y.M. wrote the paper.

**Funding:** This research was funded by the Siberian Branch of the Russian Academy of sciences, Program of Interdisciplinary Studies, grant number 64 (project 303).

**Conflicts of Interest:** The authors declare no conflict of interest. The funding sponsor had no role in the design of the study; in the collection, analyses, or interpretation of data; in the writing of the manuscript, and in the decision to publish the results.

## References

1.  Davidson, D.F. *Selenium in Some Epithermal Deposits of Antimony, Mercury and Silver and Gold*; Geol. Survey Bull: Reston, VA, USA, 1960; Volume 1112-A, pp. 1–16.

2.  Barton, M.D. The Ag–Au–S system. *Econom. Geol.* **1980**, *75*, 303–316. [CrossRef]

3.	Liu, J.; Liu, J.; Zheng, M.; Liu, X. Au–Se paragenesis in Cambrian stratabound gold deposits, Western Qinling Mountains, China. *Int. Geol. Rev.* **2000**, *42*, 1037–1045. [CrossRef]

4.	Bindi, L.; Cipriani, C. Structural and physical properties of fischesserite, a rare gold–silver selenide from the De Lamar mine, Owyhee County, Idaho, USA. *Can. Mineral.* **2004**, *42*, 1733–1737. [CrossRef]

5.	Pal'yanova, G.A.; Kokh, K.A.; Seryotkin, Y.V. Formation of gold and silver sulfides in the system Ag–Au–S. *Russ. Geol. Geophys.* **2011**, *52*, 443–449. [CrossRef]

6.	Cocker, H.A.; Mauk, J.L.; Rabone, S.D.C. The origin of Ag–Au–S–Se minerals in adularia-sericite epithermal deposits: Constraints from the Broken Hills deposit, Hauraki Goldfield, New Zealand. *Miner. Depos.* **2013**, *48*, 249–266. [CrossRef]

7.	Seryotkin, Y.V.; Pal'yanova, G.A.; Savva, N.E. Sulfur–selenium isomorphous substitution and morphotropic transition in the $Ag_3Au(Se,S)_2$ series. *Russ. Geol. Geophys.* **2013**, *54*, 646–651. [CrossRef]

8.	Palyanova, G.; Karmanov, N.; Savva, N. Sulfidation of native gold. *Am. Mineral.* **2014**, *99*, 1095–1103. [CrossRef]

9.	Pal'yanova, G.A.; Kravtsova, R.G.; Zhuravkova, T.V. $Ag_2(S,Se)$ solid solutions in the ores of the Rogovik gold–silver deposit (northeastern Russia). *Russ. Geol. Geophys.* **2015**, *56*, 1738–1748. [CrossRef]

10.	Bindi, L.; Stanley, C.J.; Seryotkin, Y.V.; Bakakin, V.V.; Pal'yanova, G.A.; Kokh, K.A. The crystal structure of uytenbogaardtite, $Ag_3AuS_2$, and its relationships with gold and silver sulfides–selenides. *Mineral. Mag.* **2016**, *80*, 1031–1040. [CrossRef]

11.	Palyanova, G.A.; Seryotkin, Y.V.; Kokh, K.A.; Bakakin, V.V. Isomorphism and solid solutions among Ag- and Au-selenides. *J. Solid State Chem.* **2016**, *241*, 157–163. [CrossRef]

12.	Palyanova, G.; Seryotkin, Y.; Kokh, K.; Bakakin, V.V. Sulfur–selenium isomorphous substitution in the AgAu(Se,S) series. *J. Alloys Compd.* **2016**, *664*, 385–391. [CrossRef]

13.	Palyanova, G.A.; Savva, N.E.; Zhuravkova, T.V.; Kolova, E.E. Gold and silver minerals in low-sulfidation ores of the Dzhulietta deposit (northeastern Russia). *Russ. Geol. Geophys.* **2016**, *57*, 1171–1190. [CrossRef]

14.	Tolstykh, N.; Vymazalová, A.; Tuhý, M.; Shapovalova, M. Conditions of formation of Au–Se–Te mineralization in the Gaching ore occurrence (Maletoivayam ore field), Kamchatka, Russia. *Mineral. Mag.* **2018**, *82*, 649–674. [CrossRef]

15.	Fang, C.M.; de Groot, R.A.; Wiegers, G.A. Ab initio band structure calculations of the low-temperature phases of $Ag_2Se$, $Ag_2Te$ and $Ag_3AuSe_2$. *J. Phys. Chem. Solids* **2002**, *63*, 457–464. [CrossRef]

16.	Xiao, C.; Xu, J.; Li, K.; Feng, J.; Yang, J.; Xie, Y. Superionic phase transition in silver chalcogenide nanocrystals realizing optimized thermoelectric performance. *J. Am. Chem. Soc.* **2012**, *134*, 4287–4293. [CrossRef] [PubMed]

17.	Dalmases, M.; Ibáñez, M.; Torruella, P.; Fernàndez-Altable, V.; López-Conesa, L.; Cadavid, D.; Piveteau, I.; Nachtegaal, M.; Llorca, J.; Ruiz-González, M.L.; et al. Synthesis and thermoelectric properties of noble metal ternary chalcogenide systems of Ag–Au–Se in the forms of alloyed nanoparticles and colloidal nanoheterostructures. *Chem. Mater.* **2016**, *28*, 7017–7028. [CrossRef]

18.	Liu, M.; Zeng, H.C. General synthetic approach to heterostructured nanocrystals based on noble metals and I–VI, II–VI, and I–III–VI metal chalcogenides. *Langmuir* **2014**, *30*, 9838–9849. [CrossRef] [PubMed]

19.	Lu, Y.; Li, B.; Zheng, S.; Xu, Y.; Xue, H.; Pang, H. Syntheses and energy storage applications of $M_xS_y$ (M = Cu, Ag, Au) and their composites: Rechargeable batteries and supercapacitors. *Adv. Funct. Mater.* **2017**, *27*, 1703949. [CrossRef]

20.	Benning, L.G.; Seward, T.M. Hydrosulphide complexing of Au(I) in hydrothermal solutions from 150–400 °C and 500–1500 bar. *Geochim. Cosmochim. Acta* **1996**, *60*, 1849–1871. [CrossRef]

21.	Tagirov, B.R.; Baranova, N.N.; Zotov, A.V.; Schott, J.; Bannykh, L.N. Experimental determination of the stabilities of $Au_2S_{(cr)}$ at 25 °C and $Au(HS)_2^-$ at 25–250 °C. *Geochim. Cosmochim. Acta* **2006**, *70*, 3689–3701. [CrossRef]

22.	Pokrovski, G.S.; Tagirov, B.R.; Schott, J.; Hazemanne, J.-L.; Proux, O. A new view on gold speciation in sulfur-bearing hydrothermal fluids from in situ X-ray absorption spectroscopy and quantum-chemical modeling. *Geochim. Cosmochim. Acta* **2009**, *73*, 5406–5427. [CrossRef]

23.	Zezin, D.Y.; Migdisov, A.A.; Williams-Jones, A.E. The solubility of gold in $H_2O$–$H_2S$ vapour at elevated temperature and pressure. *Geochim. Cosmochim. Acta* **2011**, *75*, 5140–5153. [CrossRef]

24. Liu, W.; Etschmann, B.; Testemale, D.; Hazemann, J.-L.; Rempel, K.; Müller, H.; Brugger, J. Gold transport in hydrothermal fluids: Competition among the $Cl^-$, $Br^-$, $HS^-$ and $NH_{3(aq)}$ ligands. *Chem. Geol.* **2014**, *376*, 11–19. [CrossRef]

25. Trigub, A.L.; Tagirov, B.R.; Kvashnina, K.O.; Lafuerz, S.; Filimonova, O.N.; Nickolsky, M.S. Experimental determination of gold speciation in sulfide-rich hydrothermal fluids under a wide range of redox conditions. *Chem. Geol.* **2017**, *471*, 52–64. [CrossRef]

26. Ishikawa, K.; Isonaga, T.; Wakita, S.; Suzuki, Y. Structure and electrical properties of $Au_2S$. *Solid State Ion.* **1995**, *79*, 60–66. [CrossRef]

27. Osadchii, E.G.; Rappo, O.A. Determination of standard thermodynamic properties of sulfides in the Ag–Au–S system by means of a solid-state galvanic cell. *Am. Mineral.* **2004**, *89*, 1405–1410. [CrossRef]

28. Simon, G.; Essene, E.J. Phase relations among selenides, sulfides, tellurides, and oxides; I, Thermodynamic properties and calculated equilibria. *Econ. Geol.* **1996**, *91*, 1183–1208. [CrossRef]

29. Echmaeva, E.A.; Osadchii, E.G. Determination of the thermodynamic properties of compounds in the Ag–Au–Se and Ag–Au–Te systems by the EMF method. *Geol. Ore Depos.* **2009**, *51*, 247–258. [CrossRef]

30. Rabenau, A.; Rau, H.; Rosenstein, G. Phase relations in the gold-selenium system. *J. Less Common Met.* **1971**, *24*, 291–299. [CrossRef]

31. Rabenau, A.; Schulz, H. The crystal structures of α-AuSe and β-AuSe. *J. Less Common Met.* **1976**, *48*, 89–101. [CrossRef]

32. Cretier, J.E.; Wiegers, G.A. The crystal structure of the beta form of gold selenide, β-AuSe. *Mat. Res. Bull.* **1973**, *8*, 1427–1430. [CrossRef]

33. Feng, D.; Taskinen, P. Thermodynamic stability of AuSe at temperature from (400 to 700) K by a solid state galvanic cell. *J. Chem. Thermodyn.* **2014**, *71*, 98–102. [CrossRef]

34. Xu, X.L.; Chen, W.K.; Wang, X. Density functional study on adsorption of NO on AuSe (010) surface. *Chin. J. Chem.* **2008**, *26*, 107–112. [CrossRef]

35. Wagner, F.E.; Palade, P.; Friedl, J.; Filoti, G.; Wang, N. [197]Au Mössbauer study of gold selenide, AuSe. *J. Phys. Conf. Ser.* **2010**, *217*, 012039. [CrossRef]

36. Ettema, A.R.H.F.; Stegink, T.A.; Haas, C. The valence of Au in $AuTe_2$ and AuSe studied by X-ray absorption spectroscopy. *Solid State Commun.* **1994**, *90*, 211–213. [CrossRef]

37. Lee, W.R.; Jung, D. Electronic structure study of gold selenides. *Bull. Korean Chem. Soc.* **1999**, *20*, 147–150.

38. Perry, D.L. *Handbook of Inorganic Compounds*; CRC Press: Boca Raton, FL, USA, 2011; ISBN 9781439814611.

39. Nath, S.; Ghosh, S.K.; Pal, T. Solution phase evolution of AuSe nanoalloys in Triton X-100 underUV-photoactivation. *Chem. Commun.* **2004**, 966–967. [CrossRef] [PubMed]

40. Prokeš, L.; Kubáček, P.; Peña-Méndez, E.M.; Amato, F.; Conde, J.E.; Alberti, M.; Havel, J. Laser ablation synthesis of gold selenides by using a mass spectrometer as a synthesizer: Laser desorption ionization time-of-flight mass spectrometry. *Chem.-Eur. J.* **2016**, *22*, 11261–11268. [CrossRef] [PubMed]

41. Hu, B.; Cheng, R.; Liu, X.; Pan, X.; Kong, F.; Gao, W.; Xu, K.; Tang, B. A nanosensor for in vivo selenol imaging based on the formation of AuSe bonds. *Biomaterials* **2016**, *92*, 81–89. [CrossRef] [PubMed]

42. Machogo, L.F.E.; Tetyana, P.; Sithole, R.; Gqoba, S.S.; Phao, N.; Airo, M.; Shumbula, P.M.; Moloto, M.J.; Moloto, N. Unravelling the structural properties of mixed-valence α- and β-AuSe nanostructures using XRD, TEM and XPS. *Appl. Surf. Sci.* **2018**, *456*, 973–979. [CrossRef]

43. Neumann, H.; Yakushev, M.V.; Tomlinson, R.D. Diffusion effects at the Au/p–$CuInSe_2$ contact studied by XPS. *Cryst. Res. Technol.* **2003**, *38*, 676–683. [CrossRef]

44. Haldar, K.K.; Sinha, G.; Lahtinen, J.; Patra, A. Hybrid colloidal Au–CdSe pentapod heterostructures synthesis and their photocatalytic properties. *ACS Appl. Mater. Interfaces* **2012**, *4*, 6266–6272. [CrossRef] [PubMed]

45. de la Cueva, L.; Meyns, M.; Bastús, N.G.; Rodríguez-Fernández, J.; Otero, R.; Gallego, J.M.; Alonso, C.; Klinke, C.; Juárez, B.H. Shell or dots-precursor controlled morphology of Au−Se depositson CdSe nanoparticles. *Chem. Mater.* **2016**, *28*, 2704–2714. [CrossRef]

46. Jia, J.; Bendounan, A.; Kotresh, H.M.N.; Chaouchi, K.; Sirotti, F.; Sampath, S.; Esaulov, V.A. Selenium adsorption on Au(111) and Ag(111) surfaces: Adsorbed seleniumand selenidefilms. *J. Phys. Chem. C* **2013**, *117*, 9835–9842. [CrossRef]

47. Ruano, G.; Tosi, E.; Sanchez, E.; Abufager, P.; Martiarena, M.L.; Grizzi, O.; Zampieri, G. Stages of Se adsorption on Au (111): A combined XPS, LEED, TOF-DRS, and DFT study. *Surf. Sci.* **2017**, *662*, 113–122. [CrossRef]

48. Yee, C.K.; Ulman, A.; Ruiz, J.D.; Parikh, A.; White, H.; Rafailovich, M. Alkyl selenide- and alkyl thiolate-functionalized gold nanoparticles: Chain packing and bond nature. *Langmuir* **2003**, *19*, 9450–9458. [CrossRef]

49. Hohman, J.N.; Thomas, J.C.; Zhao, Y.; Auluck, H.; Kim, M.; Vijselaar, W.; Kommeren, S.; Terfort, A.; Weiss, P.S. Exchange reactions between alkanethiolates and alkaneselenols on Au{111}. *J. Am. Chem. Soc.* **2014**, *136*, 8110–8121. [CrossRef] [PubMed]

50. Averitt, R.D.; Sarkar, D.; Halas, N.J. Plasmon resonance shifts of Au-coated $Au_2S$ nanoshells: Insight into multicomponent nanoparticle growth. *Phys. Rev. Lett.* **1997**, *78*, 4217–4220. [CrossRef]

51. Morris, T.; Copeland, H.; Szulczewski, G. Synthesis and characterization of gold sulfide nanoparticles. *Langmuir* **2002**, *18*, 535–539. [CrossRef]

52. Majimel, J.; Bacinello, D.; Durand, E.; Vallee, F.; Treguer-Delapierre, M. Synthesis of hybrid gold–gold sulfide colloidal particles. *Langmuir* **2008**, *24*, 4289–4294. [CrossRef] [PubMed]

53. Kuo, C.-L.; Huang, M.H. Hydrothermal synthesis of free-floating $Au_2S$ nanoparticle superstructures. *J. Phys. Chem. C* **2008**, *112*, 11661–11666. [CrossRef]

54. Schwartzberg, A.M.; Grant, C.D.; van Buuren, T.; Zhang, J.Z. Reduction of $HAuCl_4$ by $Na_2S$ revisited: The case for Au nanoparticle aggregates and against $Au_2S$/Au core/shell particles. *J. Phys. Chem. C* **2007**, *111*, 8892–8901. [CrossRef]

55. Mikhlin, Y.; Likhatski, M.; Karacharov, A.; Zaikovski, V.; Krylov, A. Formation of gold and gold sulfide nanoparticles and mesoscale intermediate structures in the reactions of aqueous $HAuCl_4$ with sulfide and citrate ions. *Phys. Chem. Chem. Phys.* **2009**, *11*, 5445–5454. [CrossRef] [PubMed]

56. Mikhlin, Yu.; Likhatski, M.; Tomashevich, Ye.; Romanchenko, A.; Erenburg, S.; Trubina, S. XAS and XPS examination of the Au–S nanostructures produced via the reduction of aqueous gold (III) by sulfide ions. *J. Electron Spectrosc. Relat.* **2010**, *177*, 24–29. [CrossRef]

57. Mikhlin, Y.; Karacharov, A.; Likhatski, M.; Podlipskaya, T.; Zizak, I. Direct observation of liquid pre-crystallization intermediates during the reduction of aqueous tetrachloroaurate by sulfide ions. *Phys. Chem. Chem. Phys.* **2014**, *16*, 4538–4543. [CrossRef] [PubMed]

58. Likhatski, M.; Karacharov, A.; Kondrasenko, A.; Mikhlin, Y. On a role of liquid intermediates in nucleation of gold sulfide nanoparticles in aqueous media. *Faraday Discuss.* **2015**, *179*, 235–245. [CrossRef] [PubMed]

59. Dhayagude, A.C.; Maiti, N.; Debnath, A.K.; Joshi, S.S.; Kapoor, S. Metal nanoparticle catalyzed charge rearrangement in selenourea probed by surface-enhanced Raman scattering. *RSC Adv.* **2016**, *6*, 17405–17414. [CrossRef]

60. Mikhlin, Y.L.; Nasluzov, V.A.; Romanchenko, A.S.; Shor, A.M.; Pal'yanova, G.A. XPS and DFT studies of the electronic structures of AgAuS and $Ag_3AuS_2$. *J. Alloys Compd.* **2014**, *617*, 314–321. [CrossRef]

61. Mikhlin, Y.L.; Pal'yanova, G.A.; Tomashevich, Y.V.; Vishnyakova, E.A.; Vorobyev, S.A.; Kokh, K.A. XPS and Ag $L_3$-edge XANES characterization of silver- and silver-gold sulfoselenides. *J. Phys. Chem. Solids* **2018**, *116*, 292–298. [CrossRef]

62. Hough, R.M.; Noble, R.R.P.; Reich, M. Natural gold nanoparticles. *Ore Geol. Rev.* **2011**, *42*, 55–61. [CrossRef]

63. Saunders, J.A.; Burke, M. Formation and aggregation of gold (electrum) nanoparticles in epithermal ores. *Minerals* **2017**, *7*, 163. [CrossRef]

 *minerals*

*Article*

# Composition and Ligand Microstructure of Arsenopyrite from Gold Ore Deposits of the Yenisei Ridge (Eastern Siberia, Russia)

Anatoly M. Sazonov [1,*], Sergey A. Silyanov [1], Oleg A. Bayukov [2], Yuriy V. Knyazev [2], Yelena A. Zvyagina [1] and Platon A. Tishin [3]

[1]   Institute of Mining, Geology and Geotechnology, Siberian Federal University, 79 pr. Svobodny, 660041 Krasnoyarsk, Russia; silyanov-s@mail.ru (S.A.S.); elena_zv@mail.ru (Y.A.Z.)
[2]   Kirensky Institute of Physics, Federal Research Center Krasnoyarsk Scientific Center of the Siberian Branch of the Russian Academy of Sciences, 50 Bld. 38 Akademgorodok, 660036 Krasnoyarsk, Russia; helg@iph.krasn.ru (O.A.B.); yuvknyazev@mail.ru (Y.V.K.)
[3]   Faculty of Geology and Geography, Tomsk National Research State University, 36 Lenina, 634050 Tomsk, Russia; tishin_pa@mail.ru
*   Correspondence: sazonov_am@mail.ru; Tel.: +7-902-923-5177

Received: 13 October 2019; Accepted: 26 November 2019; Published: 28 November 2019

**Abstract:** The Mössbauer spectroscopy method was used to study the ligand microstructure of natural arsenopyrite (31 specimens) from the ores of the major gold deposits of the Yenisei Ridge (Eastern Siberia, Russia). Arsenopyrite and native gold are paragenetic minerals in the ore; meanwhile, arsenopyrite is frequently a gold carrier. We detected iron positions with variable distribution of sulfur and arsenic anions at the vertexes of the coordination octahedron {6S}, {5S1As}, {4S2As}, {3S3As}, {2S4As}, {1S5As}, {6As} in the mineral structure. Iron atoms with reduced local symmetry in tetrahedral cavities, as well as iron in the high-spin condition with a high local symmetry of the first coordination sphere, were identified. The configuration {3S3As} typical for the stoichiometric arsenopyrite is the most occupied. The occupation degree of other configurations is not subordinated to the statistic distribution and varies within a wide range. The presence of configurations {6S}, {3S3As}, {6As} and their variable occupation degree indicate that natural arsenopyrites are solid pyrite {6S}, arsenopyrite {3S3As}, and loellingite {6As} solutions, with the thermodynamic preference to the formation of configurations in the arsenopyrite–pyrite–loellingite order. It is assumed that in the variations as part of the coordination octahedron, the iron output to the tetrahedral positions and the presence of high-spin Fe cations depend on the physical and chemical conditions of the mineral formation. It was identified that the increased gold concentrations are typical for arsenopyrites with an elevated content of sulfur or arsenic and correlate with the increase of the occupation degree of configurations {5S1As}, {4S2As}, {1S5As}, reduction of the share of {3S3As}, and the amount of iron in tetrahedral cavities.

**Keywords:** arsenopyrite; crystal lattice; ligand surroundings; non-equal positions; Mössbauer Effect; gold; "invisible" gold; gold ore deposits

---

## 1. Introduction

Arsenopyrite is a quasistoichometric mineral with the chemical formula FeAsS. The chemical composition of the mineral is sensitive to the physical and chemical conditions of formation and is used for studying the thermodynamic parameters of hydrothermal ore formation [1–6]. In most gold ore deposits, arsenopyrite is present in significant quantities. The interrelation of the S/As ratio with the gold content in arsenopyrite ores has been noted; however, unambiguous dependence has not been demonstrated [6–10]. It is assumed that gold in arsenopyrite can be present in a metal and/or chemically bound states [11–22].

The crystal structure of stoichiometric arsenopyrite (FeAsS) demonstrates that the mineral is of monoclinic symmetry (space group $P2_1/c$); the parameters of the elementary cell have been identified as $a = 5.7612(8)$, $b = 5.6841(7)$, $c = 5.7674(8)$ Å, $\beta = 111.72°$ [23–25]. Iron in the crystal lattice of arsenopyrite is located in the center of distorted octahedrons formed by sulfur and arsenic atoms. Meanwhile, the S and As atoms occupy the vertexes of the opposite planes of the octahedron [26,27].

Moreover, the Mössbauer spectroscopy of natural arsenopyrites [4,26] shows three quadrupole doublets, the chemical shift (IS = 0.34 mm/s), and quadrupole splitting (QS = 1.05 mm/s) of which is close to the parameters of pyrite—$FeS_2$ (IS = 0.314 mm/s, QS = 0.614 mm/s); marcasite—$FeS_2$ (IS = 0.27 mm/s, QS = 0.50 mm/s); arsenopyrite—FeAsS (IS = 0.34 mm/s, QS = 1.05 mm/s); loellingite—$FeAs_2$ (IS = 0.39 mm/s, QS = 1.68 mm/s) [28,29]. Iron in these structural motives is in the low-spin condition [26,30,31]. Some arsenopyrite specimens contain additional iron positions explained by the occurrence of high-spin cations of $Fe^{2+}$ and the formation of the nearest ligand surroundings (2S + 4As). Recently, an attempt was made to explain the occurrence of non-equivalent iron positions in arsenopyrites with different sulfur and arsenic distribution by vertexes of the coordination octahedron [6,25]. Natural arsenopyrite is heterogeneous in terms of the sulfur and arsenic content, which can be observed even within a single crystal grain [6,9,13]. This agrees with the variations of the ligand surroundings of iron and can cause the occurrence of gold in the arsenopyrite structure [6,9].

Arsenopyrite is characterized by a wide range of chemical compositions and variations in the crystalline structure, which explains presence of "invisible" gold in it. Despite a large number of publications dedicated to the arsenopyrite structure, such studies have been performed for a limited number of specimens. In this paper, the results of studies of the chemical and structural heterogeneity of arsenopyrite associated with gold and as one of the main carriers of gold are given. The representative selection of the samples consisting of 31 arsenopyrite specimens from the ores of 13 gold deposits of the Yenisei Ridge (Siberia, Russia) was studied.

## 2. General Data on Geology of Gold-Arsenopyrite Mineralization of the Yenisei Ridge

The Yenisei Ridge, as a unique geological province in terms of gold reserves, is located in Eastern Siberia (Russia) and is confined to the junction of the West Siberian Plateau and the Siberian Platform (Figure 1). Two geological-structural elements, Anagara–Kan Archaean and Lower Proterozoic offset (interfluve of the Kan and Angara Rivers) and the Zaangarskoye folding structure of the Baikal age (from the lower reaches of the Angara River to the estuary of the Podkamennaya Tunguska River), form the structure of the Yenisei Ridge. The boundary between the structural elements of the ridge coincide with the valley of the Angara River, which is confined to the broadly West–East fault zone. The multiple gold ore units are grouped in the eastern part of the structure of the ridge in the form of an extended (350 km) belt called the Eastern gold-bearing belt of the Yenisei Ridge.

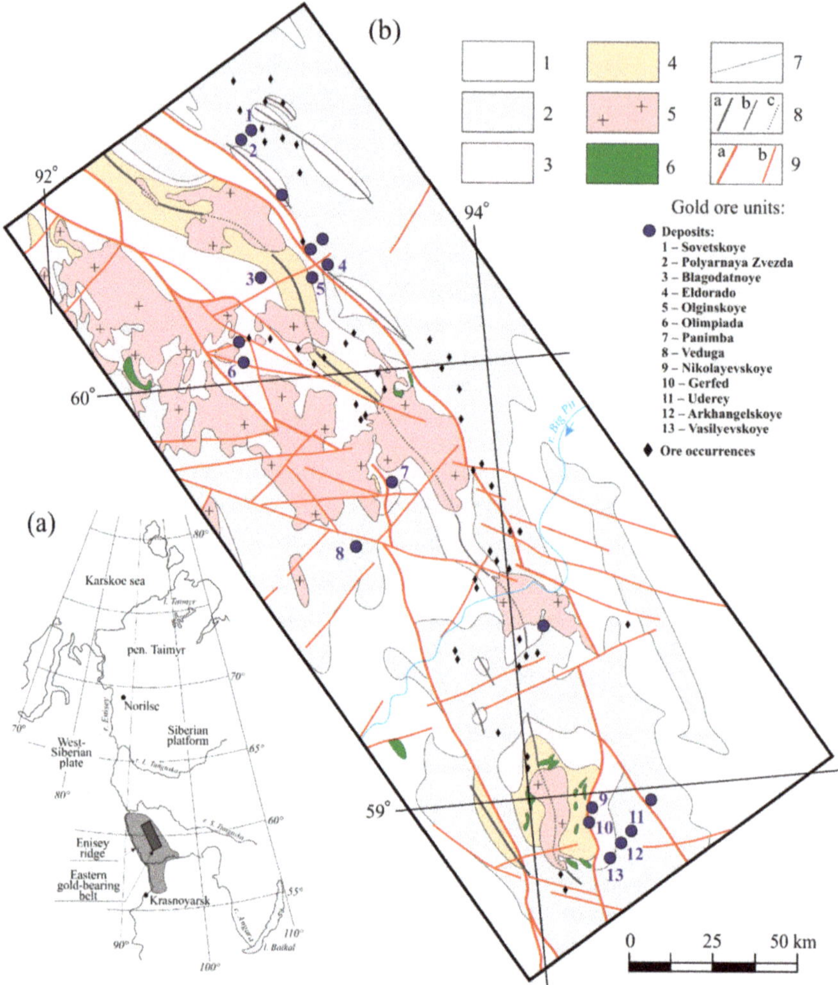



**Figure 1.** (**a**) Geographic position of the Yenisei Ridge (gray) and Eastern gold-bearing belt (dark gray); (**b**) map of geological structure of the Eastern gold-bearing belt of the Yenisei Ridge (compiled by A.M. Sazonov): 1—stratified Upper Riphean-Cambrian deposits; 2—Middle Riphean. Aladyinskaya, Pogoryuy, Uderey, and Gorbilok suites (dolomites, clay slates, siltstone, sandstone, carboniferous phyllites, chlorite–micaceous slates, and phyllites); 3—Lower Riphean. Kordinskaya suite—quartzite-like two-mica slates, gravelites, limestones; 4—Lower Proterozoic. Penchenginskaya suite and suite of the Karpinsky Ridge (quartzites, amphibolites, aluminous slates, and gneiss); 5—Proterozoic granitoid; 6—Proterozoic intrusions of mafic composition; 7—geological borders; 8—anticline axes of: a—Teya series; b—Kordinskaya suite; c—anticline axes within granite intrusions; 9—faults: a—major, b—other.

The gold ore deposits are located in the metamorphosed deposits of the Sukhopitskaya Riphean series. The thickness of the stratified layers in the region was 17.25 km at the final stages of ore formation (approximately 600 Ma). The vertical span of mineralization distribution is about 3.25 km in the lower part of the section of Vendian–Riphean deposits. The lithostatic pressure in this interval of the ore-hosting stratum was 3.6 kbar in the upper part of the ore shoot and 4.4 kbar in the lower part, and the temperature was 450 and 560 °C, respectively, at a gradient of 33 °/km [32]. The host rocks of the deposits are represented by meta-aleuropelites, less often carbonate rocks and metasandstones metamorphosed in the conditions of the greenschist up to epidote–amphibolite facies. Granite intrusions, which the genesis of the deposits is associated with, are located at a distance of 1.5 to 18 km. The wide range of distribution of deposits in relation to the granitoid massifs and a significant distance of some deposits from granites defines the negotiable nature of the genetic relationship between granitoid magmatism and mineralization. The deposits are classified as a gold–quartz–sulfide type. The ore bodies are represented by quartz veins with variable thickness, with sulfide mineralization and zones of sulfidized, silicified, and sericite-altered schists. The ore bodies have been formed in various temporal stages: quartz-vein (I), gold-arsenopyrite (II), sulfide-polymetal (III), ulmanite–gersdorffite (IV), and stibnite (V) [33–37]. The main volume of gold in the ores was formed in connection with the occurrence of the arsenopyrite stage. In the later stages of the hydrothermal process, as a result of metamorphism of early formations and telescoping of ore deposition, arsenopyrite, gold, and other early minerals of the ores were recrystallized and redeposited with consolidation or disintegration of particles. Functioning of the hydrothermal system, which led to the formation of gold ore deposits, is estimated to last about 150 Ma [37,38].

## 3. Preparations and Research Methods

In this work, we studied the composition, ligand surroundings of iron, and features of microparticle and gold distribution in 31 arsenopyrite specimens from the ores of 13 gold deposits at the Yenisei Ridge.

Arsenopyrite was collected from ores of the Sovetskoye, Polyarnaya Zvezda, Blagodatnoye, Eldorado, Olginskoye, Olimpiada, Panimba, Veduga, Nikolayevskoye, Gerfed, Uderey, Arkhangelskoye, and Vasilyevskoye deposits (Figure 1). The features of geology, ore composition, and formation conditions have been discussed in the early works of the authors [37,39]. The ore samples containing arsenopyrite were exposed to soft crushing in the metal mortar up to the fraction size of −1 to +0.5 mm. This fraction further was processed with the water-gravity method. The selection of arsenopyrite grains was carried out manually under the binocular magnifier. Twenty mineral grains were selected and sealed into epoxy pellets (each) from the obtained arsenopyrite weighted portions for the electron microscopic study and determination of the Fe, As, and S. The remaining part of the weighted portions was ground in the agate mortar for the Mössbauer study, as well as for determination of concentrations of gold and other impurities with the ICP-MS method.

Monomineral arsenopyrite powders with a particle size of no more than 10 microns were used for the Mössbauer study. The purity of the samples analyzed was controlled by a parallel X-ray analysis, which did not demonstrate the presence of other mineral phases (e.g., pyrite, pyrrhotite, and lollingite) [6,25]. The thickness of the specimens was 5–10 mg/cm$^2$. The Mössbauer spectra of arsenopyrites were measured at room temperature in the MC-1104Eмspectrometer in the transmission geometry with the Mössbauer source of Co$^{57}$(Cr), at the natural iron content. The analysis was performed at the Institute of Physics, Siberian Branch of the Russian Academy of Sciences, Krasnoyarsk.

Determination of the main mineral-forming elements (Fe, As, S, Au) in the local areas of mineral grains was performed using a TESCAN VEGA II LMU scanning electron microscope with an integrated system of X-ray energy-dispersive microanalysis OXFORD INCA ENERGY 350. Minerals were analyzed at 20 kV and 1.2 nA. Pure elements (Au, Ag, Cu), as well as FeS$_2$ (for S and Fe) and InAs (for As), were used as standards.

Impurity elements (Au, Ag, Ru, Pd, Pt) in arsenopyrite were determined with the ICP-MS method in the Agilent 7500cx device manufactured by Agilent Technologies. The sulfide sample weighted portion was preliminarily transferred by progressive digestion into nitric acid and aqua regia solutions, which allowed it to be kept in the liquid phase and for the maximum possible set of elements to be analyzed. The quality of the obtained results was estimated on the basis of rock and ore standards BCR-2, BHWO, SSL-1. The analyses to determine the main mineral-forming elements and impurities were performed in the common-use center "Analytical Center of Geochemistry of Natural Systems" of the Tomsk State University, Tomsk, Russia.

## 4. Results and Discussion

### 4.1. Mössbauer Spectroscopy

The Mössbauer arsenopyrite spectra represent quadrupole doublets with asymmetric absorption lines. This asymmetry indicates that the spectra can be the sum of several doublets (Figure 2a). The spectra were interpreted in two stages. First, the distribution of the probability of quadrupole splittings in the experimental spectrum was determined. This made it possible to determine probable non-equivalent crystallographic positions of iron. For this purpose, a model spectrum was formed, which is the sum of three groups of doublets with a natural line of absorption. These groups represented a set of quadrupole doublets with the quadrupole splitting (QS) values from 0.0 to 2.0 mm/s, with a space of 0.01 mm/s. Each group of doublets has a certain chemical shift (IS) (Figure 2b). Then, the isomer shift and intensity of Mössbauer lines were varied. As a result, we obtained a set of intensities corresponding to each group. These data conform to the probability of each doublet's existence in the experimental spectrum. The result of processing at the first stage was distributions of probability P(QS) for each group.

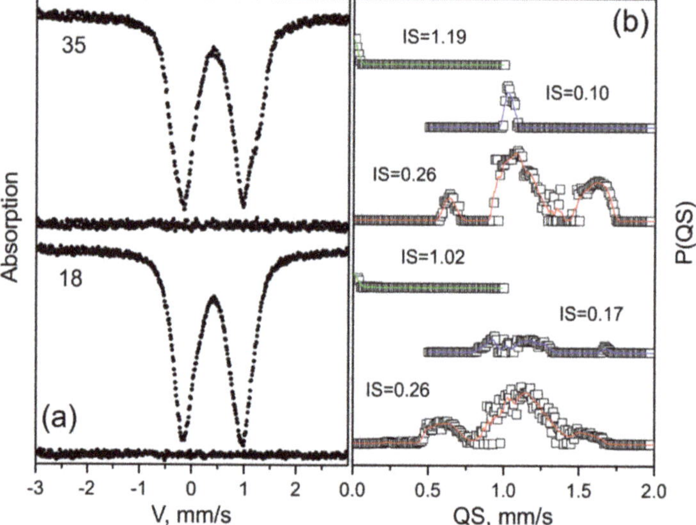


**Figure 2.** (**a**) Mössbauer spectra of arsenopyrites of the Sovetskoye deposit (18) and Blagodatnoye deposit (35), the difference between the theoretical and experimental spectra is shown under the spectra; (**b**) distribution of the probability of quadrupole splittings (QS) in the spectra of these specimens. The solid lines show the approximation of the calculated data.

The peaks and features of the P(QS) distribution obtained are the occurrences of possible nonequivalent iron positions and/or states with different distortion of the local surroundings and electronic density. However, the occurrence of false peaks is possible.

The second stage of spectra interpretation was performed taking into account the detected peculiarities of the P(QS) distribution. A preliminary spectrum contained a set of Mössbauer doublets, corresponding to possible nonequivalent positions and modeling as a group of the analytical functions. Mössbauer absorption lines were represented by the pseudo-Voigt function, following the Equation

$$I = I \frac{W^2}{W^2 + 4 \cdot (IS - i)^2} \tag{1}$$

Here, $x_i = 2 \cdot \left(\frac{IS-i}{W}\right)$, $i$ is a channel number. $I$, $IS$, and $W$ are hyperfine parameters (line intensity, isomer shift, and line-width, respectively).

The model spectrum was fit to the experimental spectrum with variation of the whole set of hyperfine parameters. Meanwhile, in the course of spectra fitting, the amplitudes and widths of false doublets tended to zero, and the parameters of actual doublets were consolidated.

To reveal the nature of occurrence of non-equivalent iron positions in arsenopyrites, the preliminary evaluation of the main component of the electric field gradient (EFG) was performed in the pyrite–arsenopyrite–loellingite series with the use of the X-ray structural data on inter-ion distances [27,40,41]. The observed quadrupole splittings for pyrite, arsenopyrite, and loellingite "phases" marked in Figure 3 with crosses reflect the linear dependence.

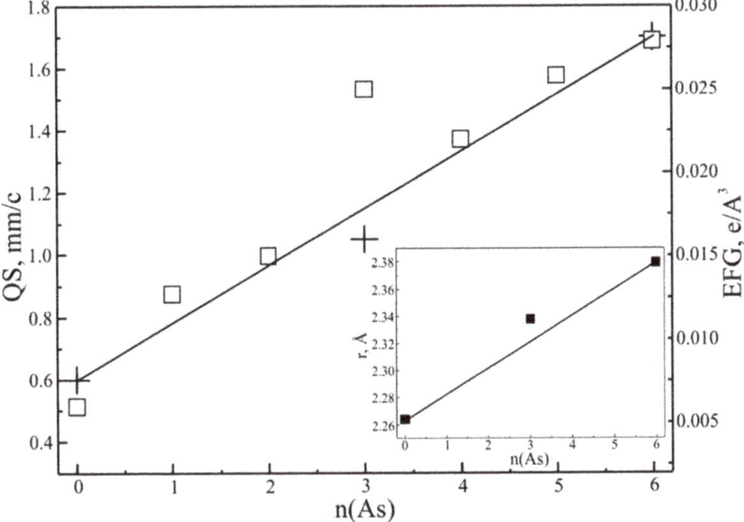

**Figure 3.** Dependence of quadrupole splitting (QS) and electric field gradient (EFG) on the number of arsenic atoms in the coordination ligand octahedron around the iron cation. Crosses are experimental QS values for pyrite, arsenopyrite, and loellingite [28,30]. n(As) is the number of As atoms in the coordination octahedron; r, Å is the average interionic distance.

The electronic envelope of the low-spin $Fe^{2+}$ cation has spherical symmetry similar to the S ion; therefore, it is assumed that the EFG at the core of this cation is preconditioned mainly by the lattice contribution. The result of the EFG evaluation at the iron cation formed by ligands of the coordination octahedron at a different number of S and As atoms in the octahedron is shown by squares (Figure 3). In the EFG evaluation, only the change of the Fe–S and Fe–As inter-ion distances at the replacement of S with As was taken into account. The calculated EFG values repeat the linear dependence of quadrupole splitting (degree of octahedron deviation) on the ligand composition. The noticeable deviation of the EFG calculated value on the linear dependence for n(As) = 3 is probably associated with the features of the ligand ordering and change in the covalent contribution to EFG at the change of inter-ion distances for this composition. The dependence of the average inter-ion distance on n(As) is shown in the insert in Figure 3. The monotonic dependence of the quadrupole splitting on the As content in the nearest surroundings of the iron cation was used for the identification of ligand configurations.

According to the spot microprobe tests, the contents of sulfur and arsenic in the studied arsenopyrites varies. The S/As ratio varies from 0.79 to 1.24 both for different deposits and for grains in one sample. The estimated distribution of P(QS) using a single group of doublets with a general chemical shift is shown in Figure 2b and demonstrates the main peak, the position of which corresponds to the quadrupole splitting of arsenopyrite. However, in addition to the main peak, there are also additional peaks; the quadrupole splittings of which are close to the parameters of pyrite $FeS_2$ and loellingite $FeAs_2$.

Processing of experimental spectra of the majority of the studied arsenopyrites using three doublets gives an unsatisfactory result, which testifies to the presence of additional unaccounted iron positions. The analysis of the spectrum—taking into account the three electronic iron conditions (three chemical shifts)—helped to reveal the additional nonequivalent iron positions both as a part of the main arsenopyrite phase leading to the smearing of the P(QS) distribution of main peaks, and the iron position with a minor and major value of chemical shifts not typical for arsenopyrites (Figure 2b). The result of the two-stage interpretation of the arsenopyrite spectra is given in Table 1.

**Table 1.** Mössbauer parameters of the studied specimens of arsenopyrite.

| Specimen Number, Deposit, Morphology of Arsenopyrite Grains (Mineralization Stage) | IS, mm/s (±0.005) | QS, mm/s (±0.02) | W, mm/s (±0.02) | A, % (±3) | Configuration of Ligands |
|---|---|---|---|---|---|
| 1. Olimpiada, isometric pseudopyramidal grains (IV) | 0.26 | 0.58 | 0.33 | 8 | {6S} |
|  | 0.25 | 1.11 | 0.33 | 68 | {3S3As} |
|  | 0.25 | 1.20 | 0.27 | 15 | {2S4As} |
|  | 0.11 | 0.97 | 0.23 | 3 | Fe-LC-1 |
|  | 0.21 | 1.56 | 0.28 | 4 | Fe-LC-2 |
|  | 1.00 | 0.11 | 0.24 | 2 | Fe-HS |
| 2. Olimpiada, isometric pseudopyramidal grains (IV) | 0.29 | 0.58 | 0.28 | 8 | {6S} |
|  | 0.25 | 1.14 | 0.35 | 69 | {3S3As} |
|  | 0.28 | 1.28 | 0.36 | 18 | {2S4As} |
|  | 0.13 | 0.94 | 0.26 | 2 | Fe-LC-1 |
|  | 0.15 | 1.28 | 0.22 | 3 | Fe-LC-2 |
| 3. Olimpiada, prismatic grains (IV) | 0.26 | 0.64 | 0.36 | 13 | {6S} |
|  | 0.24 | 1.11 | 0.33 | 66 | {3S3As} |
|  | 0.26 | 1.30 | 0.26 | 8 | {2S4As} |
|  | 0.26 | 1.66 | 0.40 | 11 | {6As} |
|  | 0.90 | 0.11 | 0.22 | 2 | Fe-HS |
| 4. Olimpiada, needle-like grains (IV) | 0.28 | 0.53 | 0.32 | 14 | {6S} |
|  | 0.25 | 0.75 | 0.24 | 6 | {5S1As} |
|  | 0.25 | 1.10 | 0.35 | 65 | {3S3As} |
|  | 0.26 | 1.27 | 0.23 | 5 | {2S4As} |
|  | 0.24 | 1.58 | 0.24 | 3 | {6As} |
|  | 0.15 | 1.14 | 0.28 | 5 | Fe-LC-1 |
|  | 1.00 | 0.14 | 0.22 | 2 | Fe-HS |
| 5. Olimpiada, prismatic grains (III) | 0.26 | 0.62 | 0.38 | 18 | {6S} |
|  | 0.26 | 1.10 | 0.29 | 54 | {3S3As} |
|  | 0.25 | 1.22 | 0.26 | 11 | {2S4As} |
|  | 0.24 | 1.55 | 0.29 | 7 | {6As} |
|  | 0.16 | 1.04 | 0.29 | 9 | Fe-LC-1 |
|  | 0.92 | 0.15 | 0.28 | 2 | Fe-HS |
| 6. Olimpiada, isometric pseudopyramidal grains (V) | 0.25 | 0.58 | 0.33 | 10 | {6S} |
|  | 0.26 | 1.10 | 0.27 | 63 | {3S3As} |
|  | 0.25 | 1.32 | 0.23 | 8 | {2S4As} |
|  | 0.25 | 1.64 | 0.36 | 6 | {6As} |
|  | 0.15 | 1.06 | 0.22 | 8 | Fe-LC-1 |
|  | 0.20 | 1.61 | 0.23 | 3 | Fe-LC-2 |
|  | 0.92 | 0.16 | 0.22 | 2 | Fe-HS |

**Table 1.** *Cont.*

| Specimen Number, Deposit, Morphology of Arsenopyrite Grains (Mineralization Stage) | IS, mm/s (±0.005) | QS, mm/s (±0.02) | W, mm/s (±0.02) | A, % (±3) | Configuration of Ligands |
|---|---|---|---|---|---|
| 7. Olimpiada, prismatic grains (II) | 0.27 | 0.61 | 0.32 | 8 | {6S} |
| | 0.25 | 1.10 | 0.29 | 70 | {3S3As} |
| | 0.23 | 1.21 | 0.24 | 12 | {2S4As} |
| | 0.20 | 1.58 | 0.23 | 2 | {6As} |
| | 0.18 | 0.95 | 0.22 | 4 | Fe-LC-1 |
| | 0.18 | 1.36 | 0.21 | 2 | Fe-LC-2 |
| | 1.04 | 0.10 | 0.29 | 2 | Fe-HS |
| 18. Sovetskoye, aggregates of coalescent pseudopyramidal grains (III) | 0.31 | 0.60 | 0.31 | 12 | {6S} |
| | 0.26 | 0.81 | 0.29 | 10 | {4S2As} |
| | 0.26 | 1.13 | 0.33 | 59 | {3S3As} |
| | 0.23 | 1.56 | 0.22 | 8 | {6As} |
| | 0.12 | 0.80 | 0.23 | 5 | Fe-LC-1 |
| | 0.15 | 1.40 | 0.28 | 4 | Fe-LC-2 |
| | 1.02 | 0.09 | 0.22 | 2 | Fe-HS |
| 19. Sovetskoye, aggregates of coalescent pseudopyramidal grains (II) | 0.29 | 0.58 | 0.33 | 14 | {6S} |
| | 0.27 | 1.11 | 0.28 | 58 | {3S3As} |
| | 0.26 | 1.15 | 0.22 | 8 | {2S4As} |
| | 0.29 | 1.58 | 0.22 | 6 | {6As} |
| | 0.14 | 1.05 | 0.26 | 13 | Fe-LC-1 |
| | 0.97 | 0.20 | 0.22 | 1 | Fe-HS |
| 28. Eldorado, aggregates of coalescent pseudopyramidal grains (II) | 0.28 | 0.61 | 0.33 | 13 | {6S} |
| | 0.25 | 1.12 | 0.34 | 70 | {3S3As} |
| | 0.17 | 0.99 | 0.36 | 14 | Fe-LC-1 |
| | 0.92 | 0.12 | 0.24 | 2 | Fe-HS |
| 29. Eldorado, aggregates of coalescent pseudopyramidal grains (III) | 0.27 | 0.62 | 0.37 | 15 | {6S} |
| | 0.30 | 1.11 | 0.32 | 66 | {3S3As} |
| | 0.12 | 1.06 | 0.28 | 7 | Fe-LC-3 |
| | 0.19 | 1.11 | 0.22 | 8 | Fe-LC-1 |
| | 0.19 | 1.70 | 0.24 | 4 | Fe-LC-2 |
| 30. Eldorado, aggregates of coalescent pseudopyramidal grains (II) | 0.29 | 0.62 | 0.30 | 12 | {6S} |
| | 0.25 | 1.14 | 0.30 | 68 | {3S3As} |
| | 0.27 | 1.58 | 0.27 | 6 | {6As} |
| | 0.18 | 0.78 | 0.34 | 12 | Fe-LC-1 |
| | 0.87 | 0.25 | 0.23 | 2 | Fe-HS |
| 31. Olginskoye, aggregates of coalescent pseudopyramidal grains (III) | 0.28 | 0.58 | 0.32 | 12 | {6S} |
| | 0.25 | 1.08 | 0.33 | 71 | {3S3As} |
| | 0.26 | 1.25 | 0.22 | 6 | {2S4As} |
| | 0.17 | 1.03 | 0.31 | 8 | Fe-LC-1 |
| | 0.92 | 0.18 | 0.28 | 2 | Fe-HS |
| 32. Olginskoye, aggregates of coalescent pseudopyramidal grains (III) | 0.27 | 0.58 | 0.32 | 12 | {6S} |
| | 0.25 | 1.09 | 0.34 | 70 | {3S3As} |
| | 0.28 | 1.27 | 0.25 | 10 | {2S4As} |
| | 0.18 | 1.61 | 0.26 | 5 | {6As} |
| | 0.85 | 0.41 | 0.22 | 2 | Fe-HS |
| 35. Blagodatnoye, disseminated aggregates of pseudopyramidal grains (III) | 0.27 | 0.63 | 0.26 | 8 | {6S} |
| | 0.28 | 1.12 | 0.30 | 54 | {3S3As} |
| | 0.26 | 1.56 | 0.27 | 23 | {6As} |
| | 0.17 | 1.14 | 0.26 | 12 | Fe-LC-1 |
| | 1.03 | 0.07 | 0.23 | 3 | Fe-HS |
| 37. Blagodatnoye, disseminated aggregates of pseudopyramidal grains (III) | 0.29 | 0.64 | 0.33 | 13 | {6S} |
| | 0.28 | 0.96 | 0.25 | 8 | {4S2As} |
| | 0.27 | 1.12 | 0.29 | 50 | {3S3As} |
| | 0.28 | 1.40 | 0.22 | 8 | {2S4As} |
| | 0.26 | 1.67 | 0.28 | 9 | {6As} |
| | 0.18 | 0.80 | 0.25 | 7 | Fe-LC-1 |
| | 0.17 | 1.33 | 0.26 | 5 | Fe-LC-2 |

**Table 1.** *Cont.*

| Specimen Number, Deposit, Morphology of Arsenopyrite Grains (Mineralization Stage) | IS, mm/s (±0.005) | QS, mm/s (±0.02) | W, mm/s (±0.02) | A, % (±3) | Configuration of Ligands |
|---|---|---|---|---|---|
| 40. Gerfed, aggregates of coalescent pseudopyramidal grains (III) | 0.26 | 0.61 | 0.32 | 13 | {6S} |
| | 0.27 | 1.12 | 0.32 | 69 | {3S3As} |
| | 0.26 | 1.59 | 0.26 | 8 | {6As} |
| | 0.14 | 1.07 | 0.22 | 8 | Fe-LC-1 |
| | 0.97 | 0.09 | 0.22 | 2 | Fe-HS |
| 41. Arkhangelskoye, aggregates of coalescent pseudopyramidal grains (III) | 0.29 | 0.56 | 0.33 | 18 | {6S} |
| | 0.27 | 0.97 | 0.33 | 33 | {4S2As} |
| | 0.27 | 1.23 | 0.31 | 36 | {3S3As} |
| | 0.26 | 1.66 | 0.29 | 7 | {6As} |
| | 0.10 | 1.08 | 0.21 | 6 | Fe-LC-1 |
| 42. Nikolayevskoye, flattened pseudopyramidal grains (III) | 0.28 | 0.57 | 0.31 | 10 | {6S} |
| | 0.27 | 1.12 | 0.32 | 76 | {3S3As} |
| | 0.25 | 1.52 | 0.27 | 6 | {1S5As} |
| | 0.16 | 1.08 | 0.24 | 8 | Fe-LC-1 |
| 43. Nikolayevskoye, flattened pseudopyramidal grains (III) | 0.27 | 0.65 | 0.34 | 15 | {6S} |
| | 0.24 | 1.11 | 0.28 | 70 | {3S3As} |
| | 0.26 | 1.33 | 0.23 | 9 | {2S4As} |
| | 0.24 | 1.56 | 0.24 | 6 | {6As} |
| 44. Veduga, aggregates of coalescent pseudopyramidal grains (III) | 0.26 | 0.66 | 0.34 | 18 | {6S] |
| | 0.25 | 1.10 | 0.34 | 59 | {3S3As} |
| | 0.28 | 1.44 | 0.35 | 22 | {2S4As} |
| 45 Polyarnaya Zvezda, aggregates of coalescent pseudopyramidal grains (II) | 0.27 | 0.62 | 0.31 | 16 | {6S} |
| | 0.25 | 1.09 | 0.37 | 66 | {3S3As} |
| | 0.28 | 1.31 | 0.26 | 11 | {2S4As} |
| | 0.28 | 1.57 | 0.24 | 6 | {6As} |
| 46. Uderey, needle-like grains (III) | 0.29 | 0.57 | 0.31 | 19 | {6S} |
| | 0.25 | 0.98 | 0.31 | 36 | {4S2As} |
| | 0.26 | 1.20 | 0.26 | 30 | {3S3As} |
| | 0.26 | 1.52 | 0.30 | 15 | {1S5As} |
| 47. Uderey, needle-like grains (II) | 0.29 | 0.54 | 0.28 | 11 | {6S} |
| | 0.26 | 0.95 | 0.36 | 45 | {4S2As} |
| | 0.25 | 1.18 | 0.23 | 22 | {3S3As} |
| | 0.27 | 1.37 | 0.22 | 14 | {2S4As} |
| | 0.25 | 1.64 | 0.32 | 8 | {6As} |
| 49. Uderey, needle-like grains (V) | 0.29 | 0.59 | 0.32 | 22 | {6S} |
| | 0.24 | 0.93 | 0.31 | 26 | {4S2As} |
| | 0.25 | 1.13 | 0.23 | 22 | {3S3As} |
| | 0.26 | 1.33 | 0.29 | 26 | {2S4As} |
| | 0.26 | 1.66 | 0.24 | 5 | {6As} |
| 56. Vasilyevskoye, prismatic grains (III) | 0.29 | 0.56 | 0.34 | 21 | {6S} |
| | 0.25 | 0.90 | 0.30 | 23 | {4S2As} |
| | 0.25 | 1.16 | 0.28 | 32 | {3S3As} |
| | 0.27 | 1.42 | 0.31 | 20 | {2S4As} |
| | 0.22 | 1.71 | 0.38 | 4 | {6As} |
| 57. Vasilyevskoye, prismatic grains (III) | 0.29 | 0.50 | 0.34 | 19 | {6S} |
| | 0.28 | 0.79 | 0.24 | 13 | {5S1As} |
| | 0.28 | 1.00 | 0.22 | 12 | {4S2As} |
| | 0.28 | 1.21 | 0.28 | 34 | {3S3As} |
| | 0.26 | 1.57 | 0.31 | 16 | {6As} |
| | 0.12 | 1.01 | 0.24 | 6 | Fe-LC-1 |
| 58. Vasilyevskoye, prismatic grains (III) | 0.28 | 0.51 | 0.35 | 18 | {6S} |
| | 0.28 | 0.80 | 0.28 | 14 | {5S1As} |
| | 0.28 | 0.99 | 0.23 | 14 | {4S2As} |
| | 0.28 | 1.21 | 0.29 | 32 | {3S3As} |
| | 0.26 | 1.59 | 0.36 | 16 | {6As} |
| | 0.11 | 1.01 | 0.22 | 6 | Fe-LC-1 |
| 61. Panimba, isometric pseudopyramidal grains (III) | 0.28 | 0.58 | 0.33 | 16 | {6S} |
| | 0.25 | 1.1 | 0.33 | 74 | {3S3As} |
| | 0.27 | 1.43 | 0.27 | 10 | {2S4As} |

**Table 1.** *Cont.*

| Specimen Number, Deposit, Morphology of Arsenopyrite Grains (Mineralization Stage) | IS, mm/s (±0.005) | QS, mm/s (±0.02) | W, mm/s (±0.02) | A, % (±3) | Configuration of Ligands |
|---|---|---|---|---|---|
| 62. Panimba, isometric pseudopyramidal grains (III) | 0.32 | 0.67 | 0.39 | 21 | {6S} |
|  | 0.26 | 1.13 | 0.32 | 66 | {3S3As} |
|  | 0.33 | 1.35 | 0.22 | 5 | {2S4As} |
|  | 0.06 | 1.08 | 0.28 | 8 | Fe-LC-3 |
| 63. Panimba, isometric pseudopyramidal grains (II) | 0.28 | 0.60 | 0.32 | 14 | {6S} |
|  | 0.25 | 1.09 | 0.33 | 70 | {3S3As} |
|  | 0.27 | 1.32 | 0.31 | 16 | {2S4As} |

Notes: IS, mm/s is a chemical isomeric shift; QS, mm/s is quadrupole splitting; W, mm/s is the width of absorption line; A, % is a fraction of iron population of a particular crystallographic position; I–IV are stages of mineral formation.

The last column in Table 1 provides the identification of ligand surroundings of iron cations on the basis of the monotonic dependence of quadrupole splitting on the number of arsenic atoms in the coordination ligand octahedron (see Figure 3), i.e., reference to a specific configuration of ligands was made on the basis of the values of quadrupole splittings detected in the iron positions spectrum. The determination of configurations is qualitative because the value of quadrupole splitting is influenced specifically by the nature of atom distribution in the following coordination spheres. Table 1 includes the iron positions with an occupation of over 1% of the total iron content in the specimen.

In some specimens, there are iron positions marked in Table 1 as Fe-HS (high spin), with a high value of the chemical shift ≈1 mm/s typical for the high-spin state of $Fe^{2+}$. The occupation of these positions is minor and constitutes < 2%. These positions can be referred to as local areas in the mineral with increased sulfur concentrations. Meanwhile, it is worth noting that the presence of the subspectrum of these positions in the general spectrum of the specimen increases the area of the right line of the doublet. The Goldansky–Karyagin effect can also lead to the doublet asymmetry [42].

The specimens of some deposits contained iron positions marked in the Table 1 as Fe-LC (low-coordinated iron), with a minor chemical shift typical for iron cations with reduced coordination by ligands. It is known that the elementary arsenopyrite cell contains 24 free tetrahedron positions (Figure 4); therefore, it can be assumed that they can be occupied by iron at specific crystal formation conditions. The maximum iron output to tetrahedral positions was detected in arsenopyrites of the Blagodatnoye and Sovetskoye deposit was up to 13%, and the Eldorado was up to 19% of the total iron content.

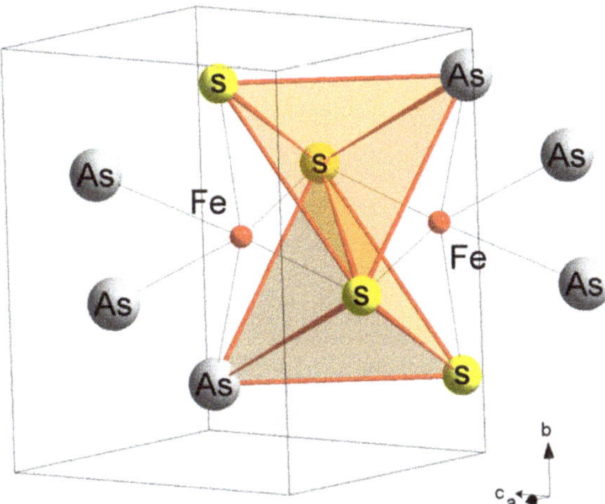


**Figure 4.** Tetrahedral cavities in the arsenopyrite structure. Thick lines show two of 24 tetrahedral iron positions in the elementary cell of arsenopyrite.

Therefore, in the studied arsenopyrites, the ligand configurations of the nearest surroundings are formed at almost equal contents of sulfur and arsenic. The maximum occupied configuration is {3S3As}. The occurrence of other configurations testifies to the concentration heterogeneity of ligands with variable composition in the specimen and suggests that arsenopyrites are solid solutions of pyrite, arsenopyrite, and loellingite.

Figure 5 shows diagrams of occupation of ligand configurations in arsenopyrite of different deposits, where the preference of the configuration {3S3As} in the studied mineral samples is evident. The continuous lines show the distribution of configurations expected upon the chaotic distribution in the crystal of ligands of two structural motives accounting for the span of the S/As ratio in the specimens of this deposit. The evident deviation of the real situation from the statistic distribution of structural motives {6S}, {3S3As}, {6As} in the studied specimens speaks to the fact that the arsenopyrite lattice is thermodynamically more favorable as compared to pyrite or loellingite lattices.

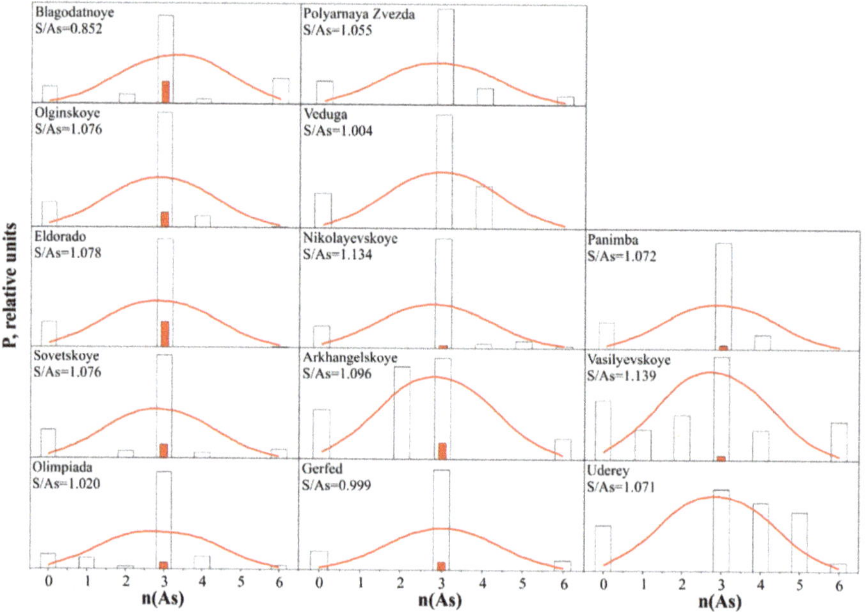


**Figure 5.** Occupation of ligand configurations in arsenopyrites from different deposits.

Excess of occupation of the ligand configuration {6S} over occupation of the {6As} configuration in arsenopyrite from almost all deposits draws attention. The pyrite "phase" is probably more favorable thermodynamically than the loellingite "phase" for the studied mineral selection. The red column in the background of the configuration {3S3As} of the diagrams (Figure 5) shows the population of low-coordinated iron (Fe-LC). The average ratio of the S/As grades is indicated by the numbers of the deposits in brackets. As is seen, there is no strict correlation between the S/As ratio and the occupation of ligand configurations. The physicochemical conditions of mineral formation probably significantly influence the set of the implemented ligand configurations of a specific structural motive.

The impact of the geological environment of ore formation on the ligand microstructure of the studied arsenopyrites shall be summarized as follows. The Olimpiada and Panimba deposits located in close proximity to the parent granitoid intrusions (1.5–2.0 km) and occupying the lower part of the ore shoot have the most ordered arsenopyrites ({3S3As} > 70%). They are characterized by a frequent predominance of arsenic ligands ({6As}, {2S4As}, {1S5As}) over sulfuric ligands ({6S}, {5S1As}, {4S2As}) in the crystal lattice. Arsenopyrites of low structural ordering ({3S3As} = 36–22%) are developed at the Vasilievskoye and Uderey deposits, which are the most distant ones from granitoids (16–18 km) and occupy the upper part of the ore shoot. Sulfuric and arsenic ligands distributed in approximately equal quantities predominate in the crystal lattice of the mineral from these deposits, and there is no Fe in tetrahedral positions (see Table 1). The presence of such regularities may be associated with an inhomogeneous temperature field in the mineral formation environment, contributing to the formation of chemical and structural varieties of the mineral. The reflection of regular variability of the ligand microstructure in arsenopyrite generations has not been revealed in this study, which can probably be explained by the abrupt change in the parameters of formation and recrystallization of the mineral in the ore formation process.

### 4.2. Arsenopyrite Composition and Distribution of Impurities of Precious Metals

Concentrations of the main mineral-forming and impurity elements—Au, Ag, Ru, Pd, Pt—were determined in arsenopyrite samples (Table 2). The content of mineral-forming elements in arsenopyrite varies within the limits (wt. %): Fe—(32.8–34.7); As—(43.3–49.1); S—(17.8–21.9) at the theoretical composition of the mineral Fe—34.3; As—46.0; S—19.7. Arsenopyrites under study are deficient in iron ((S + As)/Fe > 2), with a variable ratio of 0.85 ≤ S/As ≤ 1.18 (Table 2). The observed variability in concentrations of the main mineral-forming elements is reflected in the physicochemical conditions of mineral crystallization [1,5,6,43].

The highest concentrations of precious elements are characteristic of gold and silver. The platinoid grade does not exceed 1 ppm. The optical and electron microscope study of the samples showed the presence of native gold inclusions of various fineness and aurostibite in arsenopyrite. No Ag, Ru, Pd, and Pt minerals were found in arsenopyrite. Gold particles are observed on the surface of arsenopyrite crystals, in fractures, in the form of isometric inclusions, confined to fractures and microdroplet particles outside the cataclasis fractures (Figure 6).

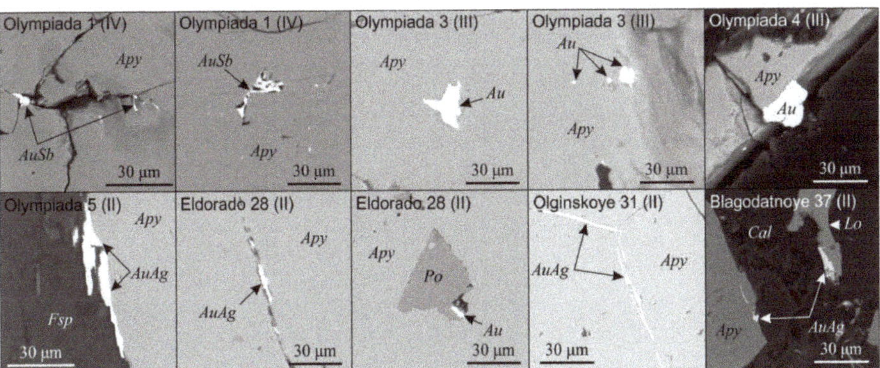


**Figure 6.** Back-scattered electron (BSE) images of distribution of gold mineral particles in arsenopyrite. Arsenopyrite associations (stages): (II)—sulfide–polymetallic; (III)—ullmannite–gersdorffite; (IV)—stibnite. Abbreviations of minerals: Apy—arsenopyrite; AuSb—aurostibite; Au—native gold (fineness ≥ 900‰); Fsp—alkaline feldspar; AuAg—native gold (fineness 800‰–899‰); Po—pyrrhotite; Cal—calcite; Lo—loellingite.

The observed particles of gold minerals were found in samples with minimal concentrations of the element in the samples, and, on the contrary, in some arsenopyrite samples, no gold mineral inclusions were found at high concentrations of gold, according to the ICP tests. Usually, visible gold and aurostibite particles occur in cataclased and recrystallized arsenopyrite grains in association with minerals of the stage of mineral formation following the arsenopyrite stage. The concentration of gold reveals a nonlinear dependence on the composition of the mineral (S/As) (Figure 7).

**Table 2.** Composition and parameters of arsenopyrite formation.

| No. | Fe | As | S | $T\,^\circ$C | $\log fS_2$ | Au | Ag | Ru | Pd | Pt | $\Sigma$ | S/As | (S + As)/Fe |
|---|---|---|---|---|---|---|---|---|---|---|---|---|---|
| 1 | 33.9 | 46.4 | 19.7 | 300 | −14.6 | 1.94 | 0.11 | 0.001 | 0.03 | 0.01 | 2.09 | 0.99 | 2.03 |
| 2 | 34.1 | 45.5 | 20.4 | 460 | −5.9 | 0.88 | 3.21 | 0.001 | 0.09 | 0.01 | 4.19 | 1.05 | 2.03 |
| 3 | 33.6 | 47.1 | 19.3 | 340 | −12.8 | 0.36 | 2.18 | 0.001 | 0.07 | 0.00 | 2.61 | 0.96 | 2.05 |
| 4 | 34.0 | 44.7 | 21.2 | 430 | −6.5 | 93.67 | 1.13 | 0.001 | 0.08 | 0.02 | 94.90 | 1.11 | 2.07 |
| 5 | 34.1 | 45.4 | 20.5 | 450 | −5.9 | 12.52 | 1.05 | 0.002 | 0.14 | 0.02 | 13.73 | 1.06 | 2.04 |
| 6 | 33.7 | 46.7 | 19.5 | 320 | −13.4 | 4.13 | 6.35 | 0.002 | 0.09 | 0.00 | 10.57 | 0.98 | 2.04 |
| 7 | 33.7 | 46.4 | 19.9 | 300 | −14.6 | 2.22 | 0.55 | 0.001 | 0.09 | 0.02 | 2.87 | 1.00 | 2.05 |
| 18 | 33.9 | 45.5 | 20.6 | 460 | −5.9 | 20.26 | 46.69 | 0.003 | 0.22 | 0.05 | 67.23 | 1.06 | 2.06 |
| 19 | 34.0 | 44.9 | 21.0 | 440 | −6.4 | 0.38 | 0.85 | 0.001 | 0.02 | 0.00 | 1.25 | 1.09 | 2.06 |
| 28 | 34.4 | 44.5 | 21.2 | 420 | −6.7 | 3.63 | 14.16 | 0.003 | 0.10 | 0.57 | 18.46 | 1.11 | 2.04 |
| 29 | 34.4 | 44.8 | 20.8 | 440 | −6.4 | 0.32 | 22.98 | 0.003 | 0.10 | 0.49 | 23.90 | 1.09 | 2.02 |
| 30 | 34.2 | 45.6 | 20.2 | 460 | −5.9 | 0.15 | 0.84 | 0.001 | 0.02 | 0.01 | 1.01 | 1.03 | 2.02 |
| 31 | 34.0 | 45.4 | 20.5 | 460 | −5.9 | 1.40 | 3.44 | 0.002 | 0.32 | 0.05 | 5.22 | 1.06 | 2.05 |
| 32 | 34.2 | 44.9 | 21.0 | 440 | −6.4 | 1.56 | 2.72 | 0.001 | 0.15 | 0.03 | 4.46 | 1.09 | 2.05 |
| 35 | 32.8 | 49.1 | 17.8 | 530 | −7.4 | 9.44 | 6.03 | 0.013 | 0.17 | 0.04 | 15.69 | 0.85 | 2.06 |
| 37 | 32.9 | 49.1 | 18.0 | 510 | −7.5 | 38.45 | 63.66 | 0.001 | 0.06 | 0.65 | 102.82 | 0.86 | 2.06 |
| 40 | 33.9 | 46.2 | 19.8 | 300 | −14.6 | 6.14 | 49.92 | 0.002 | 0.16 | 0.73 | 56.96 | 1.00 | 2.03 |
| 41 | 34.3 | 44.7 | 21.0 | 430 | −6.5 | 40.56 | 15.27 | 0.001 | 0.07 | 0.12 | 56.02 | 1.10 | 2.04 |
| 42 | 34.4 | 44.1 | 21.4 | 410 | −6.9 | 22.98 | 28.53 | 0.002 | 0.12 | 0.69 | 52.33 | 1.14 | 2.04 |
| 43 | 34.3 | 44.2 | 21.4 | 415 | −6.8 | 10.90 | 25.81 | 0.003 | 0.11 | 0.00 | 36.83 | 1.13 | 2.05 |
| 44 | 33.8 | 46.2 | 19.8 | 300 | −14.6 | 5.40 | 23.52 | 0.002 | 0.03 | 0.00 | 28.96 | 1.00 | 2.04 |
| 45 | 34.2 | 45.3 | 20.4 | 455 | −6.0 | 29.70 | 26.00 | 0.003 | 0.08 | 0.00 | 55.79 | 1.05 | 2.03 |
| 46 | 34.3 | 45.2 | 20.5 | 450 | −6.2 | 36.18 | 18.54 | 0.002 | 0.26 | 0.02 | 55.00 | 1.06 | 2.02 |
| 47 | 34.4 | 45.0 | 20.7 | 445 | −6.4 | 87.41 | 378.71 | 0.001 | 0.09 | 0.02 | 466.22 | 1.07 | 2.02 |
| 49 | 34.3 | 44.9 | 20.7 | 445 | −6.4 | 34.11 | 21.61 | 0.000 | 0.27 | 0.02 | 56.02 | 1.08 | 2.03 |
| 56 | 34.2 | 44.6 | 21.1 | 430 | −6.5 | 24.46 | 10.20 | 0.002 | 0.15 | 0.02 | 34.84 | 1.10 | 2.04 |
| 57 | 34.7 | 43.3 | 21.9 | 390 | −7.4 | 89.13 | 3.64 | 0.001 | 0.05 | 0.01 | 92.83 | 1.18 | 2.03 |
| 58 | 34.5 | 44.2 | 21.4 | 415 | −6.8 | 63.78 | 9.66 | 0.002 | 0.02 | 0.00 | 73.46 | 1.13 | 2.03 |
| 61 | 34.4 | 45.0 | 20.5 | 445 | −6.4 | 4.92 | 18.08 | 0.004 | 0.14 | 0.04 | 23.19 | 1.07 | 2.01 |
| 62 | 34.2 | 44.9 | 20.8 | 445 | −6.4 | 3.61 | 11.49 | 0.003 | 0.23 | 0.04 | 15.38 | 1.08 | 2.04 |
| 63 | 34.1 | 45.2 | 20.6 | 450 | −6.1 | 2.40 | 10.59 | 0.002 | 0.39 | 0.04 | 13.42 | 1.07 | 2.04 |

Notes: The first column contains the laboratory number corresponding to the laboratory number in Table 1; Columns 2–4 present the grades of elements, the average value in the sample, in wt. % Columns 5–6 contain $T\,^\circ$C and $\log fS_2$ determined by [1,43]; Columns 7–12 present concentrations of precious metals and their total grade, in ppm; Columns 13 and 14 provide the indicator ratios of S, As, and Fe, calculated on the basis of atomic quantities.

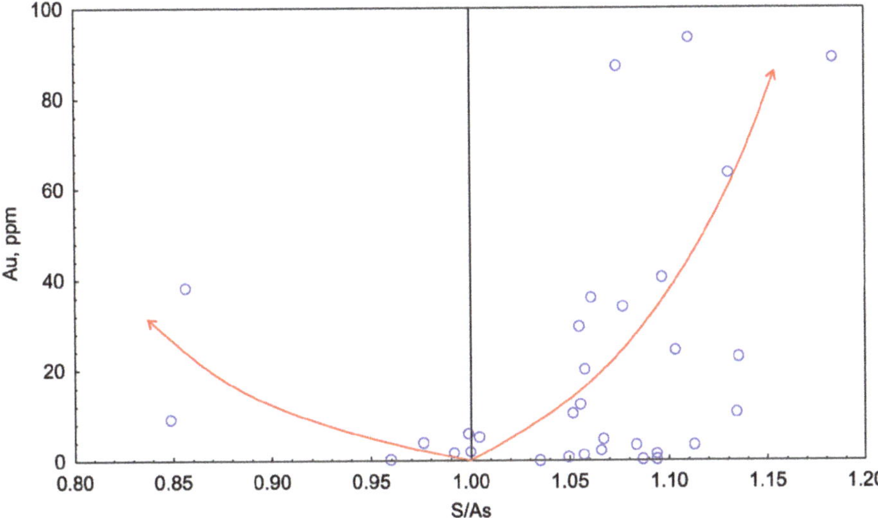

**Figure 7.** Distribution of gold in arsenopyrite depending on S/As ratio.

The revealed tendency of correlation between gold, sulfur, and arsenic concentrations suggests the presence of gold in arsenopyrite in the form of isomorphic impurity or solid solution. Thermodynamic calculations [44] confirm this assumption, since the fields of existence of complex gold with sulfur and arsenic ions in solutions coincide with the fields of pyrite and arsenopyrite stability. In this case, it is assumed that there are atoms of gold introduction in the mineral structure along with the observed crypto-inclusions of native gold.

Crystallization of arsenopyrite of the studied samples took place in the range of 300–520 °C and sulfur activity $\log fS_2 = -5.8$–$-14.6$ [32]. Increased concentrations of precious metals were revealed in nonstoichiometric mineral varieties formed at temperatures 385–410, 430, 445–450, and 510 °C and sulfur activity $\log fS_2 = -7.4$–$-7.0$, $-6.5$, $-6.3$–$-6.1$ and $-8.0$, respectively. The previous work on studying the stability of minerals of the Fe–As–S system, as well as the behavior of gold and silver in hydrothermal solutions, indicate that the formation of various parageneses and stability of Au and Ag complexes depend on many factors—partial pressure of sulfur and arsenic, acidity–alkalinity and redox potential of the medium, temperature, and pressure [44–55].

Gold concentrations (<1 wt. %) are not recorded in the Mössbauer spectra. In the studied arsenopyrite samples, the gold content does not exceed 0.01 wt. %, which does not allow the Mössbauer spectra to confirm the possibility of the element entering the crystal lattice in the form of isomorphic impurity or introduction atoms. A collection of a series of arsenopyrite samples with discrete values of the content of "invisible" gold over 1 wt. % is required to determine the forms of gold and the nature of the bond in the crystal lattice. The conclusion made earlier by Genkin et al. [9,17] on the presence of chemically bound gold in arsenopyrite suggests it is located in the coordination octahedra instead of iron. High symmetry of gold allowed Genkin [9] to refer the observed singlet characterizing high local symmetry to gold. Only the octahedron has high symmetry among tetrahedral and octahedral structural chains in arsenopyrite, but the explanation of the charge state of gold and its covalent bond with S and As remains to be determined conclusively. Moreover, the study of gold forms in arsenopyrite with the Mössbauer spectroscopy method should be carried out on the [197]Au nuclei more diligently than in the paper [17]. Without the knowledge of the charge state of gold and the nature of covalent effects, it is impossible to draw conclusions about the method of chemical binding of gold in arsenopyrite. The results of the study of the mineral ligand microstructure show regularities, which indicate that the increased gold grades in arsenopyrite are typical upon the reduction of its structural stoichiometry and are accompanied by the occurrence of the configurations {5S1As}, {4S2As}, {1S5As}. The tendency of the gold grade to increase upon the reduction of the share of the {3S3As} configurations and iron in the low-coordinated condition (Fe-LC) was also identified.

Trigub et al. [56,57] demonstrated the possibility of replacing iron atoms in the structure of arsenopyrite and pyrite with gold atoms using synthetic minerals of the Fe–As–S system. However, in our case, there is no correlation between the iron and gold grades, which may indicate a different mechanism of gold entry into the mineral structure. On the other hand, the same paper [56,57] demonstrates the extraction of isomorphic gold in pyrite into metallic, nanoscale gold, upon thermal metamorphism of the mineral. Recently, we [58,59] found gold enrichment of pyrite varieties with an increase and decrease of the S/Fe ratio from the stoichiometric one—S/Fe = 2. D. Fougerouse et al. [60] showed the possibility of remobilization and redeposition of gold from arsenopyrite under the influence of late fluids. It is possible that some of the gold in the arsenopyrite samples we have determined was redeposited due to thermal metamorphism or late fluid impact, and hides the existing correlations. In this case, the occurrence of the later pyrite–pyrrhotite–chalcopyrite–sphalerite–galena mineral association in the ores has a significant impact on the morphological features of gold particles, with their usual significant increase in sizes (Figure 8). Observations indicate a late generation of visible gold, and early gold, which is in association with early arsenopyrite, pyrrhotite, and pyrite, is an invisible form representing isomorphic gold, or a group of atoms of introduction [6,59,61,62]. Arsenopyrite in association with Cu, Pb, and Zn sulfide minerals is characterized by the poikiloblastic structure (Figure 8a), with numerous inclusions of galena among non-ore minerals. Native gold forms continuous edges around arsenopyrite. Gold inclusions in pyrite are sometimes geometrically ordered, reminiscent of the structure of decay or helicite structure, which occurs in the form of linear (in the polished section) distribution of gold particles in the host mineral (see Figure 8c). The spine-like-grained gold aggregate in cataclased vein quartz (see Figure 8d) is spatially distant from the sulfide aggregates and occupies the intergranular space in vein quartz and thin wedge-shaped cataclasis fractures, characterizing the final stage of gold deposition [6,62].

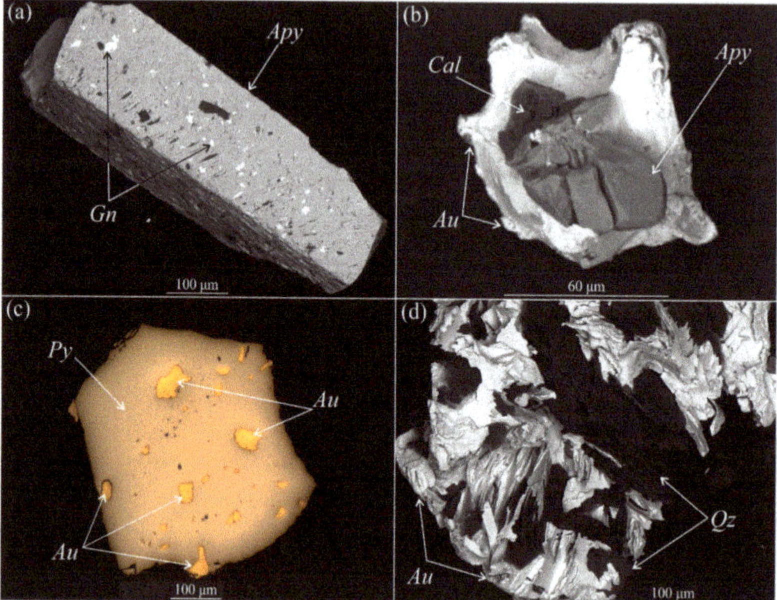


**Figure 8.** BSE (a,b,d) and optical (c) photo of structural and textural features of associations of the sulfide–polymetallic ore-formation stage: (**a**) arsenopyrite (Apy) poikiloblast with galena (Gn) inclusions;

(**b**) margin of native gold (Au) growth around arsenopyrite with calcite (Cal) inclusion; (**c**) strip distribution of native gold (Au) inclusions in pyrite (Py) crystalloblast; (**d**) morphology of intergranular-fractured native gold (Au) aggregate in cataclased quartz (Qz).

## 5. Conclusions

The Mössbauer studies of arsenopyrite specimens from gold ore deposits of the Yenisei Ridge were carried out. In the mineral, the ligand configurations of the coordination octahedron around the iron cation, differing from the configuration {3S3As} typical for the stoichiometric arsenopyrite FeAsS, were detected. The occupation of configurations with other compositions of the ligand surroundings is not subordinated to the statistic distribution. Natural arsenopyrites can be considered a solid pyrite, loellingite, and arsenopyrite solution with the physicochemical preference to the formation in the arsenopyrite–pyrite–loellingite order.

Iron output to tetrahedral positions was detected in arsenopyrites of some deposits. It is assumed that the nature of ligand disordering and the occupation of tetrahedral iron positions are preconditioned by the features of thermodynamic and geological conditions of mineral formation.

The composition of arsenopyrite is characterized by variability in the concentrations of main and impurity elements in different parts of grains, neighboring grains of disseminated aggregates, and in the ores of different deposits. The elevated gold and other precious elements' concentrations are typical for the mineral, with the predominance of sulfur or arsenic in relation to the stoichiometric composition ($0.85 \leq$ S/As $\leq 1.18$), reduced structural stoichiometry for ligand configurations {5S1As}, {4S2As}, {1S5As} and reduction of the share of the configurations {3S3As} and iron in the low-coordinated condition (Fe-LC). The structure of arsenopyrite assumes that gold atoms have been incorporated.

**Author Contributions:** Conceptualization, A.M.S. and S.A.S.; Data curation, S.A.S., O.A.B. and Y.V.K.; Investigation, A.M.S., S.A.S., O.A.B., Y.V.K., Y.A.Z., and P.A.T.; Writing—Original draft, A.M.S., S.A.S., O.A.B., Y.V.K., Y.A.Z., and P.A.T.

**Funding:** The work was supported by the Russian Foundation for Basic Research (project number 19-35-90017) and the Government of the Russian Federation (project 14.Y26.31.0012).

**Acknowledgments:** We thank Ye.V. Korbovyak for X-ray energy dispersive microanalysis; Ye.V. Rabtsevich and Ye.I. Nikitin for ICP-MS study. We are grateful to G.A. Palyanova, three anonymous reviewers, and the Editorial Board members, for their comments and improvements.

**Conflicts of Interest:** The authors declare no conflicts of interest.

## References

1. Kretschmar, U.; Scott, S.D. Phase relations involving arsenopyrite in the system Fe–As–S and their application. *Can. Mineral.* **1976**, *14*, 364–386.

2. Sharp, Z.D.; Essene, E.J.; Kelly, W.C. A reexamination of the arsenopyrite geothermometer: Pressure considerations and applications to natural assemblages. *Can. Mineral.* **1985**, *23*, 517–534.

3. Sustavov, S.G.; Murzin, V.V.; Ivanov, O.K.; Litoshko, D.N. Arsenopyrite. In *Ural Mineralogy: Arsenides and stibnides. Tellurides. Selenides. Fluorides. Chlorides and Bromides*; Ural Branch of the Russian Academy of Sciences: Sverdlovsk, Russia, 1991; pp. 4–12. (In Russian)

4. Murzin, V.V.; Semenkin, V.A.; Pikulev, A.I.; Mil'der, O.B.; Sustavov, S.G.; Krinov, D.I. Mössbauer spectroscopic evidence for nonequivalent sites of iron atoms in gold-bearing arsenopyrite. *Geochem. Int.* **2003**, *41*, 812–820.

5. Tyukova Ye, E.; Voroshin, S.V. *Composition and Parageneses of Arsenopyrite in Deposits and Hosting Rocks of the Upper Kolyma Region (for Interpretation of Sulfide Associations Genesis)*; SVKNII of the Far-Eastern Branch of the Russian Academy of Sciences: Moscow, Russia, 2007; p. 107. (In Russian)

6. Sazonov, A.M.; Kirik, S.D.; Silyanov, S.A.; Bayukov, O.A.; Tushin, P.A. Typomorphism of arsenopyrite from Blagodatnoye and Olimpiada gold ore deposits (Yenisei Ridge). *Mineralogy* **2016**, *3*, 52–70. (In Russian)

7. Stankovič, J.; Jančula, D. K typochemizmu arzenopyritu z niektorých ložísk na južných svahoch Nízkych Tatier. *Miner. Slovaca* **1981**, *13*, 373–377.

8. Boiron, M.C.; Cathelineau, M.; Trescases, J.J. Conditions of gold-bearing arsenopyrite crystallization in the Villeranges Basin, Marche-Combrailles shear zone, France: A mineralogical and fluid inclusion study. *Econ. Geol.* **1989**, *84*, 1340–1362. [CrossRef]

9. Genkin, A.D. Gold-bearing arsenopyrite from gold ore deposits: Internal structure of grains, composition, growth mechanism and state of gold. *Geol. ore Depos.* **1998**, *40*, 551–557. (In Russian)

10. Kovalev, K.R.; Kalinin, Y.A.; Naumov, Y.A.; Kolesnikova, M.K.; Korolyuk, V.N. Gold-bearing arsenopyrite in eastern Kazakhstan gold-sulfide deposits. *Russ. Geol. Geophys.* **2011**, *52*, 178–192. [CrossRef]

11. Fuess, H.; Kratz, T.; Töpel-Schadt, J.; Miehe, G. Crystal structure refinement and electron microscopy of arsenopyrite. *Z. Krist.* **1987**, *179*, 335–346. [CrossRef]

12. Wagner, F.E.; Marion, P.; Regnard, J.-R. $^{197}$Au Mössbauer study of gold ores, mattes, roaster products, and gold minerals. *Hyperfine Interact.* **1988**, *41*, 851–854. [CrossRef]

13. Johan, Z.; Marcoux, E.; Bonnemaison, M. Arsenopyrite aurifere: Mode de substitution de Au dans la structure de FeAsS. *Comptes Rendus Acad. Sci.* **1989**, *308*, 185–191.

14. Friedl, J.; Wagner, F.E.; Sawicki, J.A.; Harris, D.C.; Mandarino, J.A.; Marion, P. $^{197}$Au, $^{57}$Fe and $^{121}$Sb Mössbauer study of gold minerals and ores. *Hyperfine Interact.* **1992**, *70*, 945–948. [CrossRef]

15. Mumin, A.H.; Fleet, M.E.; Chryssoulis, S.L. Gold mineralization in As-rich mesothermal gold ores of the Bogosu–Prestea mining district of the Ashanti Gold Belt, Ghana: Remobilization of "invisible" gold. *Miner. Depos.* **1994**, *29*, 445–460. [CrossRef]

16. Jiuling, L.; Daming, F.; Jeng, Q.; Guilan, Z. The existence of the negative valence state of gold in sulfide minerals and its formation mechanism. *Acta Geol. Sin.* **1995**, *69*, 67–77.

17. Genkin, A.D.; Bortnikov, N.S.; Cabri, L.J.; Wagner, F.E.; Stanley, C.J.; Safonov, Y.G.; McMahon, G.; Friedl, J.; Kerzin, A.L.; Gamyanin, G.N. A multidisciplinary study of invisible gold in arsenopyrite from four mesothermal gold deposits in Siberia, Russian Federation. *Econ. Geol.* **1998**, *93*, 463–487. [CrossRef]

18. Cabri, L.J.; Newville, M.; Gordon, R.A.; Crozier, E.D.; Sutton, S.R.; McMahon, G.; Jiang, D.-T. Chemical speciation of gold in arsenopyrite. *Can. Mineral.* **2000**, *38*, 1265–1281. [CrossRef]

19. Sung, Y.-H.; Brugger, J.; Ciobanu, C.L.; Pring, A.; Skinner, W.; Nugus, M. Invisible gold in arsenian pyrite and arsenopyrite from a multistage Archaean gold deposit: Sunrise Dam, Eastern Goldfields Province, Western Australia. *Miner. Depos.* **2009**, *44*, 765–775. [CrossRef]

20. Zhu, Y.; Fang, A.; Tan, J. Geochemistry of hydrothermal gold deposits: A review. *Geosci. Front.* **2011**, *2*, 367–374. [CrossRef]

21. Silva, J.C.M.; De Abreu, H.A.; Duarte, H.A. Electronic and structural properties of bulk arsenopyrite and its cleavage surfaces—A DFT study. *RSC Adv.* **2015**, *5*, 2013–2023. [CrossRef]

22. Kravtsova, R.G.; Tauson, V.L.; Nikitenko, E.M. Modes of Au, Pt, and Pd occurrence in arsenopyrite from the Natalkinskoe deposit, NE Russia. *Geochem. Int.* **2015**, *53*, 964–972. [CrossRef]

23. Buerger, M.J. The symmetry and crystal structure of the minerals of the arsenopyrite group. *Z. Krist.* **1936**, *95*, 83–113. [CrossRef]

24. Morimoto, N.; Clark, L.A. Arsenopyrite crystal-chemical relations. *Am. Mineral.* **1961**, *46*, 1448–1469.

25. Kirik, S.D.; Sazonov, A.M.; Silyanov, S.A.; Bayukov, O.A. Investigation of disordering in natural arsenopyrite by X-ray powder crystal structure analysis and nuclear gamma resonance. *J. Sib. Fed. Univ. Eng. Technol. Ser.* **2017**, *10*, 578–592. (In Russian) [CrossRef]

26. Nickel, E.H. Structural stability of minerals with the pyrite, marcasite, arsenopyrite and loellingite structures. *Can. Mineral.* **1968**, *9*, 311–323.

27. Bindi, L.; Moëlo, Y.; Léone, P.; Suchaud, M. Stoichiometric arsenopyrite, FeAsS, from La Roche-Balue Quarry Loire-Atlantique, France: Crystal structure and Mössbauer study. *Can. Mineral.* **2012**, *50*, 471–479. [CrossRef]

28. Kerler, W.; Neuwirth, W.; Fluck, E.Z. Isomerieverschiebung und Quadrupolaufspaltung beim Mößbauer-Effekt von Fe$^{57}$ in Eisenverbindungen. *Z. Phys.* **1963**, *175*, 200–220. [CrossRef]

29. Evans, B.J.; Johnson, R.G.; Senftle, F.E.; Blaine Cecil, C.; Dulong, F. The $^{57}$Fe Mössbauer parameters of pyrite and marcasite with different provenances. *Geochim. Cosmochim. Acta* **1982**, *46*, 761–775. [CrossRef]

30. Imbert, P.; Gerard, A.; Wintenberger, M. Study of naturally occurring sulfides, sulfarsenides, and arsenides of iron by Mössbauer effect. *Comptes Rendus* **1963**, *256*, 4391–4393.

31. Kjekshus, A.; Nicholson, D.G. The significance of π Back-Bonding in compounds with pyrite, marcasite and arsenopyrite Type structures. *Acta Chem. Scand.* **1971**, *25*, 866–876. [CrossRef]

32. Sazonov, A.M.; Zvyagina Ye, A.; Silyanov, S.A.; Lobanov, K.V.; Leontyev, S.I.; Kalinin Yu, A.; Savichev, A.A.; Tishin, P.A. Ore genesis of the Olimpiada gold deposit (Yenisei Ridge, Russia). *Geosph. Res.* **2019**, 17–43. (In Russian) [CrossRef]

33. Sazonov, A.M.; Bernatonis, V.K. Features of formation of gold-bearing quartz-vein zones in crystalline slates. In *Geological and Geochemical Criteria of Gold Mineralization*; Nauka: Novosibirsk, Russia, 1990; pp. 44–57. (In Russian)

34. Sazonov, A.M.; Ananiev, A.A.; Vlasov, V.S. On the conditions of spatial combination of gold ore and antimony mineralization in slate strata of one of the Siberian regions. In *Geological and Geochemical Criteria of Gold Mineralization*; Nauka: Novosibirsk, Russia, 1990; pp. 84–96. (In Russian)

35. Sazonov, A.M.; Sarayev, V.A.; Ananiev, A.A. Sulfide and quartz gold deposits in metamorphic strata of the Yenisei Ridge. *Geol. Geophys.* **1991**, *5*, 28–37. (In Russian)

36. Sazonov, A.M.; Zvyagina, E.A.; Romanovsky, A.E.; Shvedov, G.I.; Leontiev, S.I. Aposlate gold-bearing metasomatites of beresite formation (Yenisei Ridge). *Geol. Geophys.* **1995**, *36*, 53–61. (In Russian)

37. Sazonov, A.M.; Ananyev, A.A.; Poleva, T.V.; Khokhlov, A.N.; Vlasov, V.S.; Zvyagina, E.A.; Fedorova, A.V.; Tishin, P.A.; Leontyev, S.I. Gold metallogeny of the Yenisei Ridge: Geological and structural position, structural types of ore fields. *J. Sib. Fed. Univ. Ser. Eng. Technol.* **2010**, *3*, 371–395. (In Russian)

38. Gibsher, N.A.; Sazonov, A.M.; Travin, A.V.; Tomilenko, A.A.; Ponomarchuk, A.V.; Sil'yanov, S.A.; Nekrasova, N.A.; Shaparenko, E.O.; Ryabukha, M.A.; Khomenko, M.O. Age and duration of the formation of the Olimpiadinski gold deposit (Yenisei ridge, Russia). *Geochem. Int.* **2019**, *57*, 593–599. [CrossRef]

39. Sazonov, A.M. *Geochemistry of Gold in Metamorphic Strata*; Publishing House of TPU: Tomsk, Russia, 1998; p. 168. (In Russian)

40. Brostigen, G.; Kjekshus, A. Redetermined Crystal Structure of FeS$_2$ (Pyrite). *Acta Chem. Scand.* **1969**, *23*, 2186–2188. [CrossRef]

41. Lutz, H.D.; Jung, M.; Waschenbach, G. Kristallstrukturen des lollingits FeAs$_2$ und des pyrits RuTe$_2$. *Z. Anorg. Allg. Chem.* **1987**, *554*, 87–91. [CrossRef]

42. Goldansky, V.I. *Chemical Applications of Moessbauer Spectroscopy*; Mir: Moscow, Russia, 1970; p. 504. (In Russian)

43. Scott, S.D. Chemical behaviour of sphalerite and arsenopyrite in hydrothermal and metamorphic environments. *Mineral. Mag.* **1983**, *47*, 427–435. [CrossRef]

44. Pavlov, A.L.; Pavlova, L.K. Elements of thermodynamics of gold behavior in ore formation sample. In *Physics and Physicochemistry of Ore-Forming Processes*; Nauka: Novosibirsk, Russia, 1971; pp. 121–147. (In Russian)

45. Kolonin, G.R.; Palyanova, G.A.; Shironosova, G.P. Stability and solubility of arsenopyrite in hydrothermal solutions. *Geochemistry* **1988**, *6*, 843–855. (In Russian)

46. Palyanova, G.A.; Kolonin, G.R. Arsenopyrite-containing mineral associations as indicators of physical and chemical conditions of hydrothermal ore formation. *Geochemistry* **1991**, *10*, 1481–1492. (In Russian)

47. Reich, M.; Kesler, S.; Utsunomiya, S.; Palenik, C.S.; Chryssoulis, S.L.; Ewing, R.C. Solubility of gold in arsenian pyrite. *Geochim. Cosmochim. Acta* **2005**, *69*, 2781–2796. [CrossRef]

48. Pokrovski, G.S.; Roux, J.; Ferlat, G.; Jonchiere, R.; Seitsonen, A.P.; Vuilleumier, R.; Hazemann, J.-L. Silver in geological fluids from in situ X-ray absorption spectroscopy and first-principles molecular dynamics. *Geochim. Cosmochim. Acta* **2013**, *106*, 501–523. [CrossRef]

49. Pokrovski, G.S.; Tagirov, B.R.; Schott, J.; Hazemann, J.-L.; Proux, O. A new view on gold speciation in sulfurbearing hydrothermal fluids from in situ X-ray absorption spectroscopy and quantum-chemical modeling. *Geochim. Cosmochim. Acta* **2009**, *73*, 5406–5427. [CrossRef]

50. Pokrovski, G.S.; Tagirov, B.R.; Schott, J.; Bazarkina, E.F.; Hazemann, J.-L.; Proux, O. An in situ X-ray absorption spectroscopy study of gold–chloride complexing in hydrothermal fluids. *Chem. Geol.* **2009**, *259*, 17–29. [CrossRef]

51. Mikhlin, Y.; Likhatski, M.; Tomashevich, Y.; Romanchenko, A.; Erenburg, S.; Trubina, S. XAS and XPS examination of the Au–S nanostructures produced via the reduction of aqueous gold(III) by sulfide ions. *J. Electron Spectrosc. Relat. Phenom.* **2010**, *177*, 24–29. [CrossRef]

52. Pokrovski, G.S.; Kara, S.; Roux, J. Stability and solubility of arsenopyrite, FeAsS, in crustal fluids. *Geochim. Cosmochim. Acta* **2002**, *66*, 2361–2378. [CrossRef]

53. Morey, A.A.; Tomkins, A.G.; Bierlein, F.P.; Weinberg, R.F.; Davidson, G.J. Bimodal Distribution of Gold in Pyrite and Arsenopyrite: Examples from the Archean Boorara and Bardoc Shear Systems, Yilgarn Craton, Western Australia. *Econ. Geol.* **2008**, *103*, 599–614. [CrossRef]

54. Xing, Y.; Brugger, J.; Tomkins, A.; Shvarov, Y. Arsenic evolution as a tool for understanding formation of pyritic gold ores. *Geology* **2019**, *47*, 335–338. [CrossRef]

55. Vilor, N.V.; Pavlova, L.A.; Kaz'min, L.A. Aresenopyrite–pyrite paragenesis in gold deposits (thermodynamic modeling). *Russ. Geol. Geophys.* **2014**, *55*, 824–841. [CrossRef]

56. Trigub, A.L.; Tagirov, B.R.; Kvashnina, K.O.; Lafuerza, S.; Filimonova, O.N.; Nickolsky, M.S. Experimental determination of gold speciation in sulfide-rich hydrothermal fluids under a wide range of redox conditions. *Chem. Geol.* **2017**, *471*, 52–64. [CrossRef]

57. Trigub, A.L.; Tagirov, B.R.; Kvashnina, K.O.; Chareev, D.A.; Nickolsky, M.S.; Shiryaev, A.A.; Baranova, N.N.; Kovalchuk, E.V.; Mokhov, A.V. X-ray spectroscopy study of the chemical state of "invisible" Au in synthetic minerals in the Fe-As-S system. *Am. Mineral.* **2017**, *102*, 1057–1065. [CrossRef]

58. Sazonov, A.M.; Ananyev, A.A.; Ananyeva, T.A.; Leontiev, S.I. Reflection of gold-bearing content of low-sulphide quartz veins in typomorphism of constituent minerals. *ZVMO* **1991**, *2*, 17–24. (In Russian)

59. Sazonov, A.M.; Zvyagina, E.A.; Krivoputskaya, L.M. Structural and chemical heterogeneity of pyrite of Sarala deposit (Kuznetsky Alatau). *Geol. Geophys.* **1992**, *8*, 87–95. (In Russian)

60. Fougerouse, D.; Micklethwaite, S.; Tomkins, A.G.; Mei, Y.; Kilburn, M.; Guagliardo, P.; Fisher, L.A.; Halfpenny, A.; Gee, M.; Paterson, D.; et al. Gold remobilisation and formation of high grade ore shoots driven by dissolution–reprecipitation replacement and Ni substitution into auriferous arsenopyrite. *Geochim. Cosmochim. Acta* **2016**, *178*, 143–159. [CrossRef]

61. Sazonov, A.M.; Onufrienok, V.V.; Kolmakov, Y.V.; Nekrasova, N.A. Pyrrhotite of gold-bearing ores: Composition, spot defects, magnetic properties, gold distribution. *J. Sib. Fed. Univ. Eng. Technol.* **2014**, *7*, 717–737. (In Russian)

62. Palyanova, G.A.; Sazonov, A.M.; Zhuravkova, T.V.; Silyanov, S.A. Composition of pyrrhotite as an indicator of mineralization conditions at the Sovetskoye gold deposit. *Russ. Geol. Geophys.* **2019**, *60*, 735–751. [CrossRef]

*Article*

# Distribution of "Invisible" Noble Metals between Pyrite and Arsenopyrite Exemplified by Minerals Coexisting in Orogenic Au Deposits of North-Eastern Russia

Vladimir Tauson, Sergey Lipko *, Raisa Kravtsova, Nikolay Smagunov, Olga Belozerova and Irina Voronova

A.P.Vinogradov Institute of Geochemistry, Siberian Branch of Russian Academy of Sciences, Irkutsk 664033, Russia; vltauson@igc.irk.ru (V.T.); krg@igc.irk.ru (R.K.); nicksm@igc.irk.ru (N.S.); obel@igc.irk.ru (O.B.); ivoronova@igc.irk.ru (I.V.)
* Correspondence: slipko@yandex.ru; Tel.: +7-3952-429-967

Received: 4 October 2019; Accepted: 24 October 2019; Published: 27 October 2019

**Abstract:** The study focused on the forms of occurrence and distribution of hidden ("invisible") noble metals (NMs = Au, Ag, Pt, Pd, Ru) in the coexisting pyrites and arsenopyrites of four samples of mineral associations from three Au deposits in the north-east of Russia. The unique nature of our approach was the combination of methods of local analysis and statistics of the compositions of individual single crystals of different sizes. This allowed us to take into account the contribution of the surface component to the total NM content and to distinguish the structurally bound form of the elements. The following estimates of the distribution coefficients of the structural (str) and surficial (sur) forms of elements were obtained: $\overline{D}^{str}_{Py/Asp}$ = 2.7 (Au), 2.5 (Pd), 1.6 (Pt), 1.7 (Ru) and $\overline{D}^{sur}_{Py/Asp}$ = 1.6 (Au), 1.1 (Pd), 1.5 (Pt and Ru). The data on Ag in most cases indicated its fractionation into pyrite ($\overline{D}^{str}_{Py/Asp}$ = 17). Surface enrichment was considered as a universal factor in "invisible" NM distribution. A number of elements (i.e., Pt, Ru, Ag) tended to increase their content with a decrease in the crystallite size in pyrite and arsenopyrite. This may be due to both the phase size effect and the intracrystalline adsorption of these elements at the interblock boundaries of a dislocation nature. The excess of metal (or the presence of S vacancies) in pyrite increased Ag and Pt content in its structure and, to a lesser extent, the content of Ru, Pd and Au. Arsenopyrite exhibited a clear tendency to increase the content of Pt, Ru and Pd in samples with excess As over S. Sulphur deficiency was a favourable factor for the incorporation of Ag and platinoids into the structures of the mineral associations studied. Perhaps this was due to the lower sulphur fugacity. Pyrite with excess Fe was associated with higher contents of some NMs. The presence of other impurity elements was not an independent factor in NM concentration.

**Keywords:** noble metals; invisible species; distribution; pyrite; arsenopyrite; structural and surficial modes

---

## 1. Introduction

The distribution of noble metals (NMs) between sulphide and arsenide minerals, particularly, pyrite and arsenopyrite, is of interest from several aspects. These minerals play an important role as concentrators of platinum group elements (PGEs) and Au in magmatic and high-temperature hydrothermal systems [1,2]. Although pyrite is unstable above 742 °C and arsenopyrite above 702 °C, these minerals are, in principle, capable of forming at the magmatic stage from sulphide melts containing low-melting-point chalcophile elements. It has been experimentally proven that pyrite can crystallise as a primary magmatic phase resulting from subsolidus reactions with the participation of monosulphide solid solution (mss) and intermediate solid solution (iss), and its formation does not require obligatory participation of hydrothermal processes [3]. However, pyrite and As-containing sulphides in Ni–Cu–PGE deposits are mainly considered as substitution products that inherit the rare-element composition of primary ores [4]. In Au deposits of different types, pyrite and arsenopyrite are common ore components, but, as a rule, of NMs, only Au is usually considered and analysed in their associations. Elevated PGE contents are associated with sulphides of metal-bearing black shales [5]; however, we were unable to find data for them concerning the platinoids contents in the coexisting pyrite and arsenopyrite.

The distribution of NMs between pyrite and arsenopyrite is directly related to the issues of the forms and limits of NM incorporation into these minerals. For example, if we consider Au, then according to Reich et al. [6], the maximum Au content in pyrite is related to its As content by linear dependence. It is assumed that this dependence satisfies the low-temperature conditions of pyrite crystallisation, Au solubility in pyrite being of retrograde nature and increasing with decreasing temperature [7]. If this is true, then the low-temperature pyrite–arsenopyrite associations are of great interest for assessment of the limit of Au incorporation into pyrite. In pyrite coexisting with arsenopyrite, the chemical potential of As and its content must be extremely high for the given combination of temperature and sulphur fugacity ($fS_2$) in which this paragenesis is stable. Arsenopyrite is a mineral of variable composition, and its As content depends on these parameters. In addition, it is believed that the PGE–As bond in pyrite, which is characterised by the presence of As, is energetically more advantageous as compared to the PGE–S bond in monosulphide–pyrrhotite and pentlandite [8]. Consequently, the highest concentrations of PGE can be observed in As–pyrite, which is in association with arsenopyrite. Most recently, experimental data on the unexpectedly high isomorphic capacity of pyrite with respect to platinum have appeared: pyrite crystals synthesised at 580 and 590 °C contained up to 4 wt % Pt [9]. X-ray absorption spectroscopy demonstrated that it replaces Fe in the form of $Pt^{2+}$ in the structure of pyrite. In this regard, there arises the question of why these opportunities are not realised in the natural environment, and what parameters and conditions determine the occurrence of NMs in these minerals.

PGEs exhibit bimodal distribution, forming discrete inclusions of the metal and its compounds with chalcogenides and pnictides, as well as solid solutions in sulphides, arsenides and sulpharsenides [10,11]. The form of PGE dissolved in the crystal lattices of these minerals is called the hidden form of the element [11]. This is one of the forms of occurrence of the element that is considered "invisible", i.e., beyond the possibility of diagnosis by optical microscopy. Such forms also include submicron inclusions of inherent phases of microelements (including nanoparticles), as well as elements present in the surface layers of minerals in the composition of surface non-autonomous phases or on the surface itself in the adsorbed state [12–14]. Invisible forms are related by mutual transformations. Thus, nanoparticles of $Ag_2S$ arise during quenching from 500 °C of pyrite crystals containing a structural admixture of Ag [14]; nanoparticles of $PtS_2$ are released upon cooling pyrite crystals that contain an isomorphic admixture of Pt [9]. The same processes are likely to lead to the formation of NM nanoparticles (<100 nm) in minerals from deposits of magmatic genesis [15]. Gold nanoparticles can be formed under conditions of pyrite undersaturation with Au due to the intracrystalline reduction reactions associated with changes in the chemical state of other elements (for example, As [16]). Similarly, elements chemically bonded in surface nanoscale phases can transform into nanoparticles when external physical and chemical conditions change, which also results in bimodal distribution of the element, but only on the mineral surface. In this case, Au nanoparticles (~5 nm) on the surface of sulphide minerals (e.g., arsenopyrite and chalcopyrite) are prone to aggregation by the self-assembly mechanism with the formation of microparticles smaller than 10 μm [17].

It is easy to understand that the distribution of invisible NM forms is influenced by many factors, including various types of crystal imperfections in mineral phase-point defects, boundaries of blocks and grains in crystals and their surfaces. For example, elevated levels of Au and other NMs in arsenopyrite are associated with its non-stoichiometry, or excessive S or As [18], and in pyrite, with the formation of a block submicroscopic structure at relatively low temperatures ($\leq$300 °C) [19]. These factors might be responsible for the appearance of invisible PGE forms even under low-temperature conditions, for example, in ores formed at temperatures around 100 °C and associated with sedimentary-diagenetic processes [11].

## 2. Previous Studies and Problem Statement

It is commonly believed that arsenide minerals are the best collectors of platinoids and Au compared to sulphides. However, this applies mainly to ores of magmatic genesis [20] or ore assemblages formed as a result of remobilisation of primary late magmatic arsenide-rich ores [1]. In these cases, the distribution coefficients of platinoids and Au between arsenides and sulphides are large enough, although they vary greatly for different elements. For example, calculations performed by Pina et al. [20] for the assumed equilibrium of maucherite ($Ni_{11}As_8$) with sulphides–pyrrhotite, pentlandite and chalcopyrite yielded the following values of the $D_i^{As/sulf}$ distribution coefficient: 330 (Pt), 250 (Pd), 50 (Ru), 310 (Au), 4 (Ag). Here As denotes arsenide, sulf–sulphide phases. However, if we consider other mineral occurrences that do not have a direct connection with magmatism, the picture will be somewhat different. For example, for the mineralisation of Mo–Ni–PGE in black shales [5], the distribution coefficients of invisible forms of Pt and Pd between pyrite and gersdorffite (NiAsS), which is the Ni-analogue of arsenopyrite, on average amount to ~0.9; that is, no significant fractionation of these elements in the sulpharsenide phase is detected. On the other hand, Kravtsova et al. [21] suggested that the increased content of PGEs and Au are typical of arsenopyrites formed at the hydrothermal stage, at the deposit of metamorphogenic–hydrothermal genesis, as confirmed by the rather high content of the structural forms of Pt and Pd in arsenopyrite. However, no data on the composition of pyrite coexisting with arsenopyrite were presented in this work.

A significant number of studies are devoted to Au in pyrite and arsenopyrite. We have undertaken generalisation of the selected data on the invisible Au in the coexisting pyrite and arsenopyrite [22–35]. The data we collected are presented in Table 1. It is believed that arsenopyrite contains more chemically bound Au than the pyrite formed together with it [36]. However, there is an opinion that "...there may be an as-yet undefined partitioning of Au between the two minerals when at equilibrium" ([22], p. 1279). As we mentioned in the Introduction, some authors believe that the solubility of Au and As in pyrite is retrograde: Au and As contents increasing with decreasing temperature [7]. The same effect is assumed for Au in arsenopyrite, based on the fact that Au is released from the arsenopyrite matrix when heated to form native Au [37]. As the experiment shows, at temperatures of 450–500 °C, the solubility of Au in pyrite with low As content (30–120 ppm) is 1–3 ppm, and only at high Mn content (~3000 ppm) does it increase to 7.3 ppm, possibly due to the heterovalent substitution $2Fe^{2+} = Mn^{3+} + Au^+$ [38,39].

Table 1. Literary data on "invisible" gold in coexisting pyrite and arsenopyrite.

| No | Deposit, Location | Type * | Formation Conditions | Ore Mineral Association ** | Au Content (ppm) | | As in Py (ppm) |
|---|---|---|---|---|---|---|---|
| | | | | | Py | Asp | |
| 1 | Shuiyindong, Guizhou, China | Carlin | $220 \pm 20\,°C$, 2–4 mol.% $CO_2$, low salinity | Py, Asp, Au | 300–600 | 300–1500 | 7000–12,400 |
| 2 | Bhukia-Jagpura Rajasthan, India | Magmatic-hydrothermal | $460$–$474\,°C$, $-\log f S_2 = 4.6$–$5.8$ bar, 24–26 wt % NaCl-eq. | Py, Asp, Po | 0.3–1.6 | 0.28–10 | 143–215 |
| 3 | Neves Norte, Alentejo, Portugal | VHMS | n/a | Py, Asp, Cpy, Gn, St, Tet | 1.7 | 16–35 | <2000 |
| 4 | Hutti, Dharwar Craton, India | Orogenic | $377$–$450\,°C$, ~2 kbar | Py, Asp, Po, Sph, Cpy, Au | 0.05–0.45 | 0.43–14.18 | 400–7500 |
| 5 | Suurikuusikko, Kittilä, Finland | VHMS | n/a | Py, Asp, Tet, Bour, Gn, Cpy, Sph, Rut, Au | <22–585 | <22–964 | ~200–40,000 |
| 6 | Lodestar, Newfoundland, Canada | Magmatic-hydrothermal | ~425 °C | Py, Asp, Cpy | 0.03–0.73 0.03–1.33 | 0.12–154 1.31–152 | n/a |
| 7 | Boliden, Skellefte, Sweden | VHMSMetamorphosed | ~400–450 °C, 5 kbar | Py, Asp, Po, Cpy, Gn, Au | ≤0.22 ≤0.19 | 3.8–108 ≤2.7 | n/a |
| 8 | Kundarkocha, Singhbhum, India | Orogenic | $375$–$390\,°C$, $-\log f S_2 = 7.5$–$9.7$ | Py, Asp, Po, Cpy, Sph, Pn, Gsd, Gn, Lö, Au | 2100–2700 1300–1500 700–900 | 1600–3600 1200–1300 900–1200 | 1000–2000 |
| 9 | Roudný, Bohemia, Czech Republic | Mesothermal | 320–330 °C | Py, Asp, Mrc, Au | 3–18 | 39–152 | 4000–25,000 |

**Table 1.** *Cont.*

| No | Deposit, Location | Type * | Formation Conditions | Ore Mineral Association ** | Au Content (ppm) | | As in Py (ppm) |
|---|---|---|---|---|---|---|---|
| | | | | | Py | Asp | |
| 10 | Campbell, Ontario, Canada | Metamorphic-hydrothermal | 450–500 °C, 2–3 kbar | Py, Asp, Gh, Po, Mt, Sph, Au° | 0.1–0.3<br>0.2–8.5 | 103–5610<br>118–4522 | n/a |
| 11 | Oberon, Tanami, Australia | Orogenic | 200–430 °C, low-to moderate salinity | Asp, Py, Po, Cpy, Sph, Gn, Au, Cb, Gsd | <1–4.6 | 2–4500 | 2000–4000 |
| 12 | Orvan Brook, Rocky Turn, Murray Brook, Bathurst Mining Camp, New Brunswick, Canada | Massive sulphide Zn–Pb–Cu–Ag | 325–425 °C, 4–6 kbar | Py, Sph, Gn, Cpy, Po, Asp, Mrc, Tet | <0.53–5.6<br>16.22–31.95<br>0.56–32.04 | 2.24–9.42<br>0.11–0.26<br>1.4–10.91 | bdl EPMA-9800 |
| 13 | Anba, Qinling Orogen, China | Carlin or Orogenic | n/a | Py, Asp, Ant, Mrc, Sph | 9.7–135.9 | 17.4–461.8 | 11,962–101,332 |
| 14 | Jinlongshan, Shaanxi, China | Carlin | n/a | Py, Asp | 1.9–9.4<br>21.1–113 | 4.2–429<br>64.6–576 | bdl EPMA-62,100 |

* As indicated in the original source. ** Py—pyrite, Asp—arsenopyrite, Po—pyrrhotite, Au—native gold, Cpy—chalcopyrite, Gn—galena, St—stannite, Tet—tetrahedrite, Sph—sphalerite, Bour—bournonite, Rut—rutile, Pn—pentlandite, Gsd—gersdorffite, Lö—loellingite, Mrc—marcasite, Gh—gahnite, Mt—magnetite, Cb—cobaltite, Ant—antimonite, VHMS—volcanogenic-hosted massive sulphide, EPMA—electron probe microanalysis. Note: n/a—not available, bdl—below detection limit. References: No.1—[23]; 2—[24]; 3—[25]; 4—[26]; 5—[27]; 6—[28]; 7—[29]; 8—[30]; 9—[31]; 10—[32]; 11—[22]; 12—[33]; 13—[34]; 14—[35].

This temperature range is close to the conditions of mineral formation in a number of orogenic and volcanogenic-hosted massive sulphide (VHMS) deposits. But, if in most of these relatively high-temperature deposits the content of invisible Au in pyrite is fully consistent with the above experimental data (Nos. 2, 3, 4, 6, 7, 10, 11 in Table 1), in several others they completely contradict them (Nos. 5, 8). Unrealistically high in comparison with experimental data, Au contents in pyrite are observed for the Carlin-type deposits and even for massive sulphide, non-Au ore deposits (Nos. 1, 12–14). The occurrence of Au in pyrite is unequivocally associated with the presence of As [6,7], but the data in Table 1 do not detect such an unambiguous relationship. Very high contents of invisible Au in pyrite are registered both at ordinary contents of As (No. 8) and at elevated ones (Nos. 1, 5). On the other hand, low Au contents can also be observed with rather high As contents (Nos. 4, 9, 11). As for arsenopyrite, it does show in most cases higher levels of invisible Au compared to the coexisting pyrite. Nevertheless, there are exceptions. At high Au contents in both minerals, the coefficient of its distribution between arsenopyrite and pyrite may be close to unity (Nos. 1, 5, 8), this being the case for significantly different types of deposits (Carlin and orogenic) and the conditions of their formation. At lower Au contents, this coefficient sometimes becomes even less than unity (No. 12: Rocky Turn, Murray Brook). In other cases, we have to state very large variations of its values—from one to three and even four orders of magnitude in favour of arsenopyrite (Nos. 10, 11).

The review of published data presented above shows that NM distribution between pyrite and arsenopyrite and the factors affecting it are poorly understood. More or less certain data are presented for Au only, although they are quite contradictory. As for PGEs, the available information suffices for qualitative analysis only, at the level of trends, and is not directly related to a specific mineral assemblage. The goal of this work is to study the NM speciation and distribution in coexisting pyrite and arsenopyrite from several orogenic Au deposits. The specific feature of our approach is the combination of methods of local analysis and statistics of the compositions of individual single crystals of different sizes. This allows us to take into account the contribution of the surface component to the total NM content and to distinguish the structurally bound form of the element according to the previously proposed analytical data selections for single crystals (ADSSC) technology [13].

## 3. Objects

Pyrite and arsenopyrite crystals for the study were collected from three Au objects: the Natalkinskoe, Degdekan and Zolotaya Rechka deposits. The deposits are located in the north-east of Russia (Magadan region, Tenkin ore district), are part of the Au-bearing Yano–Kolyma fold belt and are confined to its orogenic zones of collision origin. In terms of metallogenic properties, all of them are of the same type and belong to the Au–quartz low-sulphide formation of ores deposited in the upper Permian age carbonised strata. Figure 1 shows the location of the deposits within the territory of the Magadan region together with a simplified scheme of geologic aspects of the area of investigations.

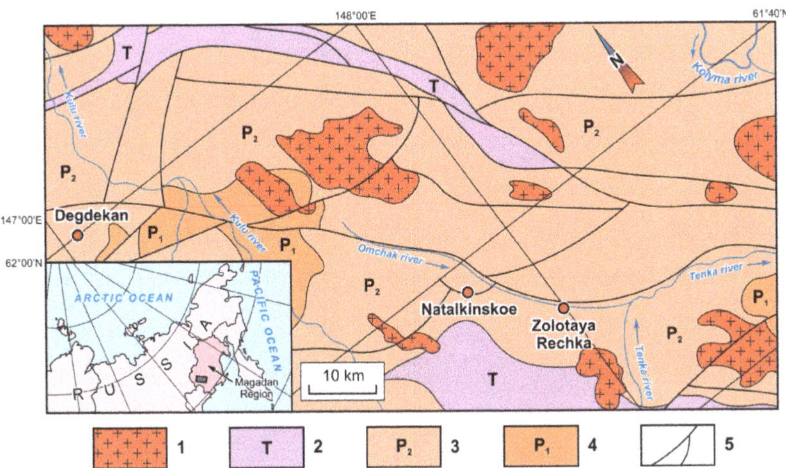



**Figure 1.** The Natalkinskoe, Degdekan and Zolotaya Rechka deposits on a simplified scheme of geologic aspects of the area of investigations. Compiled by the authors of the present paper using the data by Goncharov et al. [40]. The inset map shows the location of the deposits within the territory of the Magadan region. 1—granitoid massifs; 2—Triassic siltstones and mudstones; 3—upper Permian aleuropelitic shales with fragments of different composition rocks, aleuropelitic and carbono-argillaceous shales, carbonaceous siltstones, siltstones, sandstones, tuff sandstones and gravelites; 4—lower Permian mudstones, siltstones and sandstones; 5—faults.

Pyrite and arsenopyrite are the most common sulphide minerals in these deposits. We studied crystals from the widely distributed gold-bearing ores relatively rich in sulphides and belonging to the streaky-vein and veinlet-disseminated types (Figure 2). The mineral composition of ores was relatively constant (see below).

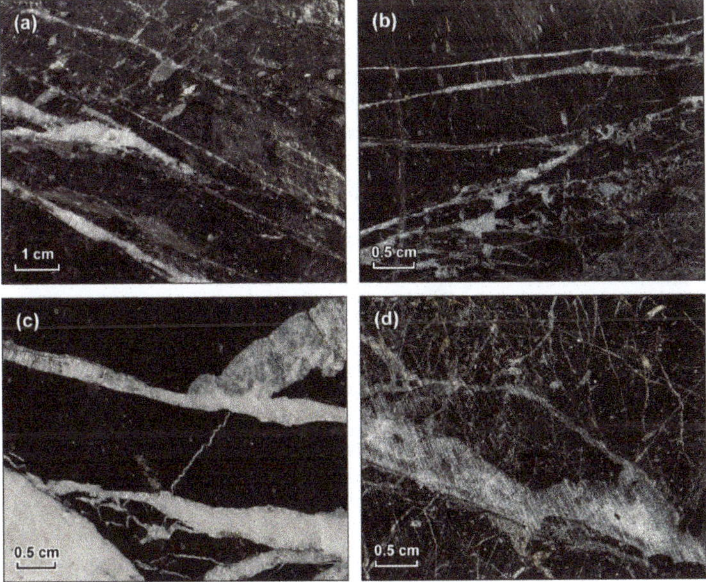



**Figure 2.** Gold-bearing streaky-vein and veinlet-disseminated ores. (**a**) Sample Nat-10 of the Natalkinskoe deposit, central area, quarry (level 787.5 m). Intensively sulphidized and carbonized aleuropelitic shales

with the inclusions of grey tufogenic material dissected with a dense network of quartz and quartz–carbonate veinlets of different widths, from hair like to 5 mm. Sulphide minerals are represented by disseminated pyrite and arsenopyrite. Pyrite crystals are cubes, pentagon dodecahedrons and cuboctahedrons, mainly up to 2 mm along an edge. Arsenopyrite crystals are represented by rhombic prisms and bipyramids, usually up to 0.5 mm along an edge. The combinations of simple forms are typical for both minerals. (**b**) Sample UV-3/13 of the Natalkinskoe deposit, south-eastern area, well DH70/5n, depth 160.1–163.1 m, level 590 m. Heavily sulphidized and carbonized siltstones and aleuropelitic shales with interlayered sandstones dissected with a network of quartz and quartz–carbonate veinlets of different width. The characteristics of ore minerals are the same as for Nat-10 except that pyrite predominates over arsenopyrite and the crystals of the later are coarser (up to 3 mm). (**c**) Sample DG-10/14 of the Degdekan deposit, Vernyi area, well DH221n, depth 231.9–233.9 m (level 470 m). Quartz veins and veinlets occur in heavily carbonized and sulphidized siltstones and characterized by a few centimetres and a few millimetres (up to 8 mm) width, accordingly. The sulphides are represented by impregnations, patches and fine stringers of pyrite. The coarse cubic crystals prevail (more than 2 mm in size). Arsenopyrite, mostly confined to veins and veinlets, is less common and represented by small rhombic-prismatic crystals (up to 0.5 mm). (**d**) Sample ZR-10/13 of the Zolotaya Rechka deposit, well DH444n, depth 307.4–310.3 m (level 310 m). Heavily fractured aleuropelitic shales dissected with a dense network of quartz veinlets of different widths, from hair like to 8–10 mm. Sulphides are represented by impregnations and patches of pyrite and arsenopyrite. Pyrite crystals are cubes and cuboctahedrons, mainly up to 2 mm along an edge. The combined forms are often observed. Arsenopyrite crystals are usually smaller (up to 0.5 mm) and have the habits of rhombic prisms and bipyramids.

The study here of PGEs, in addition to Au and Ag, the determination of their contents and forms of occurrence in ores and minerals, holds a special place in the research and is important not only theoretically but also practically. The presence in the ores of these NMs and the possibility of their extraction can supplement significantly the range of already known platinum-bearing ore formations and essentially increase the value of the extracted Au ore raw materials in the deposits where platinoids accompany Au mineralisation. The geological setting and mineralogical features of these objects are given in detail elsewhere [13,40–45]. Their brief characteristics are presented below.

*3.1. The Natalkinskoe Deposit*

The Natalkinskoe Au deposit is one of the largest in Russia by Au reserves. The deposit is embedded in stratified upper Permian sediments composed mainly of aleuropelitic shales with fragments of different composition rocks, carbono-argillaceous shales, siltstones, sandstones and gravelites. The ores belong to the Au-sulphide-impregnated type and is characterised by a complex metamorphogenic–hydrothermal genesis [40]. They make up a uniform in the internal structure load, consisting of a framework of extended linear zones of quartz, carbonate–quartz and sulphide–quartz veins and veinlets surrounded by a wide halo of sulphidized rocks. In general, the deposit is traced along the strike at a distance of 5 km, with a width of 1 km. According to K/Ar dating, the age of Au mineralisation is estimated to be from 135–130 m.y. up to 110–100 m.y., i.e., as early Cretaceous [40]. However, according to Ar–Ar sericite dating, the age of Au ores of the Natalkinskoe deposit is 135.2 m.y. [41].

Arsenopyrite and pyrite are the most common sulphide minerals in the deposit. In the ore samples Nat-10 and UV-3/13 from which they were selected, non-metallic minerals such as quartz, carbonates, feldspars, sericite and chlorite were identified. Of the ore minerals, along with arsenopyrite and pyrite, the amount of which was about 4–7%, galena, sphalerite, chalcopyrite, native Au and rutile (less than 1%) were detected. The difference between these two samples lies in the proportion of arsenopyrite and pyrite: domination of arsenopyrite over pyrite in Nat-10 and vice versa in UV-3/13. Native gold with a fineness of 750–900‰, less often low-grade (electrum), mainly coarse-grained, occurs in vein quartz and in agglomerates with sulphide minerals. Silver in its own mineral form is very rare. The most productive are pyrite–arsenopyrite mineral assemblage with Au and galena [40]. Commercial ores emerged in the interaction of host rocks with low- and moderate-salinity water–bicarbonate fluids in the salinity range of 3–12 wt % NaCl-eq., at temperatures of 360–280 °C and pressures of approximately 2.4–1.1 kbar [42]. Two samples of coexisting pyrite and arsenopyrite are considered below. The first (Nat-10) was collected in the "altarnativnyi" area of the deposit, the second (UV-3/13) in the "south-east" area. Both originated from veinlet-impregnated ores and contained from 0.5 up to 22.0 ppm of Au.

### 3.2. The Degdekan Deposit

The Degdekan Au deposit is an example of the localisation of zones of vein and veinlet-impregnated sulphide ores in stratified upper Permian sediments, composed mainly of carbonaceous siltstones and shales with a large number of interlayers of graphitised host rocks. Ore mineralisation is formed in two stages: hydrothermal–metamorphogenic and hydrothermal [43]. The absolute age of Au mineralisation by U-Pb-SHRIMP is estimated at 133–137 m.y., and by the Ar–Ar method at 137 m.y., i.e., the formation of ore refers to the beginning of the early Cretaceous [44,45]. The main vein mineral is quartz. In the host rocks, besides silification, albitisation and sericitisation are observed, carbonates (ankerite, calcite) and chlorite are found. The main ore minerals are pyrite and arsenopyrite subordinate to it, forming a small (1–2 mm) impregnation in the altered sedimentary rocks. The content of these sulphides amounts on average to about 3%. In addition to these minerals, sphalerite, galena, pyrrhotite and chalcopyrite are found, gersdorffite is rare. Native gold forms the grains varying in size from 0.1 to 0.5 mm. Its fineness varies within the range 740–800‰. Electrum and kustelite are extremely rare [43,44]. The Au–arsenopyrite–polymetallic (with pyrite) mineral assemblage stands out as a productive one [43]. The temperature interval of its formation (according to the study of fluid inclusions) is estimated at 200–230 °C, the pressure is about 1 kbar. The solutions were weakly mineralised (25 g/L) and were mostly of sodium hydrocarbonate composition, which indicates their amagmatogenic (metamorphogenic–hydrothermal) origin. The sample containing pyrite and arsenopyrite in the Degdekan deposit was collected from the veinlet-impregnated ores with an Au content from 1.4 to 15.2 ppm.

### 3.3. The Zolotaya Rechka Deposit

The Zolotaya Rechka deposit is insufficiently explored. There are no data in the published sources. The geological description and mineralogy of the deposit are presented here on the basis of the materials of unpublished geological foundations; the material collection is supplemented with data obtained by the authors of this article. The Zolotaya Rechka deposit is located in the arch of the anticline composed of weakly metamorphosed and intensely dislocated terrigenous sediments of the upper Permian age. They are dark-grey to black, aleuropelite, carbonaceous–argillaceous shales with interbeds and lenses of sandstone and sand tuffs. This deposit, similar to Natalkinskoe, is confined to the central zone of the Tenkin deep fault but is situated somewhat to the south. By the structural and morphological characteristics, the deposit belongs to the type of veinlet-mineralised zones; by genesis, as are Natalkinskoe and Degdekan, it is classed as metamorphogenic–hydrothermal. Veinlet and disseminated sulphide mineralisation (no more than 3%) are represented by pyrite and arsenopyrite; sometimes, there are occurrences of chalcopyrite. Native gold with a fineness of 750‰–800‰, mainly fine (0.1–0.8 mm), is mostly in vein quartz and in joints with sulphide minerals. Silver in its own mineral form is very rare. It is present mainly in the form of micro-inclusions of argentite (acanthite) and native silver in arsenopyrite and pyrite. The pyrite–arsenopyrite mineral assemblage with Au stands out as productive. The sample with coexisting pyrite and arsenopyrite (ZR-10/13) was selected from veinlet-mineralised zones with Au content of 1.5–13.1 ppm.

## 4. Methods

### 4.1. Electron Probe Microanalysis (EPMA)

Pyrite and arsenopyrite crystals (in well-polished epoxy pellets) were studied by EPMA using a Superprobe JXA-8200 (JEOL Ltd., Tokyo, Japan) microprobe supplied with energy-dispersive (EDS) (JEOL Ltd., Tokyo, Japan) and wavelength-dispersive spectrometers (WDS) (JEOL Ltd., Tokyo, Japan). Analysis by EDS showed qualitatively the presence of Cu, Ag, Au and Pt traces in addition to the major elements (Fe, S and As). Quantitative analyses by WDS were conducted at an accelerating voltage of 20 kV, beam current 20 nA, probe diameter 1 μm, and a counting time of 10 s for major and 30 s for trace elements. Matrix corrections (atomic number, absorption, and fluorescence) and analysed element contents were calculated using the ZAF (atomic number, absorption, and fluorescence) approach applying the software of quantitative analysis for Superprobe JXA-8200 (V01.42 Copyright (C) 2000-2007, JEOL Ltd., Tokyo, Japan). Bulk polished samples of Au–Ag alloys (fineness 700 and 750), metal Pt and minerals of known stoichiometric composition ($FeS_2$, $FeAs_{1.00}S_{0.99}$, $CuFeS_2$) were used as reference samples for WDS quantitative analysis.

Pyrite crystals were studied for homogeneity in back scattered electrons (BSE) using the scanning electron microscope of a Superprobe JXA-8200 device. The pyrite crystals displayed distinct inhomogeneity. They consisted of alternating zones of more or less bright grey colour in BSE images, corresponding to slightly higher and lower average Z.

Measurements were made at 10–20 points of each grain, depending on its homogeneity. Trace elements identified by EDS were found to be below minimum detection limits (MDLs, 0.1–0.12 wt %) and only several points showed their valuable contents. These elements were not taken into account in the final stage of data processing. Only the analyses with sums within the limits of 99.5–100.5 wt % were used. The data obtained are presented in Table 2 and analysed in detail in Section 5.1.

### 4.2. X-Ray Diffraction (XRD)

Unit cell edges and mean coherent scattering domain sizes (crystallite sizes) of powdered pyrite and arsenopyrite crystals were measured on a D8 ADVANCE diffractometer (Bruker, Karlsruhe, Germany) using CuKα radiation. Crystallite sizes were determined by the Scherrer method using EVA software (DIFFRAC Plus Evaluation package EVA; user's Manual, Bruker AXS, 2007, Karlsruhe, Germany). The results are presented in Table 3. Uncertainties in unit cell edges are at the level of $\pm 1\text{–}2 \times 10^{-5}$ nm, crystallite sizes $\pm 1\text{–}2$ nm. These values were verified in parallel measurements.

### 4.3. Atomic Absorption Spectrometry with ADSSC Data Processing (AAS-ADSSC)

The ADSSC procedure is described in detail in our recent publications [13,46]. Following this procedure, we picked up from the ore samples euhedral crystals of different sizes with clean, faultless faces. Pyrite crystals were predominantly cubes, whereas the arsenopyrite crystals used in this work were mainly of rhombic form. The crystal habit control is an important property of the ADSSC approach because the calculation of the specific surface area of an average crystal ($\bar{S}_{sp}$) in every size fraction requires information on the form coefficient k (see Equation (3) in Reference [13]). The crystal shape is approximated by a true polyhedron. Each crystal was weighed on an analytical microbalance and transferred into beakers for subsequent dissolution. We used only crystals with mass exceeding 0.1 mg.

Noble metal concentrations were determined by AAS with electrothermal element atomisation in graphite furnace (AAS-GF). The AAS measurements were performed on PerkinElmer devices (Model 503 and Analyst 800, PerkinElmer Corp., Branford, CT, USA). Gold and Ag were determined directly from the solution after acid crystal decomposition in aqua regia and creation of the required chemical medium. Measurements were accurate to ±12 and ±10% with MDLs of 0.3 μg/L (0.3 ppb) and 0.5 μg/L (0.5 ppb) for Au and Ag, respectively. Platinum, Pd and Ru were determined after preliminary extraction by concentration and separation from the matrix [13]. Tristyrylphosphine

(C$_6$H$_5$CH–CH)$_3$P was used as extracting agent. The organic phase was used to measure element concentrations. Measurements were accurate to ±10% with MDLs of 5 µg/L (5 ppb) for Pd and 50 µg/L (50 ppb) for Pt and Ru.

The data obtained were processed statistically according to the regulations of the distribution of different NM binding forms [13,46]. We divided the dataset (usually 40–60 crystals with evenly distributed NMs) into the intervals of crystal masses (sizes), chosen to be as narrow as possible although statistically representative, and determined average crystal mass in every size fraction ($\overline{m}$), average size ($\overline{r}$), specific surface area ($\overline{S}_{sp}$) and NM concentration ($\overline{C}_{NM}$) (see Table S1 in the Supplementary Materials). The more size fractions and number of crystals in the final samples, the more reliable the results obtained, unless at the expense of selection quality. When constructing the dependences $\overline{C}_{NM} = f(\overline{S}_{sp})$, we usually obtained a number of points best approximated with an exponent (Figures 5–9). The extrapolation of these curves to a zero-specific surface, i.e., to a virtual infinite crystal, gave the bulk NM content. In our model, this was equal to structurally bound NM content ($C^{str}$), because all other possible bulk modes were eliminated at the stage of initial dataset processing. The superficially bound NM content ($C^{sur}$) characterises an average crystal among all size samples, that is, the surface-related excess concentration of the element, and can be calculated with the equation given in Reference [13]. It is important to note that in that formula, the amount of the material is hereby normalised to the whole crystal, and this allows comparison of the contribution of each binding form of the element to its total concentration. The results of $C^{str}$ and $\overline{C}^{sur}$ determination and corresponding NM distribution coefficients between coexisting pyrite and arsenopyrite are shown in Table 4 and are considered in detail throughout the present article.

### 4.4. Laser Ablation Inductively Coupled Plasma Mass Spectrometry (LA-ICP-MS)

For this type of analysis, we took the same polished crystals in epoxy pellets used for the EPMA study. Measurements were performed on an Agilent 7500ce unit manufactured by Agilent Technologies with quadrupole mass analyser (Agilent Tech., Santa Clara, CA, USA) in the Center of Collective Use "Ul'tramikroanaliz" of the Limnological Institute, SB RAS (Irkutsk, Russia). The laser ablation platform of a New Wave Research UP-213 was used. Parameters of the LA-ICP-MS experiment: plasma power 1400 W, carrier gas flow rate 1 L/min, plasma forming gas flow rate 15 L/min, cooling gas 1 L/min, laser energy 0.63 mJ, frequency 20 Hz, and laser spot diameter 55 µm. Accumulation time per channel was 0.3 s and acquisition time 25 s. Measurements were made at 15 points in several (3–5) grains of each sample. Calculations of concentrations were based on the standard sample NIST 612 conformed for a number of elements with the in-house sulphide standard sample—highly homogeneous ferrous greenockite (α-CdS) crystals with many incorporated elements synthesised hydrothermally at 500 °C and 1 kbar and carefully analysed by different methods. Standard deviations are consistent with counting statistic uncertainties at around 30%. Error analysis is rated as 30%. The data obtained were processed in the manner of the ADSSC approach [38,46] to evaluate evenly distributed element concentrations identified with structurally bound impurities. The results are presented in Tables 5 and 6 for pyrite and arsenopyrite, respectively. They are considered in detail in Section 5.4 and Section 6.5.

## 5. Results

### 5.1. EPMA

The data obtained are presented in Table 2. The pyrite crystals demonstrated regions enriched with As (As–Py) and areas not containing As at the level of EPMA sensitivity, which herewith will be referred to as conventionally "pure" pyrite (Py). These areas stand out as darker sectors in the BSE images (Figure 3), but there were also cases when differences between Py and As–Py were established only by quantitative analysis. As a rule, areas of Py predominance are characterised by higher porosity (Figure 3).

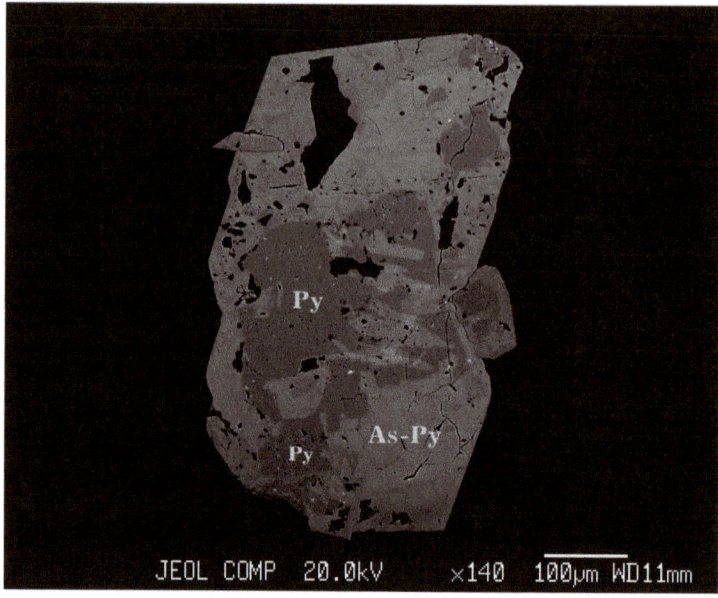



**Figure 3.** The scanning image in back scattered electrons (BSEs) of pyrite crystal with coexisting "pure" (As-free) and As–pyrite (sample ZR-10/13). Note that more dark areas corresponding to "pure" pyrite display elevated porosity.

In the mono-mineral grains of arsenopyrite of the Nat-10 sample from the Natalkinskoe deposit, the As/S ratio equalled 1.14 ± 0.01 (Table 2).

In the poly-mineral grains consisting of arsenopyrite, pyrite and As–pyrite (Asp + Py + As − Py), it was slightly higher (1.16 ± 0.02); coexisting with arsenopyrite, As–pyrite and "pure" pyrite are practically stoichiometric: Fe/(S + As) and Fe/S were 0.499 ± 0.001 and 0.496 ± 0.004 respectively. Arsenic content in As–pyrite was 0.60 ± 0.2 at.%. For arsenopyrite of the UV-3/13 sample from the south-eastern section of the Natalkinskoe deposit, As/S ratio indicated a greater deviation from stoichiometry (1.20 ± 0.02). This is most likely to indicate lower S activity during its formation, since the As content in the As–Py coexisting with arsenopyrite was practically the same as in the Nat-10 sample, equal to 0.53 ± 0.2 at. %, and Fe content in As–Py increased from 0.998 to 1.018 atoms per formula unit or from 33.28 to 33.93 at. % Fe. Consequently, As–pyrite associated with Asp is characterised by a higher ratio of Fe/(S + As) − 0.513 ± 0.001. "Pure" pyrite was not identified in this sample. The sample from the Zolotaya Rechka deposit, ZR-10/13, was similar to UV-3/13—it did not contain polyphase grains containing Asp either—and As–Py and "pure" Py were iron-redundant (0.513 ± 0.001 and 0.515 ± 0.001, respectively); As content in As–Py was 0.57 ± 0.3 at. %. This sample differed from the one mentioned above only by a more stoichiometric composition of arsenopyrite, i.e., the ratio As/S being closer to unity, at 1.14 ± 0.02. The minerals of the DG-10/14 sample of the Degdekan deposit were closest to stoichiometry: As/S in Asp was 1.08 ± 0.01, Fe/(S + As) in As–Py 0.496 ± 0.004, and Fe/S in Py 0.501. Arsenic content in As–Py was at the same level as in all studied samples at 0.47 ± 0.3 at. %. The above uncertainties were calculated for 1σ with a confidence probability of 90% for small samples and 95% for more representative datasets (arsenopyrites Nat-10, DG-10/14, Table 2).

**Table 2.** EPMA data for arsenopyrite and pyrite crystals.

| Sample | Grain No. | Arsenopyrite | | | As–Pyrite | | | | Pyrite | | |
|---|---|---|---|---|---|---|---|---|---|---|---|
| | | $n$ * | Formula | As/S | $n$ * | Formula | Fe/(S + As) | As/S | $n$ * | Formula | Fe/S |
| Nat-10 | 1 | 3 | $Fe_{0.999}As_{1.079}S_{0.922}$ | 1.17 | - | - | - | - | | - | - |
| | 2 | 4 | $Fe_{0.993}As_{1.077}S_{0.930}$ | 1.16 | - | - | - | - | | - | - |
| | 3 | 4 | $Fe_{0.987}As_{1.069}S_{0.944}$ | 1.13 | - | - | - | - | | - | - |
| | 4 | 4 | $Fe_{0.989}As_{1.065}S_{0.946}$ | 1.13 | - | - | - | - | | - | - |
| | 5 | 8 | $Fe_{0.993}As_{1.066}S_{0.941}$ | 1.13 | - | - | - | - | | - | - |
| | 6 | 7 | $Fe_{0.997}As_{1.069}S_{0.934}$ | 1.14 | - | - | - | - | | - | - |
| | 7 | 5 | $Fe_{0.987}As_{1.059}S_{0.954}$ | 1.11 | - | - | - | - | | - | - |
| | 8 | 3 | $Fe_{0.994}As_{1.060}S_{0.946}$ | 1.12 | - | - | - | - | | - | - |
| | 9 | 5 | $Fe_{0.987}As_{1.077}S_{0.936}$ | 1.15 | - | - | - | - | | - | - |
| | 10 | 5 | $Fe_{0.994}As_{1.069}S_{0.937}$ | 1.14 | - | - | - | - | | - | - |
| | 11 | 4 | $Fe_{0.992}As_{1.088}S_{0.920}$ | 1.18 | - | - | - | - | | - | - |
| | 12 | 2 | $Fe_{0.981}As_{1.098}S_{0.921}$ | 1.19 | 4 | $Fe_{0.998}S_{1.994}As_{0.008}$ | 0.499 | 0.004 | 2 | $FeS_{2.006}$ | 0.499 |
| | 13 | 2 | $Fe_{0.973}As_{1.085}S_{0.942}$ | 1.15 | 6 | $Fe_{0.998}S_{1.988}As_{0.014}$ | 0.499 | 0.007 | 1 | $FeS_{2.04}$ | 0.490 |
| | 14 | 1 | n/d | - | 3 | $Fe_{0.997}S_{1.975}As_{0.028}$ | 0.498 | 0.014 | 3 | $FeS_{2.021}$ | 0.495 |
| | 15 | 2 | $Fe_{0.991}As_{1.084}S_{0.925}$ | 1.17 | 5 | $Fe_{0.999}S_{1.984}As_{0.017}$ | 0.499 | 0.009 | - | - | - |
| | 16 | 1 | $Fe_{0.979}As_{1.076}S_{0.945}$ | 1.14 | 3 | $Fe_{0.998}S_{1.983}As_{0.019}$ | 0.499 | 0.01 | 3 | $FeS_{2.009}$ | 0.498 |
| | 17 | 4 | $Fe_{0.976}As_{1.076}S_{0.948}$ | 1.14 | 5 | $Fe_{1.000}S_{1.976}As_{0.024}$ | 0.500 | 0.012 | 3 | $FeS_{2.006}$ | 0.499 |
| UV-3/13 | 1 | 8 | $Fe_{0.982}As_{1.101}S_{0.917}$ | 1.20 | - | - | - | - | - | - | - |
| | 2 | 8 | $Fe_{0.985}As_{1.093}S_{0.922}$ | 1.19 | - | - | - | - | - | - | - |
| | 3 | 8 | $Fe_{0.986}As_{1.110}S_{0.904}$ | 1.23 | - | - | - | - | - | - | - |
| | 4 | 7 | $Fe_{0.987}As_{1.087}S_{0.926}$ | 1.17 | - | - | - | - | - | - | - |
| | 5 | 5 | $Fe_{0.991}As_{1.094}S_{0.915}$ | 1.20 | - | - | - | - | - | - | - |
| | 6 | 7 | $Fe_{0.990}As_{1.097}S_{0.913}$ | 1.20 | - | - | - | - | - | - | - |
| | 7 | - | - | - | 9 | $Fe_{1.019}S_{1.961}As_{0.020}$ | 0.514 | 0.01 | - | - | - |
| | 8 | - | - | - | 13 | $Fe_{1.015}S_{1.964}As_{0.021}$ | 0.511 | 0.01 | - | - | - |
| | 9 | - | - | - | 12 | $Fe_{1.019}S_{1.973}As_{0.007}$ | 0.515 | 0.004 | - | - | - |
| | 10 | - | - | - | 16 | $Fe_{1.018}S_{1.970}As_{0.012}$ | 0.514 | 0.006 | - | - | - |
| | 11 | - | - | - | 9 | $Fe_{1.017}S_{1.963}As_{0.020}$ | 0.513 | 0.01 | - | - | - |

**Table 2.** *Cont.*

| Sample | Grain No. | Arsenopyrite | | | As–Pyrite | | | | Pyrite | | |
| --- | --- | --- | --- | --- | --- | --- | --- | --- | --- | --- | --- |
| | | | | | Average Grain Composition (Atomic Content) | | | | | | |
| | | $n$ * | Formula | As/S | $n$ * | Formula | Fe/(S + As) | As/S | $n$ * | Formula | Fe/S |
| DG-10/14 | 1 | 7 | $Fe_{0.992}As_{1.054}S_{0.954}$ | 1.10 | - | - | - | - | - | - | - |
| | 2 | 5 | $Fe_{0.999}As_{1.016}S_{0.985}$ | 1.03 | - | - | - | - | - | - | - |
| | 3 | 5 | $Fe_{0.989}As_{1.056}S_{0.955}$ | 1.11 | - | - | - | - | - | - | - |
| | 4 | 5 | $Fe_{0.989}As_{1.037}S_{0.974}$ | 1.06 | - | - | - | - | - | - | - |
| | 5 | 4 | $Fe_{0.999}As_{1.044}S_{0.957}$ | 1.09 | - | - | - | - | - | - | - |
| | 6 | 3 | $Fe_{0.988}As_{1.050}S_{0.962}$ | 1.09 | - | - | - | - | - | - | - |
| | 7 | 4 | $Fe_{0.988}As_{1.054}S_{0.958}$ | 1.10 | - | - | - | - | - | - | - |
| | 8 | 7 | $Fe_{0.992}As_{1.034}S_{0.974}$ | 1.06 | - | - | - | - | - | - | - |
| | 9 | 7 | $Fe_{0.991}As_{1.043}S_{0.966}$ | 1.08 | - | - | - | - | - | - | - |
| | 10 | 6 | $Fe_{0.991}As_{1.041}S_{0.968}$ | 1.08 | - | - | - | - | - | - | - |
| | 11 | 5 | $Fe_{0.988}As_{1.046}S_{0.966}$ | 1.08 | - | - | - | - | - | - | - |
| | 12 | 6 | $Fe_{0.998}As_{1.049}S_{0.953}$ | 1.10 | - | - | - | - | - | - | - |
| | 13 | 6 | $Fe_{0.997}As_{1.027}S_{0.976}$ | 1.05 | - | - | - | - | - | - | - |
| | 14 | 2 | $Fe_{0.980}As_{1.050}S_{0.970}$ | 1.08 | 6 | $Fe_{0.992}S_{1.996}As_{0.011}$ | 0.494 | 0.006 | 5 | $FeS_{2.003}$ | 0.499 |
| | 15 | - | - | - | 6 | $Fe_{0.999}S_{1.984}As_{0.018}$ | 0.499 | 0.009 | - | - | - |
| | 16 | - | - | - | 5 | $Fe_{0.995}S_{1.993}As_{0.012}$ | 0.496 | 0.006 | 2 | $FeS_{1.988}$ | 0.503 |
| ZR-10/13 | 1 | 7 | $Fe_{0.997}As_{1.061}S_{0.942}$ | 1.13 | - | - | - | - | - | - | - |
| | 2 | 8 | $Fe_{0.990}As_{1.067}S_{0.943}$ | 1.13 | - | - | - | - | - | - | - |
| | 3 | 15 | $Fe_{0.994}As_{1.083}S_{0.923}$ | 1.17 | - | - | - | - | - | - | - |
| | 4 | 14 | $Fe_{0.996}As_{1.075}S_{0.929}$ | 1.16 | - | - | - | - | - | - | - |
| | 5 | 15 | $Fe_{0.992}As_{1.075}S_{0.933}$ | 1.15 | - | - | - | - | - | - | - |
| | 6 | 14 | $Fe_{0.987}As_{1.064}S_{0.949}$ | 1.12 | - | - | - | - | - | - | - |
| | 7 | - | - | - | 11 | $Fe_{1.016}S_{1.968}As_{0.016}$ | 0.512 | 0.008 | 4 | $FeS_{1.944}$ | 0.514 |
| | 8 | - | - | - | 11 | $Fe_{1.018}S_{1.972}As_{0.010}$ | 0.514 | 0.005 | 5 | $FeS_{1.944}$ | 0.514 |
| | 9 | - | - | - | 11 | $Fe_{1.017}S_{1.971}As_{0.012}$ | 0.513 | 0.006 | - | - | - |
| | 10 | - | - | - | 8 | $Fe_{1.019}S_{1.968}As_{0.013}$ | 0.514 | 0.007 | 8 | $FeS_{1.944}$ | 0.514 |
| | 11 | - | - | - | 7 | $Fe_{1.019}S_{1.969}As_{0.011}$ | 0.515 | 0.006 | 6 | $FeS_{1.935}$ | 0.517 |
| | 12 | - | - | - | 6 | $Fe_{1.015}S_{1.944}As_{0.040}$ | 0.512 | 0.021 | 4 | $FeS_{1.941}$ | 0.515 |

* Number of analysis points. Only the results of analyses with sums from 99.5 to 100.5 wt % were taken into consideration. n/d—not determined.

Thus, according to the EPMA, there were two types of As–pyrite as well as "pure" (no As) pyrite: stoichiometric pyrite with an Fe/(S + As) ratio approximately 0.5 and metal-surplus pyrite with a ratio of 0.513–0.515. It is interesting to note that coexisting minerals obeyed the principle of phase composition correlation: if As–Py was iron-excessive, "pure" pyrite was also non-stoichiometric; for their part, the stoichiometric As–Py coexisted with stoichiometric "pure" pyrite. Besides, the As distribution coefficient between arsenopyrite and As–pyrite ($D_{As}^{Asp/Py}$) was relatively constant and, on average, amounted to 66 ± 7 (in atomic concentration units) for the four associations studied.

Figure 4 exhibits two types of S-As dependence in pyrites of different samples. Negative correlation of formula units S and As (Figure 4) indicates a predominantly isomorphic substitution of S for As in the structure of As–pyrite. The dependence S (As) was approximated by a straight line: $S = a - k As$. For stoichiometric pyrite of the Natalkinskoe and Degdekan deposits $k = 1.21$ and the line crossed the x-axis at $a = 2.006$. In the case of non-stoichiometric, S-deficient pyrite of the Zolotaya Rechka and the south-eastern sector of the Natalkinskoe deposit, the line crossed the x-axis at 1.98, which does not correspond to the stoichiometry of "pure" pyrite (Figure 4). The reason for this discrepancy is not quite clear. However, our preliminary X-ray photoelectron spectroscopic data (unpublished) show that ~10–15% of As can be incorporated in pyrite as As(II) (see also Reference [47]), and this corresponds to a "stoichiometric trend" with $k = 1.21$. The "nonstoichiometric trend" corresponded to $As^{1-}$ only and was realized under lower sulphur fugacity. It might be that in the first case (DG-10/14, Nat-10), the formation of arsenic and "pure" pyrite occurred under the same conditions, whereas, in the second (ZR-10/13, UV-3/13), their formation occurred not only at lower sulphur fugacity in the system than in the first case but also at levels different from those for "pure" and "As–pyrite". It is interesting to note that different conditions and corresponding S–As trends are the characteristics of the sections of one and the same deposit (Natalkinskoe, samples Nat-10 and UV-3/13).

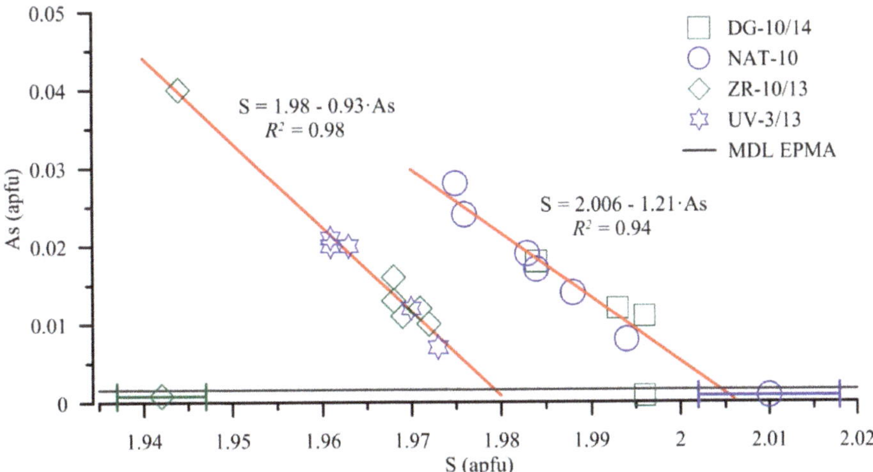



**Figure 4.** Two types of S–As correlation in As–pyrite. The "stoichiometric trend" gives Fe/S = 2.006 when extrapolated to "pure" (As-free) pyrite, whereas the "nonstoichiometric trend" gives Fe/S = 1.98 and, moreover, does not correlate with "pure" pyrite which is still more nonstoichiometric.

## 5.2. XRD

Table 3 presents X-ray data on the unit cell edges of arsenopyrite and pyrite and the average crystallite sizes or block sizes of the real mosaic structure of crystals of these minerals. The unit cell parameters differ only in the third (*a*, *c*) and fourth (*b*) decimal places for arsenopyrite (±0.0018 and ±0.0008 nm, respectively) and in the fourth decimal place for pyrite (±0.0004 nm). The average sizes of the crystallites, in contrast, differ less in arsenopyrite (within 9 nm) compared to pyrite (within 16 nm).

**Table 3.** Results of the X-ray diffraction study of arsenopyrite and pyrite crystals.

| Sample Number | Arsenopyrite (Monoclinic, P21/c) | | | | | Pyrite (Cubic, Pa3) | |
|---|---|---|---|---|---|---|---|
| | *a* | *b* | *c* | $\beta^o$ | *D* | *a* | *D* |
| Nat-10 | 0.57693 | 0.56766 | 0.57395 | 112.087 | 61 | 0.54208 | 76 |
| UV-3/13 | 0.57425 | 0.56669 | 0.57591 | 111.898 | 56 | 0.54227 | 65 |
| DG-10/14 | 0.57592 | 0.56636 | 0.57327 | 111.881 | 58 | 0.54192 | 71 |
| ZR-10/13 | 0.57357 | 0.56753 | 0.57648 | 111.979 | 54 | 0.54141 | 60 |

Note: lattice parameters (*a*, *b*, *c*) and crystallite size (*D*) are in nanometres.

### 5.3. AAS-ADSSC

The AAS data of the analysis of individual crystals processed using the ADSSC method are presented in Table S1 in the Supplementary Materials and shown graphically in Figures 5–9 in the form of dependences of the average uniformly distributed NM concentration (Au, Pd, Pt, Ru and Ag) on the specific surface of the average crystal in the size sample. In each case, the analytical form of this dependence was demonstrated: the concentration of structurally bound NM admixture was obtained as an extrapolation of the dependence on the zero-specific surface area, i.e., on a conditionally infinite crystal, and the superficially bound mode was calculated as shown above (Section 4.3, see also Reference [13]). These data, as well as NM distribution coefficients of both forms (i.e., structural and superficial), are presented in Table 4, which also shows the average compositions of Asp and As–Py for the key elements (Fe, As and S).

**Table 4.** Noble metals contents as structurally and surficially bound modes, their distribution coefficients between pyrite and arsenopyrite, and average matrix composition of minerals.

| Sample Number | Element | Element Content (ppm) | | | | D$^{Py/Asp}$ | | Matrix Composition (at. %) Asp/As–Py * | | |
|---|---|---|---|---|---|---|---|---|---|---|
| | | Pyrite | | Arsenopyrite | | | | | | |
| | | Str. | Sur. | Str. | Sur. | Str. | Sur. | As | S | Fe |
| Nat-10 | Au | 1.8 | 2.7 | 1.2 | 1.9 | 1.5 | 1.4 | | | |
| | Ag | 12 | 220 | 120 | 250 | 0.1 | 0.9 | | | |
| | Pt | 3.6 | 35 | 4.2 | 89 | 0.9 | 0.4 | 35.83/0.60 | 31.23/66.12 | 32.94/33.28 |
| | Pd | 1.1 | 4.9 | 0.4 | 8.1 | 2.8 | 0.6 | | | |
| | Ru | 4.9 | 30 | 5.7 | 79 | 0.9 | 0.4 | | | |
| UV-3/13 | Au | 0.29 | 1.4 | 0.12 | 1.3 | 2.4 | 1.1 | | | |
| | Ag | 450 | 800 | 30 | 450 | 15 | 1.8 | | | |
| | Pt | 37 | 189 | 14 | 190 | 2.6 | 1.0 | 36.57/0.53 | 30.53/65.54 | 32.90/33.93 |
| | Pd | 1.0 | 18.4 | 0.8 | 24.5 | 1.2 | 0.8 | | | |
| | Ru | 10 | 330 | 19 | 218 | 0.5 | 1.5 | | | |
| DG-10/14 | Au | 0.13 | 1.1 | 0.03 | 0.48 | 4.3 | 2.3 | | | |
| | Ag | 120 | 200 | 7 | 60 | 17 | 3.3 | | | |
| | Pt | 5.1 | 103 | 2.1 | 29 | 2.4 | 3.6 | 34.75/0.47 | 32.19/66.36 | 33.06/33.17 |
| | Pd | 1.2 | 8.2 | 0.23 | 3.7 | 5.2 | 2.2 | | | |
| | Ru | 8.8 | 97 | 3.9 | 35 | 2.3 | 2.8 | | | |
| ZR-10/13 | Au | 6.8 ** | n/d | 13.7 | 47 | 0.5 | n/d | | | |
| | Ag | 1100 | 7000 | 59 | 400 | 19 | 17.5 | | | |
| | Pt | 12 | 259 | 37 | 228 | 0.3 | 1.1 | 35.70/0.57 | 31.20/65.52 | 33.10/33.91 |
| | Pd | 3.2 | 30 | 4.1 | 38 | 0.8 | 0.8 | | | |
| | Ru | 25 | 458 | 7.7 | 369 | 3.2 | 1.2 | | | |

* Average data from the EPMA (Table 2). ** LA-ICP-MS data (Table 5). n/d—not determined.

The vast majority of dependencies for all NMs (87%) were characterised by determination coefficients $R^2 \geq 0.9$ (Figures 5–9). Less reliable are the data for Ag in Py (Figure 9): two of the four samples showed $R^2 \leq 0.7$. This may be due to the instability of the Ag solid solution in pyrite and its decomposition in the post-crystallisation period [14]. In contrast to Py, the dependencies for Ag in Asp were distinct ($R^2 \geq 0.95$) and may indicate its isomorphic incorporation into arsenopyrite at least up to 120 ppm Ag. However, in most cases (except the Nat-10 sample) the structural form of Ag was fractionated into pyrite: $D_{Ag}^{str} = 17 \pm 3$.

It should be noted that the ADSSC method failed to separate structural and surficial forms for Au in the ZR-10/13 sample, as its content size dependence was weakly determined (Table S1). For the other three samples, the Au distribution coefficients between Py and Asp were 2.7 ± 2.4 and 1.6 ± 1.0 for structural and superficial forms, respectively, i.e., both forms were distributed in favour of pyrite, although they were slightly fractionated. Here and henceforth, confidence intervals are defined for 1σ at α = 0.9. Perhaps $D_{Au}^{str}$ was lower for ZR-10/13 (0.5), but this sample contained very high levels of Ag, which could affect Au distribution. The behaviour of Pd in the arsenopyrite–pyrite system was similar to that of Au. Palladium by the level of content in pyrite and arsenopyrite was close to Au and fractionated into pyrite: $D_{Pd}^{str} = 2.5 \pm 2.4$. Palladium was concentrated in the surface layer much more than Au: $C_{Pd}^{sur} = 4.9$–30 ppm in Py (compared with $C_{Au}^{sur} = 1.1$–2.7 ppm), and the distribution coefficient between Py and Asp of its surficial form was close to unity (1.1 ± 1.0). Other platinoids have noticeably higher contents in both minerals, mainly due to the superficial forms which, in most cases, are an order of magnitude and higher than the concentration in structural forms. For Pt with structural form contents of 3.6–37 ppm in Py and 4.2–37 ppm in Asp, the Py/Asp distribution coefficient averaged 1.6 ± 1.4; for Ru with structural form contents of 4.9–25 and 3.9–19 ppm in Py and Asp, respectively, $D_{Ru}^{str} = 1.7 \pm 1.5$. That is, in both cases, as for the previously considered Au and Pd, there was a weak tendency towards NM fractionation in pyrite. This conclusion extends to superficial forms as well: the distribution coefficients $D^{sur}$ of Pt and Ru were on average the same and equal to 1.5 ± 1.4.

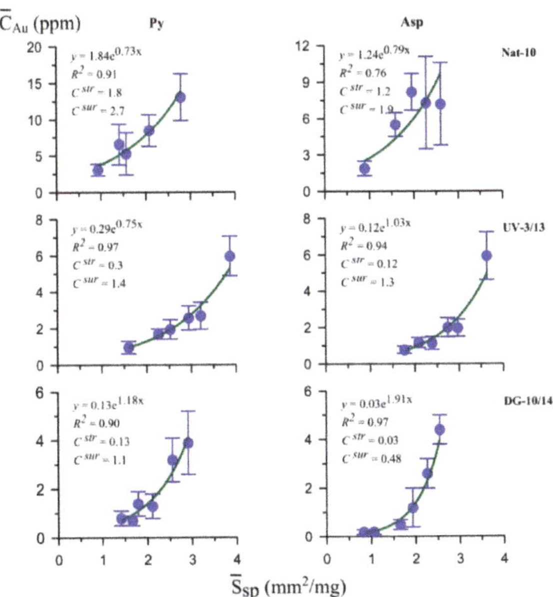



**Figure 5.** Dependence of the average concentration of evenly distributed Au in pyrite (Py) and arsenopyrite (Asp) on the specific surface area of an average crystal in size fraction. The expressions for approximate curves and concentrations of structurally and superficially bound modes are shown (see Table S1 in the Supplementary Materials for details).

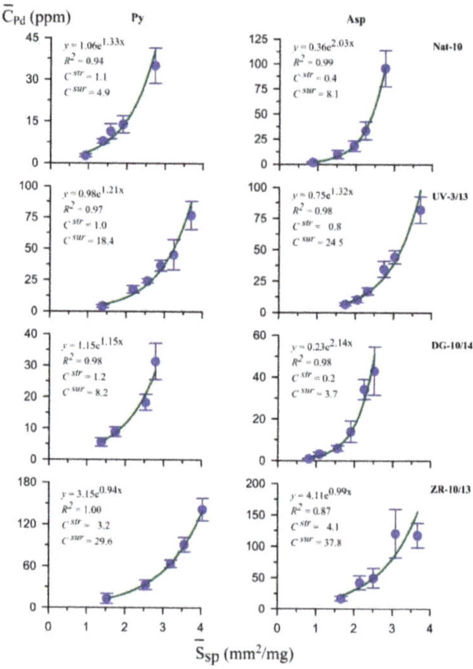

**Figure 6.** Dependence of the average concentration of evenly distributed Pd in pyrite and arsenopyrite on the specific surface area of an average crystal in size fraction. See Figure 5 for explanations.

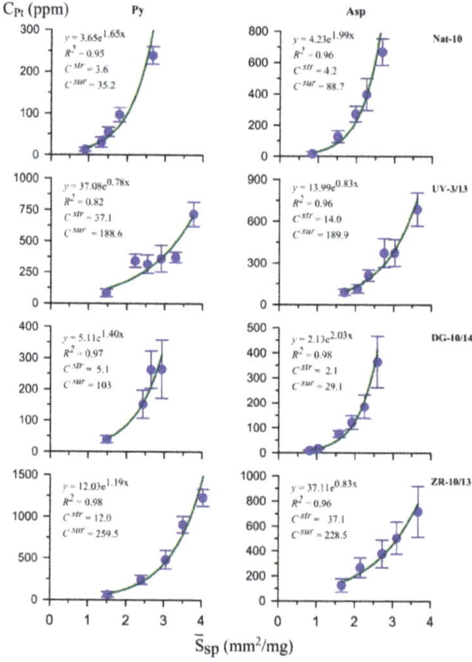

**Figure 7.** Dependence of the average concentration of evenly distributed Pt in pyrite and arsenopyrite on the specific surface area of an average crystal in size fraction. See Figure 5 for explanations.

### 5.4. LA-ICP-MS

The results are presented in Tables 5 and 6 for pyrite and arsenopyrite, respectively. It should be remembered that, according to the ADSSC method, the results that meet the distribution criteria adopted for the structurally bound form of the element are distinguished from the entire dataset for each sample (usually at least 20 points of analysis) by applying the procedure discussed in detail earlier [38,39,46]. Since grain sections were analysed, that is, only the inner regions of crystals, there is no question of the separation of structural and superficial forms. The AAS-ADSSC and LA-ICP-MS data on the Au structural forms can be compared for the Nat-10 sample (Tables 4–6): in Asp, 1.2 and 1.6 and in Py 1.8 and 1.2 ppm, respectively, i.e., the convergence of the 25 rel. % and 33 rel. %. For the sample Asp ZR-10/13, the relevant contents of $Au^{str}$ were 13.7 and 8.9 ppm ($\Delta = 35$ rel. %). This confirms the conclusion that the AAS-ADSSC method enables determination of the structural component of NM impurity with an error of ±30% [13,39]. For the UV-3/13 and DG-10/14 samples, the differences proved more significant, apparently due to the proximity to the Au detection limit (0.2 ppm) in the configuration of the LA-ICP-MS method used.

According to Tables 5 and 6, the main impurities in pyrite were Ni, Cu, Pb and Sb and in arsenopyrite Co, Ni, Cu, Pb, Sb and Se. We consider further the sum of the contents of these elements for each of the minerals and call it the Sum of Main Impurities (SMI); these values are given in the footnotes of Tables 5 and 6. It is notable that the samples of both Zolotaya Rechka minerals with the highest Ag grade contained the lowest amounts of other impurities.

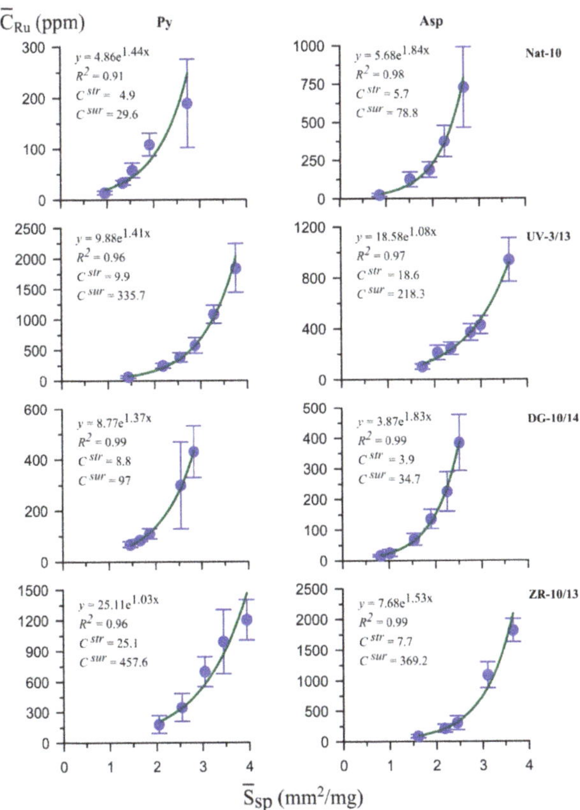

**Figure 8.** Dependence of the average concentration of evenly distributed Ru in pyrite and arsenopyrite on the specific surface area of an average crystal in size fraction. See Figure 5 for explanations.

## 6. Discussion

The results obtained allow us to consider the effect of various factors on the binding forms and ratios of NMs in the coexisting pyrite and arsenopyrite.

### 6.1. Surficial NM Accumulation

Figures 5–9 show all the NM accumulation in the surface, with the surficial forms, as a rule, exceeding by an order of magnitude the content of the element in the structurally bound forms. For Asp, in most cases, there were steeper dependences of NM content on the specific surface area than was the case for Py, except for Pt and Ag in ZR-10/13, Ru in UV-3/13 and Ag in Nat-10. This indicates somewhat greater activity of the Asp surface in the uptake of NMs. The same shape of the graphs confirms the universality of the surface enrichment factor, which is supported by the high coefficients of determination of these dependences in the vast majority of cases.

We believe that the reason for this effect is the presence of surface nanoscale formations on the surface of mineral crystals during their growth, for which, as experiments have shown [13,48,49], the distribution coefficients of elements are much higher than for the volume of the crystal. These formations are called surface non-autonomous phases (NAPs) and can be considered as primary NM concentrators [13]. The phenomenon of "latent" metal content, which was mentioned in the Introduction, is to some extent due to the presence of metals in the NAPs or in surface nano- and micro-inclusions formed as a result of their evolution. Non-autonomous phases are considered as substances regulating single-crystal growth and distribution of elements in complex multicomponent natural and experimental systems [48,49].

It should be noted that the effect of surface enrichment is often not obvious when considered in relation to the distribution of elements among coexisting minerals. The reason for this is easy to understand by referring to the data in Table 4. We see that $D^{sur}$ in some cases differs little from $D^{str}$, although $C^{sur}$ exceeds $C^{str}$. However, the excess is approximately the same for both minerals. Thus, an illusion of constancy in the distribution coefficient is created which can easily disappear with the change in the composition of the NAP, its thickness or degree of surface coverage. As can be seen from Table 4, $D^{sur}$ Py/Asp in some cases turns out slightly lower than $D^{str}$, and since $C^{sur} > C^{str}$, this will result in some decrease in the total $D^{Py/Asp}$. This factor, associated with a slightly higher activity of the Asp surface with respect to NMs, which is revealed when considering the graphs in Figures 5–9, should be taken into account (especially for small crystals). However, in general, its effect on the total distribution coefficients of invisible NM forms between pyrite and arsenopyrite is small.

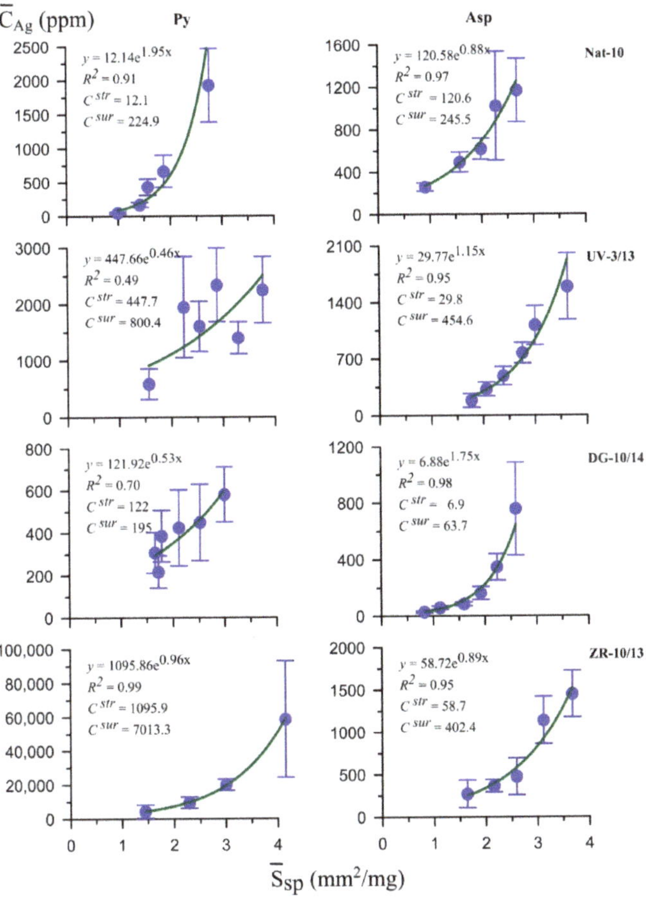



**Figure 9.** Dependence of the average concentration of evenly distributed Ag in pyrite and arsenopyrite on the specific surface area of an average crystal in size fraction. See Figure 5 for explanations.

**Table 5.** Refined LA-ICP-MS data (in ppm) for pyrite coexisting with arsenopyrite at the orogenic gold deposits located in north-eastern Russia (Magadan Region).

| Sample No. | Ti | V | Mn | Co | Ni | Cu | Pb | Bi | Sb | Se | Mo | W | Y | Nb | La | Ce | Nd | Yb | Au |
|---|---|---|---|---|---|---|---|---|---|---|---|---|---|---|---|---|---|---|---|
| Nat-10 | 8.4 | 5.3 | 6.2 | 8.0 | 56.1 | 36.4 | 23.4 | 1.1 | 11.5 | bdl | bdl | 9.7 | 2.3 | 4.2 | 1.2 | 1.6 | 4.1 | 3.6 | 1.2 |
|  | 1.8 | 2.6 | 0.8 | 1.5 | 37.1 | 15.4 | 9.8 | 0.4 | 7.4 | - | - | 9.1 | 1.7 | 2.8 | 1.0 | 1.0 | 0.9 | 2.5 | 0.9 |
| UV-3/13 | 73.9 | 3.8 | 6.7 | 7.1 | 26.0 | 57.7 | 84.5 | 2.8 | 67.2 | bdl | bdl | 5.4 | 14.4 | 1.6 | 1.5 | 2.9 | 6.2 | 4.1 | bdl |
|  | 23.8 | 2.2 | 1.8 | 2.2 | 10.3 | 20.6 | 19.5 | 1.0 | 18.1 | - | - | 2.8 | 7.4 | 0.7 | 1.0 | 2.8 | 2.5 | 1.7 | - |
| DG-10/14 | 23.6 | 4.3 | 7.9 | 9.2 | 72.3 | 18.2 | 43.6 | 4.5 | 47.5 | 68.8 | 2.8 | 5.4 | 0.7 | 7.2 | 0.4 | 0.6 | bdl | 3.3 | 0.6 |
|  | 10.3 | 3.3 | 2.7 | 2.6 | 51.5 | 9.5 | 29.3 | 2.7 | 17.0 | 12.4 | 2.7 | 3.4 | 0.5 | 2.6 | 0.2 | 0.1 | - | 0.2 | 0.2 |
| ZR-10/13 | 25.8 | 1.7 | 6.1 | 1.3 | 15.7 | 5.6 | 10.7 | 0.8 | 7.7 | bdl | bdl | 14.7 | 0.4 | 1.7 | 1.8 | 3.6 | 2.2 | 1.5 | 6.8 |
|  | 3.1 | 0.4 | 2.7 | 0.6 | 7.2 | 1.8 | 4.3 | 0.2 | 2.2 | - | - | 3.1 | 0.2 | 1.4 | 1.0 | 2.5 | 1.0 | 0.9 | 2.1 |

Note: the average concentration of structurally bound elements estimated with the ADSSC approach [38,46] (first line) and root mean square deviation (second line); bdl—below detection limit; sum of main impurities (SMI) = Ni + Cu + Pb + Sb: Nat-10, 127 ppm; UV-3/13, 235 ppm; DG-10/14, 182 ppm; ZR-10/13, 40 ppm.

**Table 6.** Refined LA-ICP-MS data (in ppm) for arsenopyrite coexisting with pyrite at the orogenic gold deposits located in north-eastern Russia (Magadan Region).

| Sample No. | Ti | V | Co | Ni | Cu | Pb | Bi | Sb | Se | Mo | W | Y | Nb | Yb | Au |
|---|---|---|---|---|---|---|---|---|---|---|---|---|---|---|---|
| Nat-10 | 3.8 | 0.7 | 10.7 | 10.7 | 10.5 | 7.7 | 1.1 | 133 | 81.2 | 1.8 | bdl | 0.6 | bdl | bdl | 1.6 |
|  | 0.6 | 0.2 | 2.4 | 3.6 | 1.8 | 4.1 | 0.5 | 45 | 26.9 | 0.8 | - | 0.6 | - | - | 0.9 |
| UV-3/13 | 6.1 | 0.6 | 4.2 | 6.1 | 3.8 | 8.5 | 0.7 | 89.7 | 105 | 2.7 | bdl | 0.3 | bdl | bdl | 0.9 |
|  | 0.4 | 0.1 | 3.3 | 1.5 | 1.3 | 1.4 | 0.2 | 12.0 | 17 | 0.5 | - | 0.1 | - | - | 0.2 |
| DG-10/14 | 71 | 1.0 | 18.8 | 24.4 | 12.4 | 16.4 | 3.5 | 146 | 187 | 2.6 | 2.5 | 0.4 | 7.2 | 1.5 | 0.6 |
|  | 50 | 0.4 | 9.6 | 8.5 | 6.5 | 6.9 | 1.5 | 27 | 26 | 0.9 | 0.8 | 0.1 | 1.2 | 0.4 | 0.2 |
| ZR-10/13 | 7.2 | 0.7 | 5.6 | 5.1 | 4.4 | 7.5 | 1.3 | 115 | 53.7 | 2.3 | bdl | 0.5 | bdl | bdl | 8.9 |
|  | 0.6 | 0.1 | 4.9 | 1.8 | 0.8 | 1.3 | 0.4 | 20 | 10.8 | 0.6 | - | 0.3 | - | - | 1.8 |

Note: the average concentration of structurally bound elements estimated with the ADSSC approach [38,46] (first line) and root mean square deviation (second line); bdl—below detection limit; SMI = Co + Ni + Cu + Pb + Sb + Se: Nat-10, 254 ppm; UV-3/13, 217 ppm; DG-10/14, 405 ppm; ZR-10/13, 193 ppm.

### 6.2. Arsenic in Pyrite: Assistance in NM Incorporation

The geochemical relation of invisible Au to As in pyrite is now generally recognised and is established at different levels of ore systems organisation—from individual crystals and micro-sized zones therein to deposits and ore provinces [7,47,50–52]. However, the nature of this relationship remains unclear as confirmed by Section 2 of this paper (Table 1). In the pyrite samples analysed, As content on average does not change much (Table 4); therefore, no direct relationships between As and NM were detected. At the same time, the state of As in pyrite may indicate indirectly the mechanism and conditions of mineral formation that lead to its enrichment with NMs. Thus, the data in Figure 4 indicate the replacement of $S^-$ with $As^-$ in the dumbbell anions of the pyrite structure which is typical of the reduction conditions of mineralisation in carbonaceous rocks. The content of As structural forms is 0.9–1.1 wt %. Substitution of $Fe^{2+}$ in pyrite for $As^{2+}$ or $As^{3+}$ is typical of oxidative hydrothermal and diagenetic conditions, including high-sulfidation epithermal deposits and shallow groundwater systems [47].

It is known from experimental studies that the solubility of As in pyrite, even at high temperatures, is limited to tenths of a percent. Clark ([53], p. 1363) shows that there is "evidence that very little arsenic is soluble in pyrite" in the range 300–700 °C. At 600 °C, pyrite solid solution may contain less than 0.53 wt % As. The hypothesis of As retrograde solubility in pyrite [7] has no ground in experiment; therefore, the formation of high-As pyrites containing more than 5 wt. % As at relatively low temperatures [7,54] remains a mineralogical mystery. However, mixed crystals $(Fe, As)S_2$ do form at low temperatures under conditions simulating deposition of authigenic pyrite in subsurface sediments [55]. Nanosized pyrite particles were synthesised at 25 and 110 °C in the presence of As(III) in solution. Research by X-ray absorption spectroscopy fine structure and modelling within density functional theory led to the establishment of the mixed nature of the As–pyrite formed, in which As, in the form of $As^{2+}$ and/or $As^{3+}$, occupies octahedral positions $Fe^{2+}$ (53–71%), and in the form of $As^-$-tetrahedral positions $S^-$ (19–47%) [55].

The supersaturated solid solutions, which are metastable in the bulk phases, often become stable in small particles due to the crystallite size effect which is the dependence of phase relationships (partially, phase boundary position) upon the crystallite size [56]. This effect was studied in detail in the $FeS_2$–$CoS_2$ system. The solubility of $CoS_2$ in the pyrite phase was shown to be 4–10 times higher for small particles (~0.1–0.2 μm) as compared to more coarse ones (1–2 μm) in the temperature interval 620–680 °C [57]. Such a mechanism can explain the formation of As-rich pyrite in small particles. Thermodynamic calculation of the change of limits of miscibility in the $FeS_2$–$FeAs_2$ system requires knowledge of the equilibrium boundaries of solid solutions in the phase diagram of the bulk phases or the dependence of miscibility limits on particle size. The mean coherent domain size along (001) for mixed crystals with 0.5 at. % As was 23 nm at 25 °C and 40 nm at 110 °C and for the composition with 1.0 at.% As, 15 and 22 nm, respectively [55]. Thus, with a decrease in crystallite size, As incorporation in pyrite increases, which is characteristic of the crystallite (phase) size effect in isomorphic mixtures [56,57]. Although a more detailed analysis is beyond the scope of the present paper, the above phenomenon may be relevant to the interpretation of our results (see the next section).

### 6.3. Crystallite Sizes

The influence of crystallite size on trace element uptake can be manifested not only in the form of a phase size effect, but also in the form of intracrystalline adsorption of these elements at the interblock boundaries of a dislocation nature. Figures 10 and 11 show the change in NM content in structural and surficial forms depending on the average size of crystallites in pyrite and arsenopyrite samples, respectively.

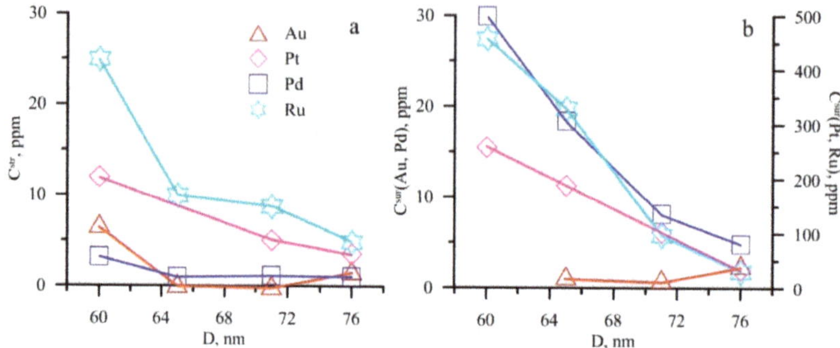

**Figure 10.** Dependences of noble metal (NM) contents in structurally (**a**) and superficially (**b**) bound modes on crystallite sizes in pyrite crystals. See Tables 3 and 4 for numerical data.

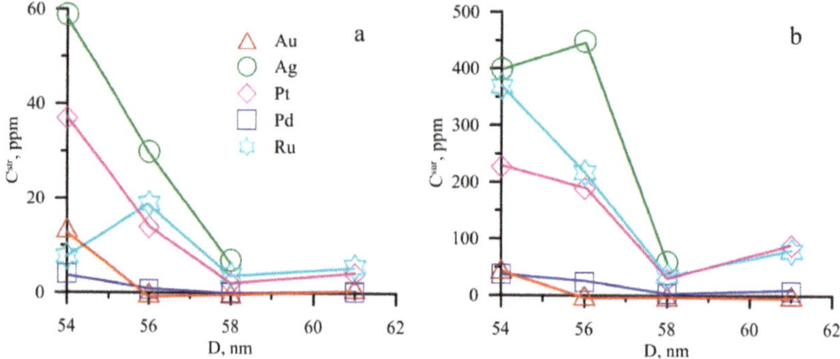

**Figure 11.** Dependences of NM contents in structurally (**a**) and superficially (**b**) bound modes on crystallite sizes in arsenopyrite crystals. See Tables 3 and 4 for numerical data.

Figure 12 shows the same data for Ag in pyrite. Most elements tend to increase their content with a decrease in the size of crystallites.

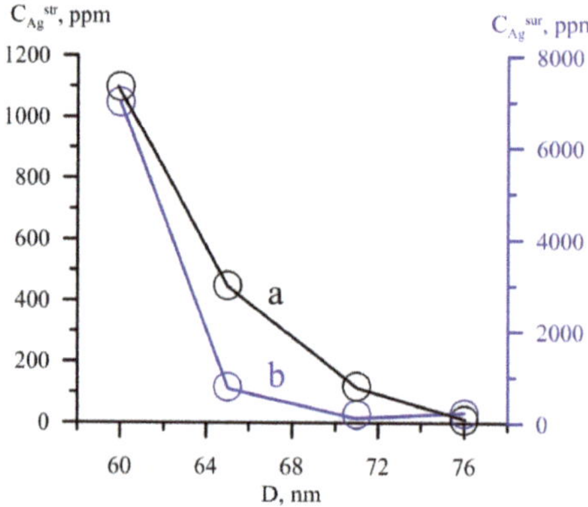

**Figure 12.** Dependences of Ag content in structurally (**a**) and superficially (**b**) bound modes on crystallite sizes in pyrite crystals. See Tables 3 and 4 for numerical data.

For structural impurities, this trend is most pronounced for Pt, Ru and Ag (Figures 10a, 11a and 12). It is interesting to note that this dependence is even more clearly manifested for these elements (and Pd) in the case of the surface-bound form of NMs (Figures 10b, 11b and 12). The reason for this phenomenon is not quite clear because the block size parameter refers to the entire crystal and not to its surface. It can be assumed that with a certain mechanism of crystal growth, namely, growth through the medium of a non-autonomous phase [13,48,49], the release of incompatible elements to the surface is more effective in the case of small blocks forming a nano-block transition zone between the substrate surface and the NAPs [48].

Intracrystalline interblock boundaries are a common defect of mineral crystals. They contain dislocation pile-ups that create a non-homogeneous elastic field around themselves. This field is able to interact with the elastic fields of point defects, such as impurity atoms. The impurities, which are larger than the matrix atoms, are concentrated in the areas of lattice extension, and smaller impurities, in the areas of compression. Regardless of the size of impurity atoms, there are always such parts of the lattice in which the interaction with the impurity is positive (impurity atoms are "attracted" to them, trying to settle in them). This effect of impurity capture by dislocations is potentially possible for any crystalline phases and impurity elements; its value depends on the density of dislocations and the nature of their distribution in the crystal volume. The effect of concentration of micro-impurities is that if there is an external source of a trace element of a sufficiently large capacity (for example, the bulk fluid phase) that is able to restore the equilibrium concentration of $C_o$ in the undistorted part of the crystal block, the total content $C_{tot}$ of trace element in the crystal will increase. The effect calculations for the case of Au in pyrite [19] have been carried out in the approximation of a simple (symmetric) tilt boundary within the formalism presented in References [58,59]. In the model considered, the block sizes were 50 and 100 nm, the distances between the dislocations in the wall were 10 and 50 nm, and $C_o$ = 1.6 ppm. The values of $C_{tot}/C_o$ at 300 °C were 49 and 2, respectively. This effect, apparently, can explain the increased content of structural Au in pyrite of the sample ZR-10/13 (Table 4), which had the smallest block size of all studied pyrites, at 60 nm (Table 3). However, due to the complexity of the analysis of multicomponent systems with a large number of dilation centres, we can only talk about the theoretical possibility of the effect on the distribution of Au and other NMs. It should be noted that for Au-REE-PGE ores of black-shale deposits, the confining of increased platinoid content to twin planes and other intracrystalline boundaries in pyrite has been previously mentioned [60].

### 6.4. Non-Stoichiometry

Figure 13 shows that the excess metal (or the presence of S vacancies) in pyrite contributes to the increase in the content of the structural forms of Ag and Pt, and to a lesser extent of Ru, Pd and possibly Au (in the latter case, the conclusion is ambiguous because of the "bounce" of point for the sample UV-3/13).

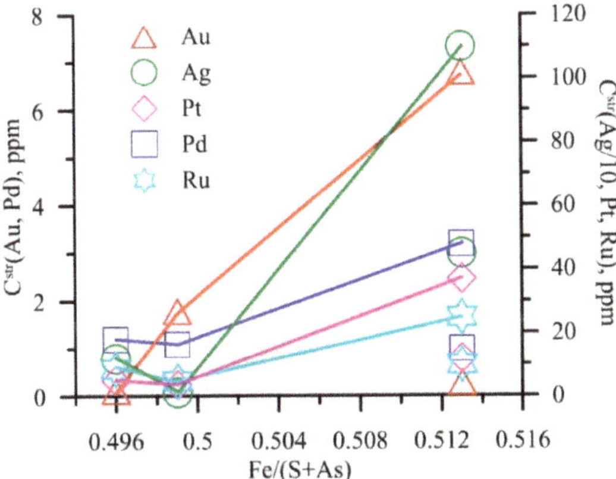


**Figure 13.** Structurally bound NMs as a function of the stoichiometric ratio in As–pyrite crystals. Note the ten-fold reduced concentration scale for Ag. See Tables 2 and 4 for numerical data.

The interaction of vacancies with impurity atoms causes the effect of impurity "trapping"—the increase of its distribution coefficient in the crystal-solution (melt) system [61] or the increase of its solubility in the solid phase [62].

The published sources seem to lack quantitative data on the effect on NM content of non-stoichiometry and point defects in natural pyrite. With regard to synthetic crystals, the following can be stated. For the pyrite crystals doped with Co, Ni and As grown by chemical vapour transport (CVT) at 700–600 °C under $Fe_{1-x}S$–$FeS_2$ buffer conditions, the S/Fe atomic ratio was found to vary from 1.95 to 2.04 [63]. The CVT-grown pyrite crystals contain $(S\text{-}Cl)^{2-}$ and $(S\text{-}Br)^{2-}$ radicals in the crystal structure [64] which substitute for the $(S_2)^{2-}$ dumbbells and act as electron donors [65]. These radical ions may be responsible for the incorporation of some impurity atoms into the crystal. Tomm et al. [66] determined Br and Au concentrations of approximately 0.2 at. % each (1094 ppm Au) in Au-doped crystals synthesised at 580–630 °C. Although the authors determined a uniform Au distribution of ~0.2 at. % in the bulk, a surface layer of about 150 Å was enhanced with Br and Au to about 1% each.

Figure 14 shows the dependence of the content of NM structural forms on the As/S ratio in arsenopyrite. An obvious trend existed expressing the elevation of Pt, Ru and Pd content for the samples with excess As (As/S ≥ ~1.14). At the same time, iron deficiency was insignificant (Table 2), and the possible concentration of vacancies in the metal position (on average ~0.3 at. %) was at the level of error of the EPMA analysis. We can assume that, for both coexisting minerals of pyrite and arsenopyrite, the sulphur deficiency is favourable for the incorporation of Ag and platinoids in their structures. Most likely, this is due to the lower sulphur fugacity under the formation of this association, which reflects two trends in the relationship between S and As in pyrite—"normal" for stoichiometric pyrite and "defective" for pyrite with excess Fe (Figure 4).

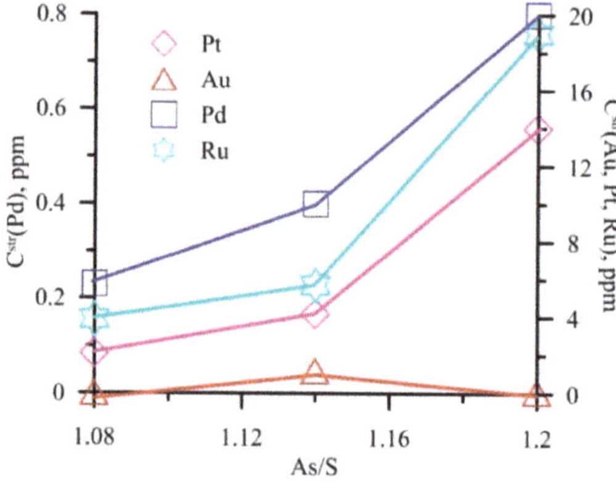


**Figure 14.** Structurally bound NM as a function of the As/S ratio in arsenopyrite crystals. See Tables 2 and 4 for numerical data.

The data for Au were not certain, as the increased content of its structural form fell to As/S = ~1.14 (samples ZR-10/13 and Nat-10, Table 4); it was not the highest of those presented in Table 2. However, for Au, there was evidence that invisible Au correlated with excess As in the most natural and synthetic arsenopyrites [54]. The authors attribute this to the uptake of Au from ore fluids by chemisorption on the surface centres containing excessive As and deficient Fe and incorporation of Au in the metastable solid solution. However, this model fails to explain why this surface structure is preserved with the continuing crystal growth and how it transfers into the crystal volume, causing zonation in the distribution of As and Au.

### 6.5. Non-Precise Metal Impurities

Figures 15 and 16 show the dependence of NM content on the sum of major non-NM impurities in the structures of pyrite and arsenopyrite, respectively. In pyrite, the highest Ag was observed for both the most "pure" (relatively non-NM impurities) sample, ZR-10/13, and the most "contaminated" with these impurities, UV-3/13. Platinum behaves in a similar way, although not in such a contrasting fashion. Gold, Pd and Ru were reduced from ZR-10/13 (40 ppm SMI) to Nat-10 (127 ppm) and then changed little, with the SMI contents being 182 and 235 ppm for DG-10/14 and UV-3/13, respectively. We believe that this dependence reflects the influence of the two above considered factors of NM concentration—fine-block substructure and non-stoichiometry defects. Both of these are better pronounced in the samples ZR-10/13 and UV-3/13 which were maximally contrasted by SMI content.

Among arsenopyrites, sample ZR-10/13 also stood out sharply with a high content of Ag (excluding Nat-10, the point of which does not fit into the scale of the graph; see Table 4), Au, Pt and Pd. It was followed by UV-3/13 with elevated Ag, Ru and Pt content. Sample DG-10/14, despite containing the highest amount of impurities (405 ppm SMI), contained less Au, Pt and Ag compared to more "pure" arsenopyrites.

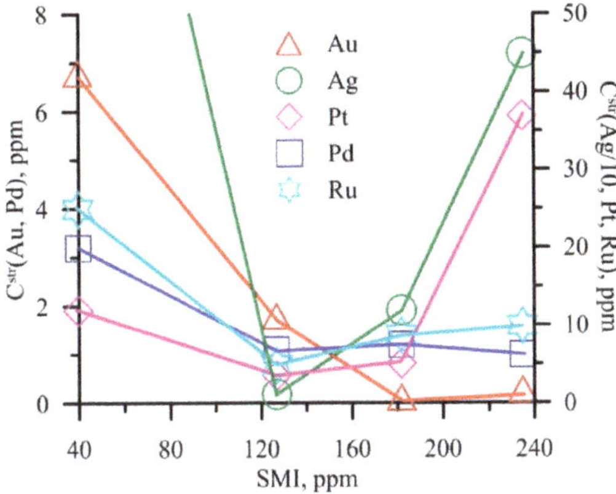


**Figure 15.** Structurally bound NM versus the SMI in pyrite crystals. Note the ten-fold reduced concentration scale for Ag. See Tables 4 and 5 for numerical data on NM and SMI, accordingly.

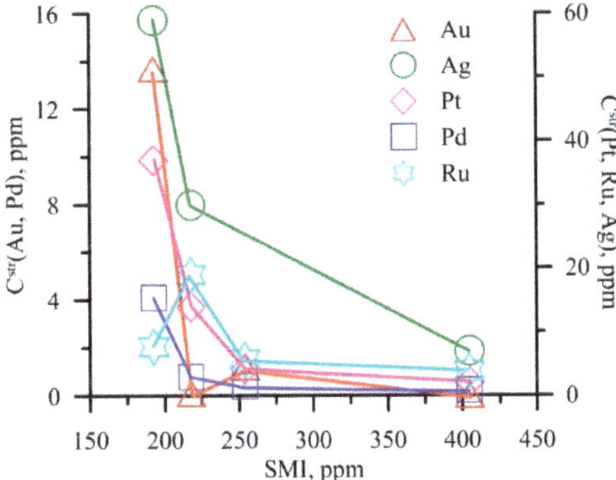


**Figure 16.** Structurally bound NM versus the SIM in arsenopyrite crystals. See Tables 4 and 6 for numerical data on NM and SMI, accordingly.

All this suggests that non-NM impurities themselves are not a factor in NM concentration in structural forms, either in pyrite or in arsenopyrite. In any case, the fine-block substructure and non-stoichiometry defects associated with the mechanism and conditions of crystallisation of minerals of this association affect NM distribution much more strongly than do these impurities.

*6.6. Effect of Crystal Inhomogeneity*

We believe that the situation reflected in Figure 3 and Table 2 could arise only as a result of the formation of the pyrite crystals by growth through the medium of a non-autonomous phase [13,48,49]. First, "pure" and As–pyrite could not be different pyrite generations because they are present in the form of a close association of irregularly located areas within one single crystal. Secondly, the situation in Figure 3 can hardly be considered characteristic of pseudomorphic pyrite substitution by its As variety [50], due to the fact that the zones of occurrence of these varieties are distributed in a complex way. Matching of compositions (non-stoichiometric indices) of coexisting Py and As–Py does not fit into this model either. The driving force of pseudomorphic substitution is the difference in chemical potentials of easily mobile components (EMCs), and the same stoichiometry of Py and As–Py indicates that there were no significant differences in the activity of EMCs (primarily $S_2$) during their formation.

It remains to be assumed that the formation of heterogeneous single crystals of pyrite (Py + As − Py) is a consequence of a special mechanism of crystal growth involving the participation of NAPs. The crystal is growing by feat of nanoscale NAPs enriched with As and NMs due to the phase size effect. Because of rather low temperatures and the relative stability of the solid solution in the process of growth, no complete transformation of As–Py into Py takes place. In fact, the solid-phase transformation of the internal NAP layer, losing the local equilibrium with a supersaturated solution [13,48], is inhibited under continuous medium conditions, since there is no free volume for the discharge of As (with the formation of the Asp phase) and other impurities. It can occur only in areas providing free space due to the presence of structural defects, such as point defects and their associates (vacancies, divacancies), dislocations, pores, etc. This leads to the observed patchy distribution of Py and As–Py sectors and the attraction of "pure" pyrite to areas with increased porosity.

## 7. Conclusions

A number of factors, the nature of which were different, influenced the distribution of hidden (invisible) forms of NMs between coexisting pyrite and arsenopyrite. The total NM content was affected by the surface accumulation effect caused by the uptake of impurities by NAPs. The surface activity of arsenopyrite was slightly higher than that of pyrite, but this had little effect on the distribution of NMs among them. The total contents increased at the expense of surficial-bound forms approximately equally for both minerals, which created the illusion of $D$ Py/Asp constancy. The distribution coefficients of structural (str) and surficial (sur) forms did not differ significantly from each other and indicated a weak tendency towards NM fractionation in pyrite in this mineral pair: $\overline{D}_{Py/Asp}^{str} = 2.7$ (Au), 2.5 (Pd), 1.6 (Pt), 1.7 (Ru); $\overline{D}_{Py/Asp}^{sur} = 1.6$ (Au), 1.1 (Pd), 1.5 (Pt), 1.5 (Ru). Data on Ag, in most cases, indicated strong fractionation in pyrite ($\overline{D}_{Py/Asp}^{str} = 17$), but they were not sufficiently reliable, probably due to the instability of the solid solution of Ag in pyrite and its decomposition in the post-crystallisation period. Surface enrichment was considered as a universal factor in "invisible" NM distribution, which was confirmed by the same shape and high coefficients of determination of the dependences of uniformly distributed concentrations of NMs on the specific surface area of the average crystal in the size fractions.

A number of NMs (Pt, Ru, Ag) tended to increase the content of structural impurities with a decrease in the size of crystallites in pyrite and arsenopyrite. This may be due to both the phase size effect and the intracrystalline adsorption of these elements at the interblock boundaries of a dislocation nature. Such dependences appear for superficially bound NM forms as well (with Pd added). Perhaps this is due to the peculiarities of the mechanism of crystal growth by means of NAPs, since the release of incompatible elements to the surface is more effective in the case of small blocks forming a nano-block transition zone between the substrate surface and the NAPs.

Excess metal (or the presence of S vacancies) in pyrite increases the content of structural forms of Ag and Pt and, to a lesser extent, those of Ru, Pd and Au. The interaction of vacancies with impurity atoms causes the effect of "trapping" the impurity, increasing its distribution coefficient in the crystal–growth medium system or increasing its solubility in the solid phase. For arsenopyrite, there was a clear trend of increasing content of Pt, Ru and Pd for the samples with an excess of As. We can assume that, for both coexisting minerals of pyrite and arsenopyrite, the deficiency in S was favourable for the incorporation of Ag and platinoids in their structures. Most likely, this was due to the fact of a lower sulphur fugacity under the formation of this association, which reflects two trends in the relationship between S and As in pyrite—"normal" for stoichiometric pyrite and "defective" for pyrite with excess Fe. The latter was associated with higher contents of some NMs.

The main non-precise metal impurities in the studied pyrite were Ni, Cu, Pb and Sb and, in arsenopyrites, Co, Ni, Cu, Pb, Sb and Se. The levels of their total contents varied quite widely—from 40 to 235 ppm in pyrite and from 193 to 405 ppm in arsenopyrite. These impurities did not seem to be a significant factor in NM concentration in structural forms; their influence was mediated by other factors, such as block substructure and non-stoichiometry defects associated with the mechanism and conditions of crystallisation of minerals of this association. These factors give rise to variations of NM accumulation and distribution characteristics even for minerals of ore deposits of the same type studied here.

The formation of heterogeneous single crystals of pyrite consisting of As-free and As-enriched areas may be a consequence of a special mechanism of crystal growth specified by the agency of NAPs.

Variations in NM distribution coefficients in the Py/Asp pair for the four samples considered from three orogenic Au deposits reached almost 100% rel. With varying degrees of influence of the abovementioned factors on the minerals of the association, we should expect significant variations of $D_{Py/Asp}$ of "invisible" NM forms which confirms a wide range of $D_{Py/Asp}$ values for "invisible" Au at various deposits.

**Supplementary Materials:** The following are available online at http://www.mdpi.com/2075-163X/9/11/660/s1, Table S1: Analysis of single-crystal size selections of coexisting pyrite and arsenopyrite of mineral samples of Natalkinskoe (Nat, UV), Degdekan (DG) and Zolotaya Rechka (ZR) deposits, north-eastern Russia.

**Author Contributions:** V.T. formulated the problem, organized the research team, and guided the study and interpretation of results. S.L. provided data handling and imaging, and organized LA-ICP-MS analysis, R.K. provided samples and contributed to interpretation, N.S. processed and rendered ADSSC data, O.B. performed EPMA data acquisition and interpretation, I.V. realized a vast amount of AAS analyses.

**Funding:** The research was performed within the frame of the state order project IX.124.3, No. 0350-2016-0025 and was partially funded by the Russian Foundation for Basic Research, grants Nos. 18-05-00077 and 17-05-00095, and integration project No.1.2 of the Irkutsk Scientific Center SB RAS.

**Acknowledgments:** We thank Ekaterina V. Kaneva for X-ray measurements and Taisa M. Pastushkova for assistance in chemical analyses. We thank the authorities of the Common Use Centers "Isotope and Geochemical Studies" of the Vinogradov Institute of Geochemistry SB RAS and "Ultra-microanalysis" of the Limnological Institute SB RAS. The authors would like to thank five anonymous reviewers for their deep insight into the problem, constructive criticism and useful comments.

**Conflicts of Interest:** The authors declare no conflicts of interest.

## References

1. Gervilla, F.; Papunen, H.; Kojonen, K.; Johanson, B. Platinum-, palladium- and gold-rich arsenide ores from the Kylmäkoski Ni-Cu deposit (Vammala Nickel Belt, SW Finland). *Mineral. Petrol.* **1998**, *64*, 163–185. [CrossRef]

2. Dare, S.A.S.; Barnes, S.-J.; Prichard, H.M. The distribution of platinum group elements and other chalcophile elements among sulfides from the Creighton Ni-Cu-PGE sulfide deposit, Sudbury, Canada, and the origin of Pd in pentlandite. *Miner. Depos.* **2010**, *45*, 765–793. [CrossRef]

3. Cafagna, F.; Jugo, P.J. An experimental study on the geochemical behavior of highly siderophile elements (HSE) and metalloids (As, Se, Sb, Te, Bi) in a mss-iss-pyrite system at 650 °C: A possible magmatic origin for Co-HSE-bearing pyrite and the role of metalloid-rich phases in the fractionation of HSE. *Geochim. Cosmochim. Acta* **2016**, *178*, 233–258. [CrossRef]

4. Duran, C.J.; Barnes, S.-J.; Corkery, J.T. Chalcophile and platinum-group element distribution in pyrites from the sulfide-rich pods of the Lac des Iles Pd deposit, Western Ontario, Canada: Implications for post-cumulus re-equilibration of the ore and the use of pyrite composition in exploration. *J. Geochem. Explor.* **2015**, *158*, 223–242. [CrossRef]

5. Pašava, J.; Zaccarini, F.; Aiglsperger, T.; Vymazalová, A. Platinum-group elements (PGE) and their principal carriers in metal-rich black shales: An overview with a new data from Mo-Ni-PGE black shales (Zunyi region, Guizhou Province, south China). *J. Geosci.* **2013**, *58*, 209–216. [CrossRef]

6. Reich, M.; Kesler, S.E.; Utsunomiya, S.; Palenik, C.S.; Chryssoulis, S.L.; Ewing, R.C. Solubility of gold in arsenian pyrite. *Geochim. Cosmochim. Acta* **2005**, *69*, 2781–2796. [CrossRef]

7. Deditius, A.P.; Reich, M.; Kesler, S.E.; Utsunomiya, S.; Chryssoulis, S.L.; Walshe, J.; Ewing, R.C. The coupled geochemistry of Au and As in pyrite from hydrothermal ore deposit. *Geochim. Cosmochim. Acta* **2014**, *140*, 644–670. [CrossRef]

8. Dare, S.A.S.; Barnes, S.-J.; Prichard, H.M.; Fisher, P.C. Chalcophile and platinum-group element (PGE) concentrations in the sulfide minerals from the McCreedy East deposit, Sudbury, Canada, and the origin of PGE in pyrite. *Miner. Depos.* **2011**, *46*, 381–407. [CrossRef]

9. Filimonova, O.N.; Nickolsky, M.S.; Trigub, A.L.; Chareev, D.A.; Kvashnina, K.O.; Kovalchuk, E.V.; Vikentyev, I.V.; Tagirov, B.R. The state of platinum in pyrite studied by X-ray absorption spectroscopy of synthetic crystals. *Econ. Geol.*. in press.

10. Gervilla, F.; Cabri, L.J.; Kojonen, K.; Oberthür, T.; Weiser, T.W.; Johanson, B.; Sie, S.H.; Campbell, J.L.; Teesdale, W.J.; Laflamme, J.H.G. Platinum-group element distribution in some ore deposits: Results of EPMA and Micro-PIXE analyses. *Microchim. Acta* **2004**, *147*, 167–173. [CrossRef]

11. Parviainen, A.; Gervilla, F.; Melgarejo, J.-C.; Johanson, B. Low-temperature, platinum-group elements-bearing Ni arsenide assemblage from the Atrevida mine (Catalonian Coastal Ranges, NE Spain). *Neues Jahrb. Mineral. Abh.* **2008**, *185*, 33–49. [CrossRef]

12. Tauson, V.L.; Parkhomenko, I.Y.; Babkin, D.N.; Men'shikov, V.I.; Lustenberg, E.E. Cadmium and mercury uptake by galena crystals under hydrothermal growth: A spectroscopic and element thermo-release atomic absorption study. *Eur. J. Mineral.* **2005**, *17*, 599–610. [CrossRef]

13. Tauson, V.L.; Lipko, S.V.; Smagunov, N.V.; Kravtsova, R.G. Trace element partitioning dualism under mineral-fluid interaction: Origin and geochemical significance. *Minerals* **2018**, *8*, 282. [CrossRef]

14. Tauson, V.L.; Lipko, S.V.; Arsent'ev, K.Y.; Mikhlin, Y.L.; Babkin, D.N.; Smagunov, N.V.; Pastushkova, T.M.; Voronova, I.Y.; Belozerova, O.Y. Dualistic distribution coefficients of trace elements in the system mineral-hydrothermal solution. IV. Platinum and silver in pyrite. *Geochem. Int.* **2017**, *55*, 753–774. [CrossRef]

15. Ballhaus, C.; Sylvester, P. Noble metal enrichment processes in the Merensky Reef, Bushveld Complex. *J. Petrol.* **2000**, *41*, 545–561. [CrossRef]

16. Morishita, Y.; Shimada, N.; Shimada, K. Invisible gold in arsenian pyrite from high-grade Hishikari gold deposit, Japan: Significance of variation and distribution of Au/As ratio in pyrite. *Ore Geol. Rev.* **2018**, *95*, 79–93. [CrossRef]

17. Akimov, V.V.; Gerasimov, I.N.; Rigin, A.V. Behavior of gold nanoparticles in formation of surficial sulfide micro- and nanophases on substrates of single crystals of arsenopyrite and chalcopyrite. *AIP Conf. Proc.* **2019**, *2064*, 030001. [CrossRef]

18. Sazonov, A.M.; Kirik, S.D.; Silyanov, S.A.; Bayukov, O.A.; Tishin, P.A. Typomorphism of arsenopyrite from the Blagodatnoe and Olimpiada gold deposits (Yenisei Ridge). *Mineralogiya* **2016**, *3*, 53–70. (In Russian)

19. Tauson, V.L.; Mironov, A.G.; Smagunov, N.V.; Bugaeva, N.G.; Akimov, V.V. Gold in sulfides: State of art of occurrence and perspectives for experimental studies. *Geol. Geofiz.* **1996**, *37*, 3–14. (In Russian)

20. Pina, R.; Gervilla, F.; Barnes, S.-J.; Ortega, L.; Lunar, R. Partition coefficients of platinum group and chalcophile elements between arsenide and sulfide phases as determined in the Beni Bousera Cr-Ni mineralization. *Econ. Geol.* **2013**, *108*, 935–951. [CrossRef]

21. Kravtsova, R.G.; Tauson, V.L.; Nikitenko, E.M. Modes of Au, Pt, and Pd occurrence in arsenopyrite from the Natalkinskoe deposit, NE Russia. *Geochem. Int.* **2015**, *53*, 964–972. [CrossRef]

22. Cook, N.J.; Ciobanu, C.L.; Meria, D.; Silcock, D.; Wade, B. Arsenopyrite-pyrite association in an orogenic gold ore: Tracing mineralization history from textures and trace elements. *Econ. Geol.* **2013**, *108*, 1273–1283. [CrossRef]
23. Su, W.; Zhang, H.; Hu, R.; Ge, X.; Xia, B.; Chen, Y.; Zhu, C. Mineralogy and geochemistry of gold-bearing arsenian pyrite from the Shuiyindong Carlin-type gold deposit, Guizhou, China: Implications for gold depositional processes. *Mineral. Depos.* **2012**, *47*, 653–662. [CrossRef]
24. Deol, S.; Deb, M.; Large, R.R.; Gilbert, S. LA-ICPMS and EPMA studies of pyrite, arsenopyrite and loellingite from the Bhukia-Jagpura gold prospect, southern Rajasthan, India: Implications for ore genesis and gold remobilization. *Chem. Geol.* **2012**, *326–327*, 72–87. [CrossRef]
25. Benzaazoua, M.; Marion, P.; Robaut, F.; Pinto, A. Gold-bearing arsenopyrite and pyrite in refractory ores: Analytical refinements and new understanding of gold mineralogy. *Mineral. Mag.* **2007**, *71*, 123–142. [CrossRef]
26. Hazarika, P.; Mishra, B.; Pruseth, K.L. Trace-element geochemistry of pyrite and arsenopyrite: Ore genetic implications for late Archean orogenic gold deposits in southern India. *Mineral. Mag.* **2017**, *81*, 661–678. [CrossRef]
27. Kojonen, K.; Johanson, B. Determination of refractory gold distribution by microanalysis, diagnostic leaching and image analysis. *Mineral. Petrol.* **1999**, *67*, 1–19. [CrossRef]
28. Hinchey, J.G.; Wilton, D.H.C.; Tubrett, M.N. A LAM-ICP-MS study of the distribution of gold in arsenopyrite from the Lodestar prospect, Newfoundland, Canada. *Can. Mineral.* **2003**, *41*, 353–364. [CrossRef]
29. Wagner, T.; Klemd, R.; Wenzel, T.; Mattsson, B. Gold upgrading in metamorphosed massive sulfide ore deposits: Direct evidence from laser-ablation-inductively coupled plasma-mass spectrometry analysis of invisible gold. *Geology* **2007**, *35*, 775–778. [CrossRef]
30. Sahoo, P.R.; Venkatesh, A.S. Constrains of mineralogical characterization of gold ore: Implication for genesis, controls and evolution of gold from Kundarkocha gold deposit, eastern India. *J. Asian Earth Sci.* **2015**, *96*, 136–149. [CrossRef]
31. Zachariáš, J.; Frýda, J.; Paterová, B.; Mihaljevič, M. Arsenopyrite and As-pyrite from the Roudný deposit, Bohemian Massif. *Mineral. Mag.* **2004**, *68*, 31–46. [CrossRef]
32. Tarnocai, C.A.; Hattori, K.; Cabri, L.J. "Invisible" gold in sulfides from Campbell mine, Red Lake greenstone belt, Ontario: Evidence for mineralization during the peak of metamorphism. *Can. Mineral.* **1997**, *35*, 805–815.
33. McClenaghan, S.H.; Lentz, D.R.; Cabri, L.J. Abundance and speciation of gold in massive sulfides of the Bathurst Mining Camp, New Brunswick, Canada. *Can. Mineral.* **2004**, *42*, 851–871. [CrossRef]
34. Liang, J.; Sun, W.; Zhu, S.; Li, H.; Liu, Y.; Zhai, W. Mineralogical study of sediment-hosted gold deposit in the Yangshan ore field, Western Qinling Orogen, Central China. *J. Asian Earth Sci.* **2014**, *85*, 40–52. [CrossRef]
35. Zhang, J.; Li, L.; Gilbert, S.; Liu, J.-J.; Shi, W.-S. LA-ICP-MS and EPMA studies on the Fe-S-As minerals from the Jinlongshan gold deposit, Qinling Orogen, China: Implications for ore-forming processes. *Geol. J.* **2014**, *49*, 482–500. [CrossRef]
36. Yang, S.; Blum, N.; Rahders, E.; Zhang, Z. The nature of invisible gold in sulfides from the Xiangxi Au-Sb-W ore deposit in northwestern Hunan, People's Republic of China. *Can. Mineral.* **1998**, *36*, 1361–1372.
37. Morey, A.A.; Tomkins, A.G.; Bierlein, F.P.; Weinberg, R.F.; Davidson, G.J. Bimodal distribution of gold in pyrite and arsenopyrite: Examples from the Archean Boorara and Bardoc shear systems, Yilgarn Craton, Western Australia. *Econ. Geol.* **2008**, *103*, 599–614. [CrossRef]
38. Tauson, V.L. Gold solubility in the common gold-bearing minerals: Experimental evaluation and application to pyrite. *Eur. J. Mineral.* **1999**, *11*, 937–947. [CrossRef]
39. Tauson, V.L.; Babkin, D.N.; Akimov, V.V.; Lipko, S.V.; Smagunov, N.V.; Parkhomenko, I.Y. Trace elements as indicators of the physicochemical conditions of mineral formation in hydrothermal sulfide systems. *Rus. Geol. Geophys.* **2013**, *54*, 526–543. [CrossRef]
40. Goncharov, V.I.; Voroshin, S.V.; Sidorov, V.A. *Natalkinskoe Gold-Ore Deposit*; North-East Complex Research Institute, Far-East Branch of RAS: Magadan, Russia, 2002. (In Russian)
41. Goryachev, N.A.; Pirajno, F. Gold deposits and gold metallogeny of Far East Russia. *Ore Geol. Rev.* **2014**, *59*, 123–151. [CrossRef]
42. Goryachev, N.A.; Vikent'eva, O.V.; Bortnikov, N.S.; Prokof'ev, V.Y.; Alpatov, V.A.; Golub, V.V. The world-class Natalka gold deposit, Northeast Russia: REE patterns, fluid inclusions, stable oxygen isotopes, and formation conditions of ore. *Geol. Ore Depos.* **2008**, *50*, 362–390. [CrossRef]

43. Mikhailov, B.K.; Struzhkov, S.F.; Natalenko, M.V.; Cimbalyuk, N.V. Multifactor model of large-volume gold-ore deposit Degdekan (Magadan region). *Otechestvennaya Geol.* **2010**, *2*, 20–31.

44. Khanchuk, A.I.; Plyusnina, L.P.; Nikitenko, E.M.; Kuzmina, T.V.; Barinov, N.N. The noble metal distribution in the black shales of the Degdekan gold deposit in Northeast Russia. *Russ. J. Pac. Geol.* **2011**, *5*, 89–96. [CrossRef]

45. Akinin, V.V.; Voroshin, S.V.; Gelman, M.L.; Leonova, V.V.; Miller, E.L. Shrimp-dating of metamorphic xenolyths of lamprophyre at gold deposit Degdekan: To the history of continental earth crust transformations at Ayan-Uriakh anticlinorium (Yano-Kolyma fold system). In *Geodynamics, Magmatism and Minerageny of Continental Margins of North Pacific*; North-East Complex Research Institute, Far-East Branch of RAS: Magadan, Russia, 2003; Volume 2, pp. 142–146. (In Russian)

46. Tauson, V.L.; Lustenberg, E.E. Quantitative determination of modes of gold occurrence in minerals by the statistical analysis of analytical data sampling. *Geochem. Int.* **2008**, *46*, 423–428. [CrossRef]

47. Deditius, A.P.; Utsunomiya, S.; Renock, D.; Ewing, R.C.; Ramana, C.V.; Becker, U.; Kesler, S.E. A proposed new type of arsenian pyrite: Composition, nanostructure and geological significance. *Geochim. Cosmochim. Acta* **2008**, *72*, 2919–2933. [CrossRef]

48. Tauson, V.L.; Lipko, S.V.; Smagunov, N.V.; Kravtsova, R.G.; Arsent'ev, K.Y. Distribution and segregation of trace elements during the growth of ore mineral crystals in hydrothermal systems: Geochemical and mineralogical implications. *Rus. Geol. Geophys.* **2018**, *59*, 1718–1732. [CrossRef]

49. Tauson, V.L.; Lipko, S.V.; Arsent'ev, K.Y.; Smagunov, N.V. Crystal growth through the medium of nonautonomous phase: Implications for element partitioning in ore systems. *Cryst. Rep.* **2019**, *64*, 496–507. [CrossRef]

50. Xing, Y.; Brugger, J.; Tomkins, A.; Shvarov, Y. Arsenic evolution as a tool for understanding formation of pyritic gold ores. *Geology* **2019**, *47*, 335–338. [CrossRef]

51. Kusebauch, C.; Gleeson, S.A.; Oelze, M. Coupled partitioning of Au and As into pyrite controls formation of giant Au deposits. *Sci. Adv.* **2019**, *5*, eaav5891. [CrossRef]

52. Morishita, Y.; Hammond, N.Q.; Momii, K.; Konagaya, R.; Sano, Y.; Takahata, N.; Ueno, H. Invisible gold in pyrite from epithermal, banded-iron-formation-hosted, and sedimentary gold deposits: Evidence of hydrothermal influence. *Minerals* **2019**, *9*, 447. [CrossRef]

53. Clark, L.A. The Fe-As-S system: Phase relations and applications. Part 1. *Econ. Geol.* **1960**, *55*, 1345–1381. [CrossRef]

54. Fleet, M.E.; Mumin, A.H. Gold-bearing arsenian pyrite and marcasite and arsenopyrite from Carlin Trend gold deposits and laboratory synthesis. *Am. Mineral.* **1997**, *82*, 182–193. [CrossRef]

55. Le Pape, P.; Blanchard, M.; Brest, J.; Boulliard, J.-C.; Ikogou, M.; Stetten, L.; Wang, S.; Landrot, G.; Morin, G. Arsenic incorporation in pyrite at ambient temperature at both tetrahedral $S^{-1}$ and octahedral $Fe^{II}$ sites: Evidence from EXAFS-DFT analysis. *Environ. Sci. Technol.* **2017**, *51*, 150–158. [CrossRef] [PubMed]

56. Tauson, V.L.; Akimov, V.V. Effect of crystallite size on solid state miscibility: Applications to the pyrite-cattierite system. *Geochim. Cosmochim. Acta* **1991**, *55*, 2851–2859. [CrossRef]

57. Tauson, V.L.; Akimov, V.V. Further experimental evidence for a crystallite size effect in the $FeS_2$-$CoS_2$ system. *Chem. Geol.* **1993**, *109*, 113–118. [CrossRef]

58. Abramovich, M.G.; Shmakin, B.M.; Tauson, V.L.; Akimov, V.V. Mineral typochemistry: Anomalous trace-element concentrations in solid solutions with defect structures. *Int. Geol. Rev.* **1990**, *32*, 608–615. [CrossRef]

59. Urusov, V.S.; Tauson, V.L.; Akimov, V.V. *Geochemistry of Solids*; GEOS: Moscow, Russia, 1997. (In Russian)

60. Ermolaev, N.P.; Sozinov, N.A.; Fliciyan, E.S.; Chinenov, V.A.; Khoroshilov, V.L. *New Types of Ores of Noble and Rare Elements in Carbonaceous Shales*; Beus, A.A., Ed.; Nauka: Moscow, Russia, 1992. (In Russian)

61. Urusov, V.S.; Dudnikova, V.B. The trace-component trapping effect: Experimental evidence, theoretical interpretation, and geochemical applications. *Geochim. Cosmochim. Acta* **1998**, *62*, 1233–1240. [CrossRef]

62. Tauson, V.L.; Akimov, V.V.; Parkhomenko, I.Y.; Nepomnyashchikh, K.V.; Men'shikov, V.I. A study of cadmium incorporation into pyrrhotites of different stoichiometry. *Geochem. Int.* **2004**, *42*, 115–121.

63. Lehner, S.W.; Savage, K.S.; Ayers, J.C. Vapor growth and characterization of pyrite ($FeS_2$) doped with Co, Ni, and As: Variations in semiconducting properties. *J. Cryst. Growth* **2006**, *286*, 306–317. [CrossRef]

64. Fiechter, S.; Mai, J.; Ennaoui, A.; Szacki, W. Chemical vapor transport of pyrite ($FeS_2$) with halogen (Cl, Br, I). *J. Cryst. Growth* **1986**, *78*, 438–444. [CrossRef]

65. Schieck, R.; Hartmann, A.; Fiechter, S. Electrical properties of natural and synthetic pyrite (FeS$_2$) crystals. *J. Mater. Res.* **1990**, *5*, 1567–1572. [CrossRef]
66. Tomm, Y.; Schieck, R.; Ellmer, K.; Fiechter, S. Growth mechanism and electronic properties of doped pyrite (FeS$_2$) crystals. *J. Cryst. Growth* **1995**, *146*, 271–276. [CrossRef]

 **minerals**

*Article*

# Quartz Rb-Sr Isochron Ages of Two Type Orebodies from the Nibao Carlin-Type Gold Deposit, Guizhou, China



**Lulin Zheng [1,2], Ruidong Yang [2,*], Junbo Gao [2], Jun Chen [2], Jianzhong Liu [3] and Depeng Li [2]**

[1] Mining College of Guizhou University, Guiyang 550025, China
[2] College of Resources and Environmental Engineering, Guizhou University, Guiyang 550025, China
[3] Geological Party 105, Guizhou Bureau of Geology and Mineral Exploration & Development, Guiyang 550018, China
[*] Correspondence: rdyang@gzu.edu.cn; Tel.: +86-851-8362-0551



Received: 3 May 2019; Accepted: 25 June 2019; Published: 28 June 2019



**Abstract:** The Nibao gold deposit, which includes both fault-controlled and strata-bound gold orebodies, constitutes an important part of the Yunnan–Guizhou–Guangxi "Golden Triangle" region. Defining the mineralization age of these gold orebodies may provide additional evidence for constraining the formation ages of low-temperature orebodies and their metallogenic distribution in South China. Petrographic studies of gold-bearing pyrites and ore-related quartz veins indicate that these pyrites coexist with quartz or filled in vein-like quartz, which suggests a possible genetic relationship between the two from Nibao gold deposit. Minerals chemistry shows that Rb and Sr are usually hosted in fluid inclusions in quartz ranging from 0.0786 to 2.0760 ppm and 0.1703 to 2.1820 ppm, respectively. The Rb–Sr isotopic composition of gold-bearing quartz-hosted fluid inclusions from the Nibao gold deposit were found to have Rb–Sr isochron ages of 142 ± 3 and 141 ± 2 Ma for both fault-controlled and strata-bound orebodies, respectively, adding more evidence to previous studies and thus revealing a regional gold mineralization age of 148–134 Ma. These results also confirm the Middle-Late Yanshanian mineralizing events of Carlin-type gold deposits in Yunnan, Guizhou, and Guangxi Provinces of Southwest China. In addition, previous studies indicated that antimony deposits in the region which were formed at ca. 148–126 Ma have a close affinity with gold deposits. This illustrates that the regional low-temperature hydrothermal gold mineralization is related in space and time to the Yanshanian (ca. 146–115 Ma) magmatic activity. Specifically, the large-scale gold and antimony mineralization are considered to be inherently related to mantle-derived mafic and ultramafic magmatic rocks associated with an extensional tectonic environment. Based on the initial $^{87}Sr/^{86}Sr$ ratios of 0.70844 ± 0.00022 (2σ) and 0.70862 ± 0.00020 (2σ) for gold-bearing quartz veins from fault-controlled and strata-bound gold orebodies, respectively, at the Nibao gold deposit, as well as the C, H, O, and S isotopic characteristics of gold deposits located in the Golden Triangle region, we suggest that the mantle-derived material can be involved in the formation of the Nibao gold deposit and that the ore-forming fluid can be derived from a mixed crust–mantle source.

**Keywords:** mineralization age; Nibao gold deposit; ore-forming fluid; Rb–Sr isotopic composition; Yunnan–Guizhou–Guangxi "Golden Triangle" region



---

## 1. Introduction

South China possesses a large region containing various types of low-temperature mineralization that produced many gold, antimony, mercury, thallium, and lead–zinc deposits. Southwestern Guizhou, adjacent to Yunnan and Guangxi Provinces in South China, is an important mining district for

Carlin-type gold deposits, containing several large-scale gold deposits (e.g., Shuiyindong, Lannigou, Zimudang, Getang, etc.) (Figure 1).

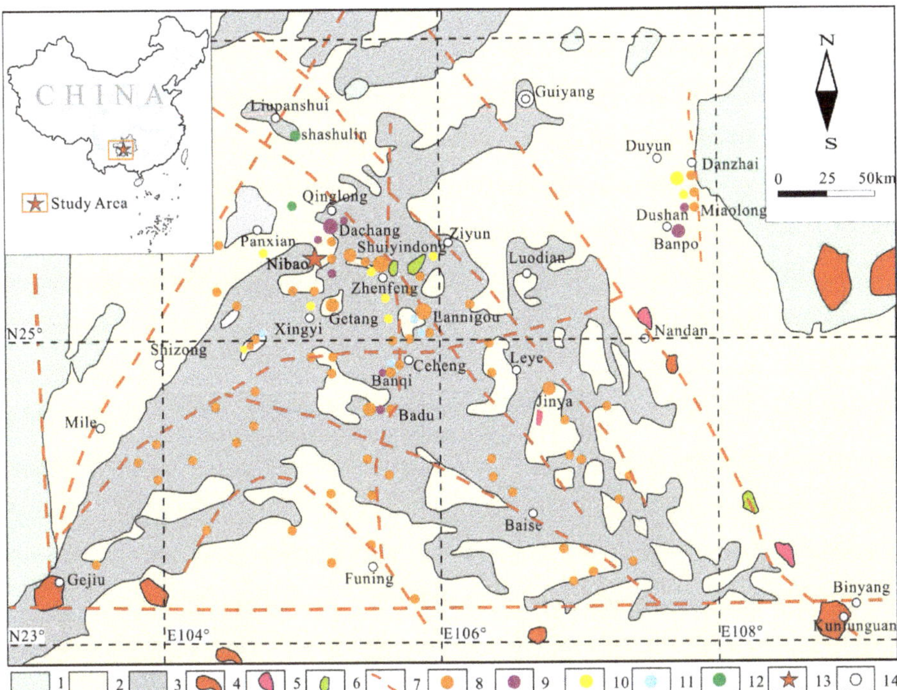



**Figure 1.** Geological map of low-temperature hydrothermal deposits distributed in the Yunnan–Guizhou–Guangxi "Golden Triangle" region: 1. Proterozoic; 2. Paleozoic; 3. Triassic; 4. Granite; 5. Quartz porphyry; 6. Alkaline mafic–ultramafic rock; 7. Fault; 8. Gold deposit; 9. Antimony deposit; 10. Mercury deposit; 11. Arsenic deposit; 12. Lead–zinc deposit; 13. Nibao gold deposit; 14. County town.

A strata-bound gold orebody was recently discovered at 1300 m below the surface in the Shuiyindong gold deposit, signifying a breakthrough in the bidimensional space prospecting in Guizhou Province. The orebody has a reserve of 263 t gold, with an average grade of approximately 5 g/t [1], making the Shuiyindong a super-large gold deposit. The newly discovered Nibao gold deposit constitutes another large-scale deposit in Guizhou Province with gold reserves of 70 t and the average grade of 2.6 g/t. Southwestern Guizhou presents extremely favorable metallogenic conditions for gold mineralization and shows a great potential for gold deposits. In addition to strata-bound gold orebodies, a large-scale fault-controlled orebody in the fault F1 has also been recently discovered in the Nibao deposit. This finding challenged the existing understanding of the Nibao gold deposit and showed that it can be not only a strata-bound-type deposits but also contains fault-controlled-type deposit [2]. Moreover, the large-scale gold deposits are mainly strata-bound or fault-controlled bodies in the Yunnan–Guizhou–Guangxi "Golden Triangle". For example, the Shuiyindong super-large gold deposit (263 t) belongs to a typical strata-bound deposit, where the orebodies occured mainly in the structure-controlled alteration zone (abbreviated as Sbt, according to its definition in Chinese) and the Longtan Formation. Whereas the Lannigou super-large gold deposit (110 t) occurred in the fault F3 and the Zimudang large gold deposit (72 t) occurred in the fault F1 are typical fault-controlled gold deposits. According to a comparison of the geological and geochemical characteristics of various typical Carlin-type gold deposits in the Yunnan–Guizhou–Guangxi Golden Triangle, it can be concluded that the Nibao gold deposit is not only unique but also general. Therefore, the Nibao gold deposit that is regarded as a representative Carlin-type gold deposit has a special and relatively high research value.

Some previous studies have focused on the genesis of the deposit, the source of ore-forming material, fluid evolution, and the metallogeny of the gold deposits in southwestern Guizhou, which have allowed a better understanding of their origin and evolution [3–13]. However, there is still considerable controversy with respect to a relatively large gold mineralization age interval (235–83 Ma) of gold deposits in this region, which makes it difficult to accurately determine the age of regional gold mineralization [8,14–18]. Defining the age of mineralizing events is considered to be the key to studying ore deposits and understanding the mechanism of ore formation and metallogeny. Therefore, in this study both the fault-controlled and strata-bound gold orebodies of the Nibao gold deposit were investigated in order to define the mineralization age by using the Rb–Sr isochron dating method of fluid inclusions in quartz, and the quartz is considered to be closely associated with gold mineralization. Moreover, the age data should also constrain the metallogenic events involved in low-temperature mineralization in the Golden Triangle region of South China.

## 2. Regional Setting and Geological Characteristics of Gold Deposits

A few Carlin-type gold deposits in Southwest China were located in the Yunnan–Guizhou–Guangxi Golden Triangle, specifically in southeastern Yunnan, southwestern Guizhou, and northwestern Guangxi. The southwestern Guizhou region sits at the junction of the Yangtze Block and the western segment of the South China Fold Belt. Yanshanian (205–66 Ma) regional tectonism involved periods of extension and rifting along the block margins alternating with episodes of extrusive magmatism, which are considered to be important geological factors for regional gold mineralization. Hu et al. [4] noted that regional gold deposits occurred in strata from the Cambrian to Cretaceous periods but mainly in Permian and Triassic strata, where gold orebodies were controlled by the strata (referred to as the strata-bound type) or the faults (referred to as the fault-controlled type). Generally, the host rocks are mainly impure carbonates, sedimentary tuffs, tuffs in Permian and siltstones (or fine sandstones) in Triassic. The ore minerals are mainly pyrite, stibnite, arsenopyrite, realgar, orpiment and cinnabar, and gangue minerals are mainly quartz, calcite, dolomite, fluorite, illite, etc. More than 200 gold deposits showing have been discovered in the Golden Triangle, making it an important gold mining district of China.

The Nibao gold deposit is an important Carlin-type gold deposit located in Pu'an County in southwestern Guizhou. Tectonically, the deposit is situated at the transition zone between the Yangtze craton and the Youjiang orogenic belt (Figure 2a). The structure of the Nibao gold mining district is relatively simple and can be represented primarily by NEE-striking faults (F1–F4), the Erlongqiangbao anticline, and NW-striking faults (F6, F8, F10, and F11). The large-scale gold orebody mentioned above occurred mainly within fault F1 (Figure 2). The stratigraphy of the deposit mainly includes the Middle Permian Maokou Formation, the Upper Permian Longtan Formation, the Lower Triassic Yongningzhen Formation, and the Middle Triassic Guanling Formation. A structure-controlled alteration zone (abbreviated to Sbt, according to its definition in Chinese) was found in the ore-bearing strata. Here, Sbt refers to altered rocks which had been generated by Yanshanian regional structures (such as large-scale decollement structures) and had experienced hydrothermal alteration near the unconformity between the Maokou Formation (shallow platform carbonates) and the Longtan Formation. A more specific description of the structure-controlled alteration zone can be found in a previous study published by Zheng et al. [2]. Field investigations have revealed that the Longtan Formation, which can be divided into three members (the first, second, and third members of the Longtan Formation), mainly includes sedimentary tuffs, claystones with a minor amount of coal, limestones, and silicified limestones. The Yongningzhen Formation and the Guanling Formation consist mainly of sedimentary carbonates. Although there are no magmatic rocks exposed in the Nibao gold mining district, small-scale Late Yanshanian mafic, ultramafic, or intermediate medium-acidic igneous bodies were found to outcrop in the surrounding area (Figure 1). Moreover, Wang et al. [19] indicated that there may be underground basic-ultrabasic rock hide in the Nibao gold deposit by using geophysical exploration, the Emeishan basaltic Formation is distributed widely in the northwestern region near Panxian County (Figure 2b).

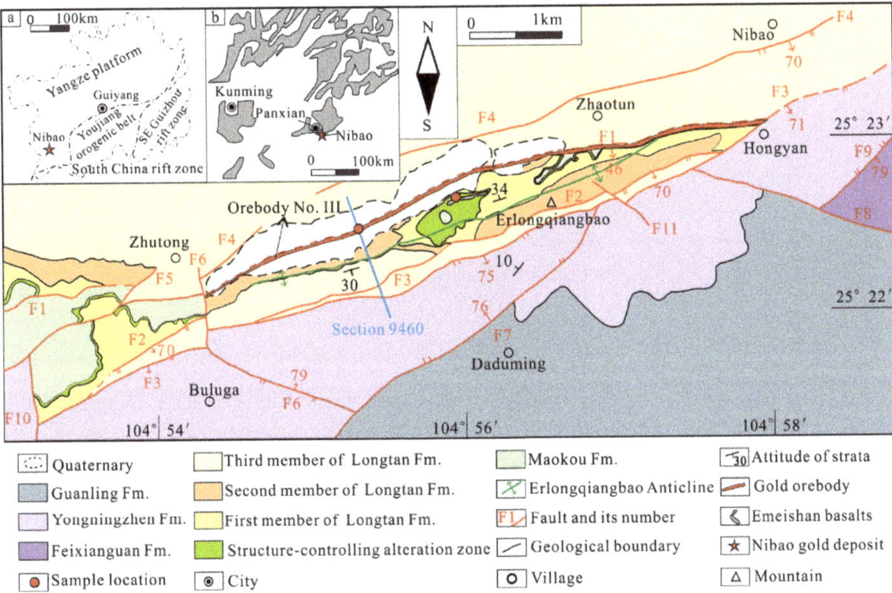


| | | | |
|---|---|---|---|
| Quaternary | Third member of Longtan Fm. | Maokou Fm. | Attitude of strata |
| Guanling Fm. | Second member of Longtan Fm. | Erlongqiangbao Anticline | Gold orebody |
| Yongningzhen Fm. | First member of Longtan Fm. | Fault and its number | Emeishan basalts |
| Feixianguan Fm. | Structure-controlling alteration zone | Geological boundary | Nibao gold deposit |
| Sample location | City | Village | Mountain |

**Figure 2.** Geological map of Nibao gold deposit: (**a**) tectonic map and (**b**) geological map of the Emeishan basalt distribution (modified from He et al. [20]).

Most of the gold orebodies in the Nibao gold deposit were found in fault F1 or in the structure-controlled alteration zones, while the rest occurred in the first or the second member of the Longtan Formation (Figure 3). Gold orebodies in the Nibao gold deposit have been classified into fault-controlled and strata-bound types, and gold orebody No. III, which occurs as stratiform-like or lenticular layers, is of the fault-controlled type that is associated with fault F1. The orebody is 4084 m along strike, with a dipping extension of 540 m, and presents an average thickness of 4.86 m and an average grade of 3.42 Au g/t. The gold reserves reach 39 t, make it the largest orebody in the deposit. Strata-bound orebody No. IV occurs in the structure-controlled alteration zone unit, while layered orebodies No. I, II, and VI are in the Longtan Formation. Among these, orebody No. IV is a relatively large orebody in the Nibao mining district.

The gold ore is principally hosted by sedimentary tuffs (Figure 4a,c–e) and tuffs (Figure 4b,f), the U–Pb zircon age of which is 251–263 Ma, as suggested by previous studies [20–23]. As revealed by this study, the gold mineralized rocks have been primarily affected by silicification (Figure 4a,b,e,f) and pyritization (Figure 4a–f), and secondly by arsenopyritization (Figure 4c), while the host rocks underwent argillization (illitization) (Figure 4c,e,f) and carbonatization (calcitization and dolomitization) (Figure 4c–e). Gold mineralization occurs as the disseminated type in sedimentary tuffs (Figure 4) and as stockwork in tuffs (Figure 4b). Ore minerals (e.g., pyrite and arsenopyrite) contain most of the gold mineralization [3,7,11,12]. The gangue minerals include quartz, clay minerals (illite), dolomite, and calcite, among others.

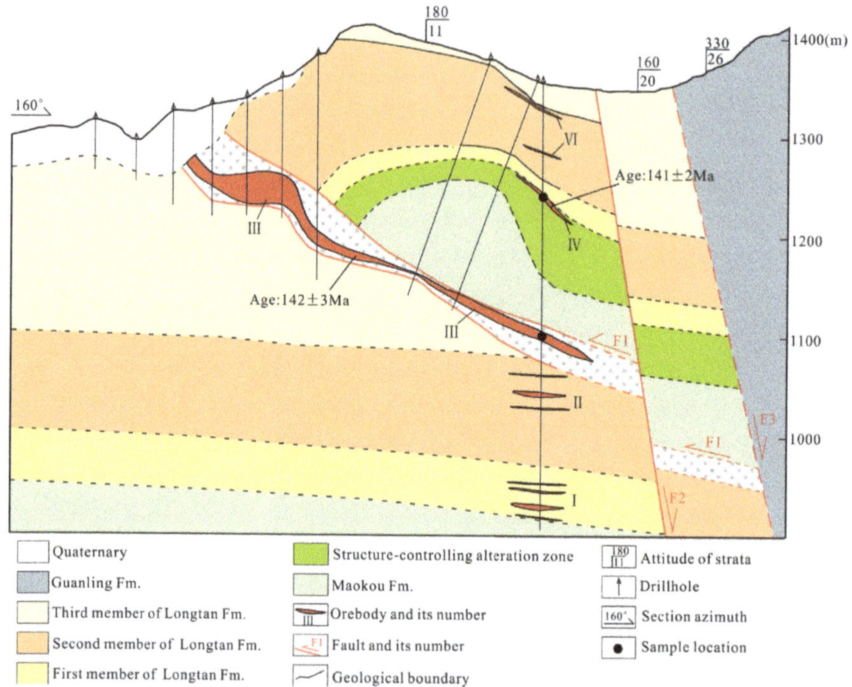



**Figure 3.** Cross section along the 9460 exploration line of the Nibao gold deposit.

**Figure 4.** Macrocharacteristics and microscopic textures of various rocks in the Nibao gold deposit: (**a**) vein (stockwork) quartz and disseminated pyrite in sedimentary tuffs; (**b**) vein quartz associated closely with pyrite in tuffs; (**c**) pyrite showing subidiomorphic–idiomorphic granular texture and arsenopyrite showing subidiomorphic texture in sedimentary tuffs (transmitted light, polarized light); (**d**) strong argillization and carbonatization in sedimentary tuffs (transmitted light, crossed polarizers); (**e**) silicification, pyritization, carbonatization, and argillization in sedimentary tuffs (BSE); (**f**) silicification, pyritization, and argillization in tuffs (BSE). Mineral abbreviations: Py = pyrite, Asp = arsenopyrite, Dol = dolomite, Cal = calcite, Clay = clay, Bx = vitric tuff, Rt = rutile, Qz = quartz, Ill = illite.

According to the ore texture characteristics and paragenesis of hydrothermal altered minerals, the Nibao gold deposit experienced three stages of formation: (1) the early quartz–pyrite stage, which involved layered quartz and quartz veinlets (stockwork) and pyrite characterized by banding or a low degree of euhedral granular texture; (2) the main stage of gold mineralization can be represented by a mineral assembly of quartz, arsenic-bearing pyrite, and arsenopyrite, in which the quartz occurred as stockwork, veinlets, and flecks along the joints and fissures. Further, there was a great amount of fine-grained disseminated pyrite and arsenic-rich pyrite showing zonal texture and arsenopyrite in angular, needle, and hair-like shapes; and (3) the later stage, when minerals such as quartz, carbonates, and clays (mostly illite) crystallized with quartz and calcite, often exhibiting a miarolitic texture.

## 3. Sampling and Analytical Methods

### 3.1. Sampling

Based on detailed field investigations of the characteristics of different orebodies (e.g., ore texture, rock alteration, etc.) in the Nibao gold deposit, a total of 10 ore samples associated with quartz veins were collected, of which 5 were collected from orebody No. III and the rest from orebody No. IV. It should be noted that samples from orebody No. III were collected from the cores of different drill holes in a shattered zone of fault F1. Quartz in these samples was mainly shaped as networks or strings, and occurred primarily in ore-bearing sedimentary tuffs (Figure 4a). Samples from orebody No. IV were all collected from an open-air profile of a structure-controlled alteration zone, the quartz of which mainly had a vein-like texture occurring in tuffs (Figure 4b). All the quartz was formed during gold mineralization and thus bore a close relationship with gold mineralization.

### 3.2. Analytical Methods

Clean quartz pieces were separated from samples by washing with distilled water, air-drying, and coarse crushing. The chosen quartz pieces were then finely crushed to less than 0.25 mm and cleaned by distilled water again. Finally, single mineral quartz grains were hand-picked under a binocular microscope, and the purity of the picked quartz was greater than 99%.

Pretreatment and isotopic analysis of all quartz samples were completed in the Isotope Open Laboratory of the Wuhan Center, China Geological Survey. Sample preparation was conducted in an ultraclean laboratory. The Rb and Sr blanks for all of the isotopic analytical procedures were 0.0005 and 0.001 ppm, respectively, and a background correction for the Rb and Sr concentrations was implemented for all analyzed samples. The Rb–Sr isotopic analysis involved the following steps: (1) Ultrapure hydrochloric acid, nitric acid, and pure water were used to clean the hand-picked quartz. (2) Cleaned quartz samples were placed in a drying oven and the temperature was between 120 to 180 °C to decrepitate and thus remove secondary fluid inclusions. (3) The quartz was then washed three to five times in purified water in an ultrasonic cleaner and dried thoroughly. (4) An appropriate amount of quartz grains was weighed and mixed with $^{85}$Rb + $^{84}$Sr mixing diluent. It was then dissolved in hydrofluoric and perchloric acids. After that, the Rb and Sr was liberated by a cationic resin (Dowex $50 \times 8$) exchange method. (5) The isotopic compositions of Rb and Sr were determined via a TRITON thermal ionization mass spectrometer (Thermo Fisher Scientific, Waltham, MA, USA). The isotopic dilution method (spiking) was employed to calculate the concentrations of Rb and Sr in the samples. During the analytical process, the $^{87}$Sr/$^{86}$Sr isotopic compositions of certified reference materials (CRM) including NBS987, NBS607, and GBW04411 were determined to validate the accuracy and precision of the methodology. As a result, NBS987 gave an $^{87}$Sr/$^{86}$Sr value of $0.71018 \pm 0.00006$ ($2\sigma$); the $^{87}$Sr/$^{86}$Sr ratio and Rb and Sr concentrations of NBS607 were $1.20037 \pm 0.00008$ ($2\sigma$), 523.50 ppm, and 65.67 ppm, respectively; and the $^{87}$Sr/$^{86}$Sr ratio and Rb and Sr concentrations of GBW04411 were $0.75985 \pm 0.00004$ ($2\sigma$), 249.10 ppm, and 158.50 ppm, respectively. The Rb and Sr concentrations and Sr isotopic ratios of these CRM within the error bound compared with the certified values. The least-squares method was applied to calculate the isochron age.

## 4. Analytical Results

### 4.1. Fluid Inclusion in Quartz

Fluid inclusion petrography studies on main-stage quartz of strata-bound and fault-controlled bodies showed that there is seldom secondary fluid inclusions in quartz, and the primary fluid inclusions in quartz are quite abundant, most of them are scattered, isolated, or clustered, with a granularity of 4–12 µm; and only a few of them are elongated, irregular, polygon, or negative crystallized (Figure 5). The fluid inclusions are mainly two-phase aqueous inclusions and a few of $CO_2$-rich inclusions. Based on their components and phase behavior at room temperature (20 °C), three types of fluid inclusions including type I—two-phase aqueous inclusions (Figure 5), type II—$CO_2$–$H_2O$ inclusions, and type III—$CO_2$ inclusions are recognized. The homogenization temperature of fluid inclusions showed that the homogenization temperatures of strata-bound bodies are mainly between 220 and 280 °C, average at 251 °C (Figure 6a), the salinity of which mainly ranged from 4 wt % to 7 wt % NaCl equivalent, average at 4.94 wt % NaCl equivalent (Figure 6b), the ore-forming fluid density of which varied from 0.74 to 0.92 $g/cm^3$, average at 0.83 $g/cm^3$, and the vapor phase of two-phase aqueous inclusions mainly contains $CO_2$ in addition to $CH_4$ and $N_2$ (the ore-forming fluid belongs to a $H_2O$–NaCl–$CO_2$–$CH_4$–$N_2$ system). In contrast, the homogenization temperatures of inclusions in fault-controlled bodies are mainly between 160 and 240 °C, average at 231 °C (Figure 6c), the salinity mainly of which ranged from 3 wt % to 5 wt % NaCl equivalent, average at 4.75 wt % NaCl equivalent (Figure 6d), the ore-forming fluid density of which varied from 0.71 to 0.93 $g/cm^3$, average at 0.85 $g/cm^3$, and the vapor phase of two-phase aqueous inclusions contains $CO_2$ in addition to $CH_4$ or $N_2$ (the ore-forming fluid belongs to a $H_2O$–NaCl–$CO_2 \pm CH_4 \pm N_2$ system).

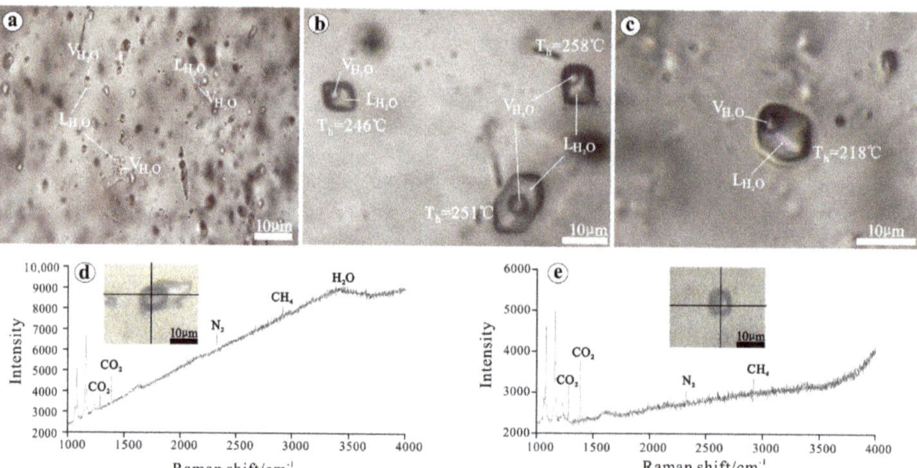

**Figure 5.** Microphotographs and laser Raman spectra of fluid inclusion in quartz from the Nibao gold deposit: (**a**,**b**) two-phase aqueous inclusions(main-stage quartz of Sbt); (**c**) two-phase aqueous inclusions (main-stage quartz of fault F1); (**d**) two-phase aqueous inclusions containing $CO_2$, $N_2$, and $CH_4$; (**e**) $CO_2$ inclusions containing $N_2$ and $CH_4$.

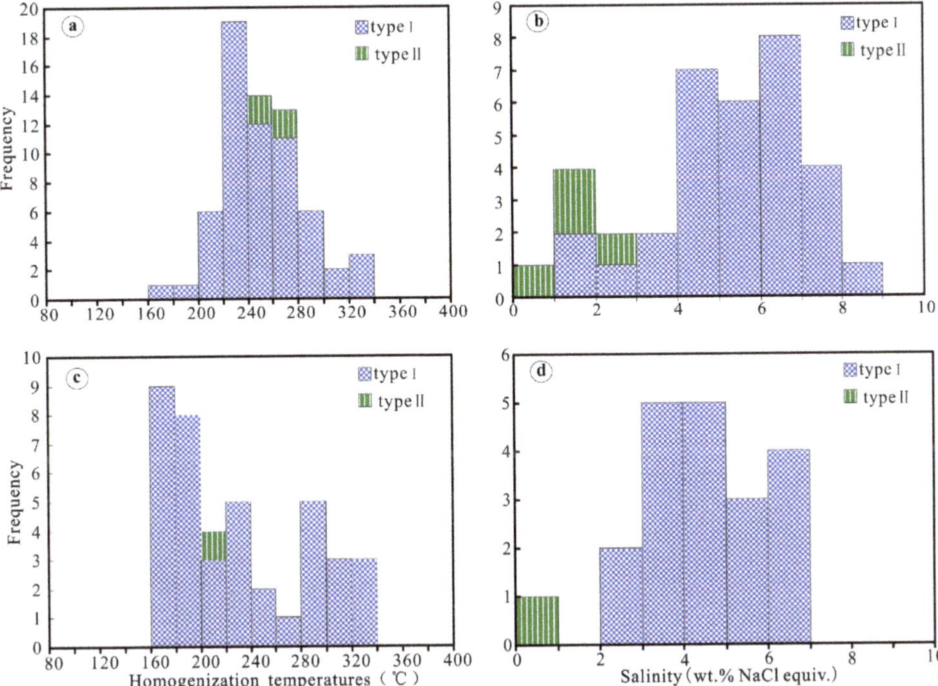

**Figure 6.** Histograms showing homogenization temperatures and salinities of fluid inclusions in two types of orebodies from the Nibao gold deposit: (**a**) histograms showing homogenization temperatures of fluid inclusions in the main-stage quartz of Sbt; (**b**) histograms showing salinities of fluid inclusions in the main-stage quartz of Sbt; (**c**) histograms showing homogenization temperatures of fluid inclusions in the main-stage quartz of fault F1; (**d**) histograms showing salinities of fluid inclusions in the main-stage quartz of fault F1.

## 4.2. Rb–Sr Isotopic Compositions

The Rb–Sr isotopic compositions of fluid inclusions in quartz collected from both the fault-controlled orebody No. III and the strata-bound orebody No. IV are presented in Table 1. For quartz from orebody No. III, the Rb content varied from 0.2921 to 2.0760 ppm, whereas the Sr concentration ranged from 0.6469 to 2.1820 ppm. Therefore, the obtained $^{87}Rb/^{86}Sr$ and $^{87}Sr/^{86}Sr$ ratios were 1.302 to 3.732 and 0.71103 ± 0.00004 to 0.71594 ± 0.00004, respectively. In the case of orebody No. IV, the Rb concentration ranged from 0.0786 to 0.3333 ppm, while the Sr content varied from 0.1703 to 0.9890 ppm. Thus, the $^{87}Rb/^{86}Sr$ and $^{87}Sr/^{86}Sr$ ratios varied from 0.491 to 1.900 and 0.70961 ± 0.00003 to 0.71246 ± 0.00004, respectively. For these two types of gold orebodies, the variation range of the Rb and Sr isotopic data of fluid inclusions in quartz was relatively broad and was sufficient for constructing valid isochrons (Figure 7). The isochron ages of the fault-controlled gold orebody No. III and the strata-bound orebody No. IV were thus determined to be 142 ± 3 Ma (95% reliability) and 141 ± 2 Ma (95% reliability), respectively. The isochron intercept gave the initial $^{87}Sr/^{86}Sr$ ratios of 0.70844 ± 0.00022 (2σ) (MSWD = 0.73) and 0.70862 ± 0.00020 (2σ) (MSWD = 0.063), respectively.

**Table 1.** The quartz fluid inclusion Rb–Sr isotopic measurement results from the Nibao gold deposit.

| Type | Position | Sequence No. | Lab. No. | Sample No. | Rb (ppm) | Sr (ppm) | $^{87}Rb/^{86}Sr$ | $^{87}Sr/^{86}Sr \pm 2\sigma$ |
|---|---|---|---|---|---|---|---|---|
| fault-controlled gold orebody No. III | fault F1 | 1 | 3015648-1 | 123-5 | 1.0450 | 1.6960 | 1.776 | 0.71198 ± 0.00005 |
| | | 2 | 3015648-2 | 543-5 | 0.2921 | 0.6469 | 1.302 | 0.71103 ± 0.00004 |
| | | 3 | 3015648-3 | 9470-5-4 | 1.0770 | 2.1820 | 1.423 | 0.71127 ± 0.00001 |
| | | 4 | 3015648-5 | 544-4 | 2.0760 | 1.6050 | 3.732 | 0.71594 ± 0.00004 |
| | | 5 | 3015648-6 | 558-4 | 0.9183 | 1.3770 | 1.923 | 0.71245 ± 0.00005 |
| strata-bound gold orebody No. IV | Sbt | 1 | 3015647-1 | NBP2-1 | 0.1197 | 0.7024 | 0.491 | 0.70961 ± 0.00003 |
| | | 2 | 3015647-2 | NBP2-2 | 0.0786 | 0.1703 | 1.332 | 0.71126 ± 0.00007 |
| | | 3 | 3015647-4 | NBP2-4 | 0.1762 | 0.9890 | 0.514 | 0.70966 ± 0.00004 |
| | | 4 | 3015647-5 | NBP2-5 | 0.3333 | 0.5059 | 1.900 | 0.71246 ± 0.00004 |
| | | 5 | 3015647-6 | NBP2-6 | 0.1179 | 0.3465 | 0.981 | 0.71060 ± 0.00004 |

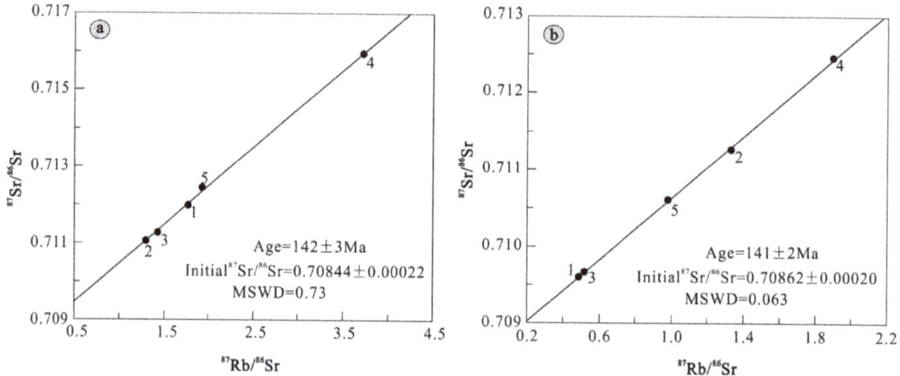


**Figure 7.** Rb–Sr isotope isochronic ages of quartz from the Nibao gold deposit for (**a**) fault-controlled gold orebodies and (**b**) strata-bound gold orebodies. The numbers (1–5) are meaning the sequence No. in Table 1.

## 5. Discussion

### 5.1. Age of Mineralization

Because quartz is extremely pure and free of Rb- and Sr-rich impurities, Rb and Sr are usually hosted in fluid inclusions, quartz is an ideal target mineral for the Rb–Sr dating method [24–28]. Therefore, the Rb–Sr isochron method for fluid inclusions in quartz has been used as a tool to reveal the mineralization age of certain gold deposits [10,28–33].

Quartz samples for Rb–Sr isotopic dating were all collected from the hydrothermal metallogenic stage that were related to gold mineralization. Five pieces of picked quartz samples were measured for gold content by the fire assay method and quantified by atomic absorption spectroscopy (Varian Spectr AA240) in ALS Chemex Co., Ltd. (Guangzhou, China). The results showed that the gold contents of these samples ranged from 0.106 to 3.640 ppm, indicating that the quartz samples were closely associated with gold mineralization. Therefore, we can conclude that the Rb–Sr isochrons of fluid inclusions in quartz from the Nibao fault-controlled and strata-bound gold orebodies have ages of 142 ± 3 and 141 ± 2 Ma, respectively, which are similar to each other within the error range. This also illustrated that these two types of orebodies were resulted from a single geological event.

Mao et al. [34] suggested that Mesozoic large-scale mineralization in South China may be related to the time taken for the lithosphere stretch in the same area. Specifically, Mesozoic mineralization occurred mainly around 170–150, 140–126, and 110–80 Ma, respectively, whereas the corresponding lithosphere stretch happened around 180–155, 145–125, and 110–75 Ma, respectively. The similarity between these values indicates that both of these events resulted from the same earth dynamic evolution, that is to say that the large-scale lithosphere stretch might have contributed to the massive mineralization. However, previous studies on the ore-forming ages of gold deposits in the Yunnan–Guizhou–Guangxi region show different results [14–16]. Age data greatly vary for different deposits due to using different dating methods, sometimes even within the same mining district (e.g., Lannigou and Shuiyindong) (Table 2), making the constraint of the mineralization age difficult. Another possible reason for this variation may be that the dating results provided by some determination methods cannot accurately represent the actual gold mineralization time. For example, since electron spin resonance spectroscopy is likely to be affected by epigenetic events, the age data are only valid before the time of the latest hydrothermal activity (i.e., the mineralization epoch's upper limit) [35]. Also, another study suggested that the accuracy of the Re–Os dating method is relatively due to the low content of arsenic pyrite, which is a target mineral for the measurement [36]. Moreover, the age of 235–193 Ma obtained by the Re–Os method is much older than those by other methods.

**Table 2.** The ages of typical gold deposits in the Golden Triangle region.

| Deposit | Stratum | Dating Methods | Dating Mineral | Ages (Ma) | Reference |
|---|---|---|---|---|---|
| Nibao | Permian | SIMS Th–Pb age | Apatite | 142 ± 3 | [16] |
| | | Fluid inclusion Rb–Sr age | Quartz | 141 ± 2~142 ± 3 | (This study) |
| Lannigou | Triassic | Fluid inclusion Rb–Sr age | Quartz, Calcite | 105.6 | [37] |
| | | Quartz electron spin resonance (ESR) age | Quartz | 82.9 ± 6.3 | [35] |
| | | Re–Os age | Arsenopyrite | 204 ± 19 | [15] |
| Suiyindong | Permian | Sm–Nd age | Calcite | 134 ± 3~136 ± 3 | [14] |
| | | Re–Os age | Arsenopyrite | 235 ± 33 | [15] |
| Zimudang | Permian | Sm–Nd age | Calcite | 148.4 ± 4.8 | [38] |
| Baidi | Triassic | Quartz electron spin resonance (ESR) age | Quartz | 87.6 ± 6.1 | [35] |
| Jinya | Triassic | Re–Os age | Arsenopyrite | 206 ± 22 | [15] |
| | | SIMS Th–Pb age | Apatite | 146 ± 6.2 | [18] |
| Badu | Permian | SIMS U–Pb age | Rutile | 137.9 ± 5.1 | [18] |
| | | SIMS U–Pb age | Zircon | 141 ± 1.9 | [18] |

In contrast, the Rb–Sr isotopic dating method for fluid inclusion in quartz and the Sm–Nd isotopic dating method for fluid inclusion in calcite are appropriate choices for determining the age of gold deposits because of the data effectiveness and relatively small error bound [14,27]. This has been verified in a number of gold deposits in the study area. For example, the determined isotopic age of the Lannigou gold deposit was 105.6 Ma by the Rb–Sr method with fluid inclusions in quartz [37], the obtained isotopic age of the Shuiyindong gold deposit ranged from 134 ± 3–136 ± 3 Ma by the Sm–Nd method with fluid inclusions in calcite [14], and the Sm–Nd method with fluid inclusions in calcite indicated that the achieved isotopic age of the Zimudang gold deposit was 148.4 ± 4.8 Ma [38].

Notably, Chen et al. [16] used the SIMS U–Pb isochron method for hydrothermal apatite to date the age of the fault-controlled orebody in the Nibao gold deposit and obtained a result of 142 ± 3 Ma, which is the same to the data in this study by the Rb–Sr method with fluid inclusion in quartz. Generally, the gold deposits mentioned above are of the same mineralization age being of different ore-bearing strata and different structures. As shown in Figure 8, the mineralization age of the Carlin-type gold deposits in the Golden Triangle is mainly 148–134 Ma. This is consistent with the structural evolution background of the study area. After the Devonian period, the Golden Triangle experienced the following four stages of continental dynamics evolution: the Hercynian basin stretch-chasmic stage (405–250 Ma), the Indo-Chinese epoch arc-rear basin development stage (250–205 Ma), the Early–Medium Yanshanian intracontinental orogenic stage (205–140 Ma), and the Late Yanshanian postorogenic and crust stretch stage (140–66 Ma) [39]. The crust stretched during the Late Yanshanian stage and produced a great amount of ore-forming materials and fluids, thus developed massive medium-low temperature hydrothermal deposits in favorable host rocks or in some secondary structures over a large area.

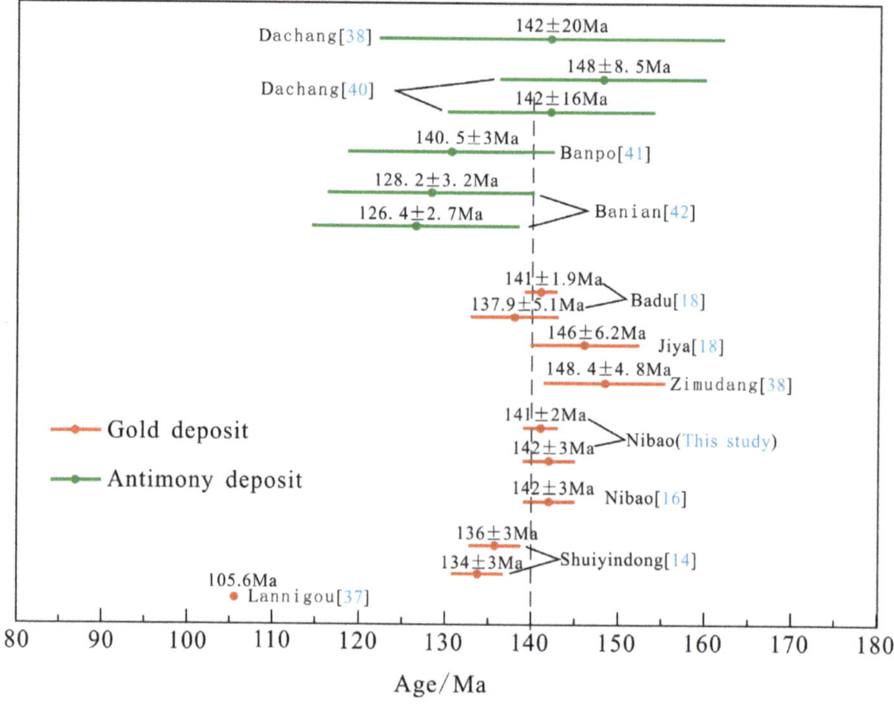



**Figure 8.** Mineralization ages of the gold and antimony deposits in the Yunnan–Guizhou–Guangxi Golden triangle region.

Furthermore, antimony deposits that are considered to be closely related to the gold deposits in the study area are of the mineralization age 148–126 Ma. For example, the fluorite Sm–Nd isochron ages of the Dachang antimony deposit in Qinglong County varied 142 ± 16–148 ± 8.5 Ma [40] and 141 ± 20 Ma [38]. Moreover, the calcite Sm–Nd isochron age of the Banpo antimony deposit is 130.5 ± 3.0 Ma [41], and the calcite Sm–Nd isochron age of the Banian deposit in the Dushan antimony ore field is 126.4 ± 2.7 to 128.2 ± 3.2 Ma [42]. Since the isochron ages of two types of orebody from the Nibao gold deposit are close to each other, we concluded that the gold and antimony deposits in the study area might have been formed around 140 Ma (i.e., in the same mineralization epoch, Middle-Late Yanshanian). This implies that both the gold and the antimony deposits were resulted from the same geological event, showing the differences only in the occurrence space, ore-bearing rocks, physicochemical conditions, fluid migration, and evolution time during the mineralization process.

Investigating the cross-cutting relationship between rock masses and orebodies is still considered to be an effective tool in terms of mineralization age research. It is actually a foundation for dating deposits when used with high-precision age determination. For example, the Shijia gold deposit located in northwestern Guangxi occurs not only in a fault-shattered zone within a diabase dyke but also at the contact zone between the diabase and the Carboniferous strata. Thus, the age of the diabase can provide the lower limit of the deposition age of the Shijia gold deposit, which is <140 Ma [43]. The muscovite $^{40}Ar/^{39}Ar$ method applied to quartz porphyry bodies located at Bama, Fengshan, Lingyun, and Liaotun in northwestern Guangxi yielded ages of 96.5–95.0 Ma. The contact relations between the porphyry and the gold deposit (Liaotun gold deposit) indicate that the latter age (95.0 Ma) represents the upper limit of the age of the gold deposit [36]. Therefore, from this perspective, it can be concluded that the mineralization of the Carlin-type gold deposits in the Yunnan–Guizhou–Guangxi Golden Triangle took place possibly around 140–95 Ma, which is basically the same as 148–134 Ma, which was obtained from the quartz Rb–Sr method and the calcite Sm–Nd method. Previous studies have shown that Mesozoic massive mineralization in South China may be related to lithosphere stretching and synchronal igneous intrusion activity; that is, there seems to exist a time–space catenation relationship between the two [4,34].

Whole-rock K–Ar isotopic analyses of intrusive rocks have yielded geochronological data that have reference value, such as the 146–115.5 Ma of diabase dykes from Pu'an and Panxian Counties [44]. Furthermore, Hu et al. [43] determined the age of the Shijia gold deposit and Badu diabase from northwest Guangxi to be 140 Ma by using the K–Ar isotopic method. Therefore, all the studies mentioned above suggest that the time of magmatic emplacement is almost the same as that of the mineralization of gold deposits (148–134 Ma) within the area of the Golden Triangle. This illustrates that the low-temperature hydrothermal deposit (i.e., gold deposit) in this area has a very close time and space relationship with Yanshanian magmatic activities. Thus, the massive gold and antimony mineralization could be attributed to late Yanshanian mantle-derived mafic–ultramafic magmatic activities that produced a stretching-structure environment.

In summary, based on the isotopic ages of two types of orebody from the Nibao gold deposit and the previous age data, we suggest that in the Yunnan–Guizhou–Guangxi Golden Triangle, the massive mineralization of low-temperature deposits represented by gold and antimony took place around 140 Ma.

### 5.2. Ore-Forming Fluids

Zheng et al. [1] studied the geochemical characteristics of gold orebodies hosted by different ore-bearing rocks in the Nibao gold deposit and found that all the orebodies possess a similar ore-forming source which had been mixed significantly by mantle-derived materials. It had been previously reported that the precipitation of quartz in a hydrothermal deposit has little influence on the Sr isotopic composition of the entire hydrothermal system, and thus, the $^{87}Sr/^{86}Sr$ ratio of fluid inclusions in quartz can reflect the properties of the ore-forming fluids [25,45]. Since quartz is closely associated with Carlin-type gold mineralization, variations of the Sr isotopic composition can provide indications for the ore-forming fluids and materials.

In the Nibao gold deposit, the initial $^{87}Sr/^{86}Sr$ ratios of fault-controlled and strata-bound gold orebodies are $0.70844 \pm 0.00022$ (2σ) and $0.70862 \pm 0.00020$ (2σ), respectively, which are distinctly lower than that of the average of the upper crust (0.71190) suggested by Palmer et al. [46], but are close to the original value of Sr (0.707) of the crust–mantle boundary reported by Faure [47]. Therefore, we can infer that ore-forming fluids possess characteristics of both mantle-derived Sr isotopes and stratigraphic Sr isotopes, suggesting that the Sr of the Nibao gold deposit may be of mixed origin. In addition, investigations of stable isotopes (C, H, and O) and in situ sulfur isotopes of the gold deposits in southwestern Guizhou also suggest that the mineralization may involve mantle-derived material. Especially, the studies on H and O isotopes of several gold deposits (e.g., Yata, Banqi, Getang, Shuiyindong, Taipingdong, and Nibao) all showed that the ore-forming fluid may be a thermal fluid with multiple origins, namely, deep magmatic water mixed with meteoric water that migrated at shallower levels. This illustrates that there may be deeply covered rock masses underground [4,8,10,13,48].

## 6. Conclusions

The Nibao gold deposit is an important part of the Yunnan–Guizhou–Guangxi Golden Triangle region. In this work, we investigated the mineralization ages of fault-controlled and strata-bound gold orebodies within the Nibao gold deposit. The key findings of this study are summarized below.

(1)   Rb–Sr isochron ages of $142 \pm 3$ and $141 \pm 2$ Ma were obtained from gold-bearing quartz vein fluid inclusions for the fault-controlled and strata-bound orebodies within the Nibao gold deposit. Since the regional age of gold mineralization varied from 148 to 134 Ma, we confirmed that the age of the Carlin-type gold deposits in the Golden Triangle is Middle-Late Yanshanian.

(2)  In the Nibao gold deposit, the initial $^{87}Sr/^{86}Sr$ ratios of fluid inclusions in gold-bearing quartz for fault-controlled and strata-bound gold orebodies are $0.70844 \pm 0.00022$ ($2\sigma$) and $0.70862 \pm 0.00020$ ($2\sigma$), respectively. Considering the isotopic data and previously reported isotopic results (C, H, O, and S) for gold deposits in this region, we infer that the mantle-derived material can be involved in the formation of gold deposits and that the ore-forming fluid was likely a mixture of crustal and mantle-derived fluids.

(3)  In South China, low-temperature gold hydrothermal deposits are spatially and temporally related to Yanshanian magmatism, suggesting that large-scale gold mineralization has geneticity with Late Yanshanian mantle-derived mafic–ultramafic magmas.

**Author Contributions:** Conceptualization, R.Y. and L.Z.; methodology, R.Y. and L.Z.; software, L.Z. and D.L.; formal analysis, L.Z. and J.G.; investigation, L.Z., J.G., J.C. and D.L.; resources, R.Y. and J.L.; data curation, L.Z.; writing—original draft preparation, L.Z.; writing—review and editing, L.Z., R.Y., J.G., J.C. and J.L.; supervision, R.Y.; funding acquisition, R.Y.

**Funding:** This research was supported by the Talent introduction research project of Guizhou University (No. [2017]36), the National Natural Science Foundation of China (No. U1812402,41802088), the National Key Research and Development Program of China (No. 2017YFC0601500) and the Project of the Scientific and Technological Innovation Team of Sedimentary Deposit in Guizhou Province (2018-5613).

**Acknowledgments:** We sincerely appreciate the editors' and anonymous reviewer's comments. And thanks the analyst (Chongfan Liu) of the Rb-Sr isotopic compositions.

**Conflicts of Interest:** The authors declare no conflict of interest.

## References


1.  Tan, Q.P.; Xia, Y.; Xie, Z.J.; Yan, J. Migration paths and precipitation mechanisms of ore-forming fluids at the Shuiyindong Carlin-type gold deposit, Guizhou, China. *Ore. Geol. Rev.* **2015**, *69*, 140–156. [CrossRef]

2.  Zheng, L.L.; Yang, R.D.; Gao, J.B.; Chen, J.; Liu, J.Z.; He, Y.N. Geochemical characteristics of the giant Nibao Carlin-type gold deposit (Guizhou, China) and their geological implications. *Arab. J. Geosci.* **2016**, *9*, 108–123. [CrossRef]

3.  Hu, R.Z.; Su, W.C.; Bi, X.W.; Tu, G.C.; Hofstra, A.H. Geology and geochemistry of Carlin-type gold deposits in China. *Miner. Depos.* **2002**, *37*, 378–392.

4.  Hu, R.Z.; Fu, S.L.; Huang, Y.; Zhou, M.F.; Fu, S.H.; Zhao, C.H.; Wang, Y.J.; Bi, X.W.; Xiao, J.F. The giant South China Mesozoic low-temperature metallogenic domain: Reviews and a new geodynamic model. *J. Asian Earth Sci.* **2017**, *137*, 9–34. [CrossRef]

5.  Liu, J.M.; Ye, J.; Ying, H.L.; Liu, J.J.; Zheng, M.H.; Gu, X.X. Sediment-hosted micro-disseminated gold mineralization constrained by basin paleo-topographic highs in the Youjiang basin, South China. *J. Asian Earth Sci.* **2002**, *20*, 517–533. [CrossRef]

6.  Peters, S.G.; Huang, J.Z.; Li, Z.P.; Jing, C.G. Sedimentary rock-hosted Au deposits of the Dian–Qian–Gui area, Guizhou and Yunnan Provinces, and Guangxi district, China. *Ore. Geol. Rev.* **2007**, *31*, 170–204. [CrossRef]

7.  Zhang, Y.; Xia, Y.; Su, W.C.; Zhang, X.C.; Liu, J.Z.; Deng, Y.M. Metallogenic model and prognosis of the Shuiyindong super-large strata-bound Carlin-type gold deposit, southwestern Guizhou province, China. *Chin. J. Geochem.* **2010**, *29*, 157–166. [CrossRef]

8.  Su, W.C.; Zhang, H.T.; Hu, R.Z.; Ge, X.; Xia, B.; Chen, Y.Y.; Zhu, C. Mineralogy and geochemistry of gold-bearing arsenian pyrite from the Shuiyindong Carlin-type gold deposit, Guizhou, China: Implications for gold depositional processes. *Miner. Depos.* **2012**, *47*, 653–662. [CrossRef]

9.  Su, W.C.; Dong, W.D.; Zhang, X.C.; Shen, N.P.; Hu, R.Z.; Hofstra, A.H.; Cheng, L.Z. Carlin-type gold deposits in the Dian-Qian-Gui "Golden Triangle" of southwest China. *Rev. Econ. Geol.* **2018**, *20*, 157–185.

10. Cline, J.S.; Muntean, J.L.; Gu, X.X.; Xia, Y. A comparison of Carlin-type gold deposits: Guizhou province, golden triangle, southwest China, and northern Nevada, USA. *Earth Sci. Front.* **2013**, *20*, 1–18.

11. Hou, L.; Peng, H.J.; Ding, J.; Zhang, J.R.; Zhu, S.B.; Wu, S.Y.; Wu, Y.; Ouyang, H.G. Textures and in situ chemical and isotopic analyses of pyrite, Huijiabao trend, Youjiang basin, China: Implications for paragenesis and source of sulfur. *Econ. Geol.* **2016**, *111*, 331–353. [CrossRef]

12. Xie, Z.J.; Xia, Y.; Cline, J.S.; Koenig, A.; Wei, D.T.; Tan, Q.P.; Wang, Z.P. Are there Carlin-type gold deposits in China? A comparison of the Guizhou, China, deposits with Nevada, USA, deposits. *Rev. Econ. Geol.* **2018**, *20*, 187–233.

13. Xie, Z.J.; Xia, Y.; Cline, J.S.; Pribil, M.J.; Koenig, A.; Tan, Q.P.; Wei, D.T.; Wang, Z.P.; Yan, J. Magmatic origin for sediment-hosted Au deposits, Guizhou province, China: In situ chemistry and sulfur isotope composition of pyrites, Shuiyindong and Jinfeng deposits. *Econ. Geol.* **2018**, *113*, 1627–1652.

14. Su, W.C.; Hu, R.Z.; Xia, B.; Xia, Y.; Liu, Y.P. Calcite Sm-Nd isochron age of the Shuiyindong Carlin-type gold deposit, Guizhou, China. *Chem. Geol.* **2009**, *258*, 269–274. [CrossRef]

15. Chen, M.H.; Mao, J.W.; Li, C.; Zhang, Z.Q.; Dang, Y. Re-Os isochron ages for arsenopyrite from Carlin-like gold deposits in the Yunnan-Guizhou-Guangxi "golden triangle", southwestern China. *Ore. Geol. Rev.* **2015**, *64*, 316–327. [CrossRef]

16. Chen, M.H.; Bagas, L.; Liao, X.; Zhang, Z.Q.; Li, Q.L. Hydrothermal apatite SIMS Th–Pb dating: Constraints on the timing of low-temperature hydrothermal Au deposits in Nibao, southwest China. *Lithos* **2019**, *324*, 418–428. [CrossRef]

17. Pi, Q.H.; Hu, R.Z.; Xiong, B.; Li, Q.L.; Zhong, R.C. In situ SIMS U–Pb dating of hydrothermal rutile: Reliable age for the Zhesang Carlin-type gold deposit in the golden triangle region, southwest China. *Miner. Depos.* **2017**, *52*, 1179–1190. [CrossRef]

18. Gao, W. Geochronology and Dynamics of Carlin-Type Gold Deposits in the Youjiang Basin (NW Guangxi). Ph.D. Thesis, Institute of Geochemistry, Chinese Academy of Sciences, Guiyang, China, 2018 (unpublished). (In Chinese)

19. Wang, L.; Long, C.L.; Liu, Y. Discussion on concealed rock mass delineation and gold source in southwestern Guizhou. *Geoscience* **2015**, *29*, 702–712. (In Chinese with English Abstract)

20. He, B.; Xu, Y.G.; Huang, L.X.; Luo, Z.Y.; Shi, Y.R.; Yang, Q.J.; Yu, S.Y. Age and duration of the Emeishan flood volcanism, southwest China: Geochemistry and SHRIMP zircon U-Pb dating of silicic ignimbrites, post-volcanic Xuanwei Formation and clay tuff at the Chaotian section. *Earth Planet. Sci. Lett.* **2007**, *255*, 306–323. [CrossRef]

21. Mundil, R.; Ludwig, K.R.; Metcalfe, I.; Renne, P.R. Age and timing of the Permian mass extinctions: U/Pb dating of closed-system zircons. *Science* **2004**, *305*, 1760–1763. [CrossRef]

22. Zhou, M.F.; Zhao, J.H.; Qi, L.; Su, W.C.; Hu, R.Z. Zircon U-Pb geochronology and elemental and Sr–Nd isotope geochemistry of Permian mafic rocks in the Funing area, SW China. *Contrib. Miner. Petrol.* **2006**, *151*, 1–19. [CrossRef]

23. Yin, H.F.; Yang, F.Q.; Yu, J.X.; Peng, Y.Q.; Wang, S.Y.; Zhang, S.X. An accurately delineated Permian-Triassic boundary in continental successions. *Sci. China: Earth Sci.* **2007**, *35*, 1281–1292. [CrossRef]

24. Shepherd, T.J.; Darbyshire, D.P.F. Fluid inclusion Rb–Sr isochrons for dating mineral deposits. *Nature* **1981**, *290*, 578–579. [CrossRef]

25. Norman, D.I.; Lands, G.P. Source of mineralizing components in hydrothermal ore fluids as evidenced by $^{87}Sr/^{86}Sr$ and stable isotope data from the Pasto Bueno deposit, Peru. *Econ Geol.* **1983**, *78*, 451–465. [CrossRef]

26. Rossman, G.R.; Weis, D.; Wasserburg, G.J. Rb, Sr, Nd and Sm concentrations in quartz. *Geochim. Cosmochim. Acta* **1987**, *51*, 2325–2329. [CrossRef]

27. Changkakoti, A.; Gray, J.; Krsticl, D.; Cumming, G.L.; Morton, R.D. Determination of radiogenic isotope (Rb/Sr, Sm/Nd, and Pb/Pb) in fluid inclusion water: An example from the Bluebell Pb–Zn deposit, British Columbia, Canada. *Geochimi. Cosmochim. Acta* **1988**, *52*, 961–967. [CrossRef]

28. Li, H.Q.; Liu, J.Q.; Wei, L. *The Chronological and Geological Application of Fluid Inclusions in Hydrothermal Deposits*; Geological Publishing House: Beijing, China, 1993; pp. 1–25. (In Chinese)

29. Mo, C.H.; Wang, X.Z.; Cheng, J.P. Auriferous quartz veins from the Dongping gold deposit, NW Hebei province and metallogenesis—Fluid inclusion Rb–Sr isochron evidence. *Chin. J. Geochem.* **1996**, *15*, 265–271.

30. Zhang, L.C.; Shen, Y.C.; Ji, J.S. Characteristics and genesis of Kanggur gold deposit in the eastern Tianshan mountains, NW China: Evidence from geology, isotope distribution and chronology. *Ore Geol. Rev.* **2003**, *23*, 71–90. [CrossRef]

31. Zhu, Y.F.; Zhou, J.; Zeng, Y.S. The Tianger (Bingdaban) shear zone hosted gold deposit, west Tianshan, NW China: Petrographic and geochemical characteristics. *Ore Geol. Rev.* **2007**, *32*, 337–365. [CrossRef]

32. Ni, P.; Wang, G.G.; Chen, H.; Xu, Y.F.; Guan, S.J.; Pan, J.Y.; Li, L. An Early Paleozoic orogenic gold belt along the Jiang–Shao Fault, south China: Evidence from fluid inclusions and Rb–Sr dating of quartz in the Huangshan and Pingshui deposits. *J. Asian Earth Sci.* **2015**, *103*, 87–102. [CrossRef]

33. Sahoo, A.K.; Krishnamurthi, R.; Vadlamani, R.; Pruseth, K.L.; Narayanan, M.; Varghese, S.; Pradeepkumar, T. Geneticpects of gold mineralization in the southern Granulite Terrain, India. *Ore Geol. Rev.* **2016**, *72*, 1243–1262. [CrossRef]

34. Mao, J.W.; Xie, G.Q.; Li, X.F.; Zhang, C.Q.; Mei, Y.X. Mesozoic large scale mineralization and multiple lithospheric extension in south China. *Earth Sci. Front.* **2004**, *11*, 45–55. (In Chinese with English Abstract)

35. Zhang, F.; Yang, K.Y. Metallogenic geochronology for the micro-grain disseminated gold deposits in southwestern Guizhou province. *Chin. Sci. Bull.* **1992**, *27*, 1593–1595. (In Chinese with English Abstract)

36. Chen, M.H.; Zhang, Y.; Meng, Y.Y.; Lu, G.; Liu, S.Q. The confirmation of the upper limit of metallogenetic epoch of Liaotun gold deposit in western Guangxi, China, and its implicationon chronology of Carlin-type gold deposits in Yunnan-Guizhou-Guangxi "golden triangle" area. *Miner. Depos.* **2014**, *33*, 1–13. (In Chinese with English Abstract)

37. Su, W.C.; Yang, K.Y.; Hu, R.Z.; Chen, F. Fluid inclusion chronological study of the Carlin-type gold deposits in southwestern China: As exemplified by the Lannigou gold deposit, Guizhou Province. *Acta Mineral. Sin.* **1998**, *18*, 359–362. (In Chinese with English Abstract)

38. Wang, Z.P. Genesis and dynamic mechanism of the epithermal ore deposits, SW Guizhou, China—A Case Study of Gold and Antimony Deposits. Ph.D. Thesis, Institute of Geochemistry, Chinese Academy of Sciences, Guiyang, China, 2013. (In Chinese)

39. Chen, B.J. Ore-Forming Mechanism and Background of Continental Dynamics of the Shuiyindong Carlin-Type Gold Deposit, Southwestern Guizhou, China. Ph.D. Thesis, Chengdu University of Technology, Chengdu, China, 2010. (In Chinese)

40. Peng, J.T.; Hu, R.Z.; Jiang, G.H. Strontium-neodymium isotope system of fluorites from the Qinglong antimony deposit, Guizhou Province: Constrains on the mineralizing age and ore-forming materials sources. *Acta Petrol. Sin.* **2003**, *19*, 785–791. (In Chinese with English Abstract)

41. Xiao, X.G. Geochronology, Ore Geochemistry and Genesis of the Banpo Antimony Deposit, Guizhou Province, China. Ph.D. Thesis, Kunming University of Seience and Technology, Kunming, China, 2014. (In Chinese)

42. Wang, J.S. Mineralization, Geochronology and Geodynamics of the Low-Temperature Epithermal Metallogenic Domain in Southwestern China. Ph.D. Thesis, Institute of Geochemistry, Chinese Academy of Sciences, Guiyang, China, 2012. (In Chinese)

43. Hu, R.Z.; Su, W.C.; Bi, X.W.; Li, Z.Q. A possible evolution way of ore-forming hydrothermal fluid for the Carlin-type gold deposits in the Yunnan-Guizhou-Guanxi triangle area. *Acta Mineral. Sini.* **1995**, *15*, 144–149. (In Chinese with English Abstract)

44. Bureau of Geology and Mineral Resources of Guizhou Province. *Guizhou Regional Geology*; Geological Publishing House: Beijing, China, 1987; pp. 1–698. (In Chinese)

45. Li, W.; Huang, Z.; Yin, M. Isotope geochemistry of the Huize Zn–Pb ore field, Yunnan Province, Southwestern China: Implication for the sources of ore fluid and metals. *Geochem. J.* **2007**, *41*, 65–81. [CrossRef]

46. Palmer, M.R.; Elderfield, H. Sr isotope composition of sea water over the past 75 Myr. *Nature* **1985**, *314*, 526–528. [CrossRef]

47. Faure, G. *Principles of Isotope Geology*, 2nd ed; Wiley: New York, NY, USA, 1986; pp. 1–589.

48. Chen, J.; Yang, R.D.; Du, L.J.; Zheng, L.L.; Gao, J.B.; Lai, C.K.; Wei, H.R.; Yuan, M.G. Mineralogy, geochemistry and fluid inclusions of the Qinglong Sb–(Au) deposit, Youjiang basin (Guizhou, SW China). *Ore Geol. Rev.* **2018**, *92*, 1–18. [CrossRef]

 *minerals*

*Article*

# Source of the Chaoyangzhai Gold Deposit, Southeast Guizhou: Constraints from LA-ICP-MS Zircon U–Pb Dating, Whole-rock Geochemistry and *In Situ* Sulfur Isotopes

Hinyuen Tsang [1,2], Jingya Cao [1,2,*] and Xiaoyong Yang [1,2,*]

[1]  CAS Key laboratory of Crust-Mantle Materials and Environments,
     University of Science and Technology of China, Hefei 230026, China; geotsang2012@yahoo.com.hk
[2]  CAS Center for Excellence in Comparative Planetology, University of Science and Technology of China,
     Hefei 230026, China
*    Correspondence: jingyacao@126.com (J.C.); xyyang@ustc.edu.cn (X.Y.);
     Tel.: +86-0551-360-6871 (J.C.); +86-0551-360-6871 (X.Y.)

Received: 19 March 2019; Accepted: 13 April 2019; Published: 16 April 2019

**Abstract:**  The Chaoyangzhai gold deposit is one of the newly discovered medium to large scale turbidite-hosted gold deposits in Southeast Guizhou, South China. In this study, laser ablation-inductively coupled plasma-mass spectrometer (LA-ICP-MS) zircon U–Pb dating on the tuffaceous- and sandy-slates of Qingshuijiang Formation, Xiajiang Group, and gold-bearing quartz vein yielded similar age distributions, indicating that zircon grains in gold-bearing quartz vein originated from the surrounding tuffaceous- and sandy-slates. In addition, the youngest weighted mean ages of the zircon grains from the tuffaceous- and sandy-slates were 775 ± 13 Ma and 777 ± 16 Ma, respectively, displaying that the tuffaceous- and sandy-slates of the Qingshuijiang Formation were likely deposited in Neoproterozoic. Based on their major and trace element compositions, the tuffaceous- and sandy-slates were sourced from a felsic igneous provenance. The sandy slates have higher contents of Au (mostly ranging from 0.019 to 0.252 ppm), than those of the tuffaceous slates (mostly lower than 0.005 ppm). The $\delta^{34}S_{V\text{-CDT}}$ values of pyrite and arsenopyrite of the gold-bearing samples range from +8.12‰ to +9.99‰ and from +9.78 to +10.78‰, respectively, indicating that the sulfur source was from the metamorphic rocks. Together with the evidence of similar geochemical patterns between the tuffaceous- and sandy-slates and gold-bearing quartz, it is proposed that the gold might be mainly sourced from sandy slates. The metamorphic devolatilization, which was caused by the Caledonian orogeny (Xuefeng Orogenic Event), resulted in the formation of the ore-forming fluid. Gold was likely deposited in the fractures due to changes of the physico-chemical conditions, leading to the formation of the Chaoyangzhai gold deposit, and the large-scale gold mineralization in Southeast Guizhou.

**Keywords:** zircon U–Pb age; in situ sulfur isotopes; gold source; ore genesis; Chaoyangzhai gold deposit; Southeast Guizhou

---

## 1. Introduction

The turbidite-hosted gold deposit, belonging to the orogenic class of gold deposits, was first explored in Australia and Canada [1–3]. Quartz vein type ore bodies of turbidite-hosted gold deposit commonly occur in the metamorphic rocks and they are related to the faults and/or shear zones [4]. In addition, these gold deposits often appear in groups and/or belts, which are characterized by abundant gold resources and reserves. Examples include Hill End in central New South Wales, Australia [3], the Bendigo-Ballarat in central Victoria, Australia [5], the Suurikuusikko in Central

Lapland Greenstone Belt, Finland [6], and Tianzhu-Jinping in southeast Guizhou, China [7–10]. The source of gold for these turbidite-hosted gold deposits has been debated and two contrasting models exist: (1) the metamorphic rocks that host the gold deposits and (2) the gold-rich fluid from the igneous intrusions [11]. However, the viewpoint that gold of these turbidite-hosted gold deposits is sourced from metamorphic rocks has been widely accepted [11–14].

The western Jiangnan Orogenic Belt, which is situated in the Xuefeng region, is one of the most important turbidite-hosted gold belts in South China, with 319 explored gold deposits/mines and an estimated gold reserve of >200 t [15]. The Southeast Guizhou province, together with the west Hunan province, are significant part of the western Jiangnan Orogenic Belt, where numerous of turbidite-hosted gold deposits were explored, such as Bake, Jinjing, Pingqiu, and Kengtou gold deposits (Figure 1) [8–10]. Although some studies have been carried out on these gold deposits, the source and genesis of these deposits are still poorly constrained [8–10]. As a newly discovered turbidite-hosted gold deposit in southeast Guizhou province, the Chaoyangzhai gold deposit has attracted little attention from the geologists. Furthermore, the source and genesis of this deposit still remain unclear. Therefore, a series of laser ablation-inductively coupled plasma-mass spectrometer (LA-ICP-MS) zircon U–Pb dating, bulk-rock major and trace element compositions, and in situ LA-multi-collector-ICP-MS (LA-MC-ICP-MS) sulfur isotopic analyses were carried out for this deposit, with the aim of providing some new insights regarding the source and genesis of this specific gold deposit and the regional style of gold mineralization.

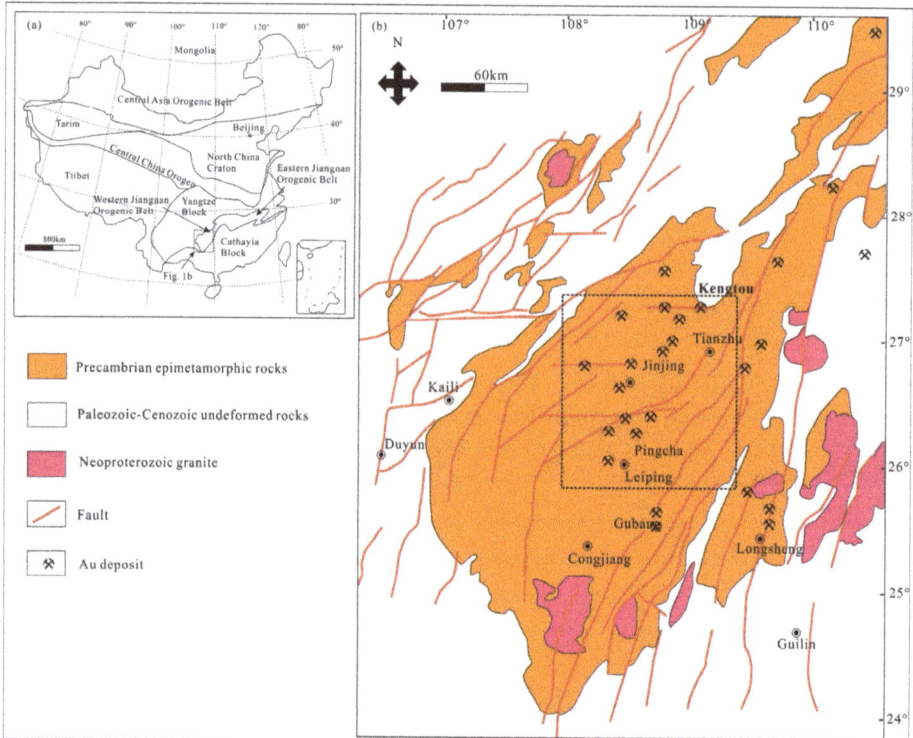


**Figure 1.** (**a**) Geological sketch map of China (modified from [8,16]); (**b**) Geological map of the western part of Jiangnan Orogenic Belt, showing the location of Precambrian rocks, granitic intrusions and gold deposits (modified from [8,17]).

## 2. Geological Setting

### *2.1. Regional Geology*

The Jiangnan Orogenic Belt, which extends from northern Guangxi in the southwest to western Zhejiang in the east, 1500 km long, and ca. 500 km wide, was proposed to be the junction zone between the Yangtze and Cathysia Blocks at Neoproterozoic (Figure 1a) [10,18]. This belt can be subdivided into eastern- and western-parts, with distinct types of mineralization. The western part is characterized by gold mineralization that is related to the orogeny, whereas the eastern part is famous for tungsten polymetallic mineralization related to the Yanshanian magmatism [7,10,19–22]. The Southeast Guizhou, which is located at the western part of the Jiangnan Orogenic Belt, is famous for the intense gold mineralization and it is one of the important gold producers in China [7–10,23,24]. The stratigraphic sequences of this area consist of the Neoproterozoic Xiajiang Group (also named Banxi Group in Hunan Province), Sinian, Carboniferous, Permian, Jurassic, Cretaceous, and Quaternary units (Figure 1b). As the basement and the gold-bearing strata, the Xiajiang Group is widespread in this area and it is subdivided into seven formations, of which four are prominent in the Southeast Guizhou Province: the Longli Formation, the Pinglue Formation, the Qingshuijiang Formation and the Fanzhao Formation (Figure 2) [4,25]. The Longli Formation, with a thickness of 1300–1700 m, is mainly composed of blastopsammite, intercalated with blastopsephitic siltstone, blastopsephitic arkose, and silty sericite slate; The Pinglue Formation (1500–2000 m thick) is mainly composed of sericite slate of silty slate, which is intercalated with a few tuffaceous slate and blastopsammite; The Qingshuijiang Formation (2300–3700 m thick) is mainly composed of bluish or gray, thin to medium banded tuffaceous slate, and laminated medium to thick tuff; The Fanzhao Formation (>1000 m thick) can be subdivided into an Upper Formation and a Lower Formation, which comprise tuffaceous slate and blastopsammite, intercalated with banded tuff, and gray slate, intercalated with blastopsammite, respectively.

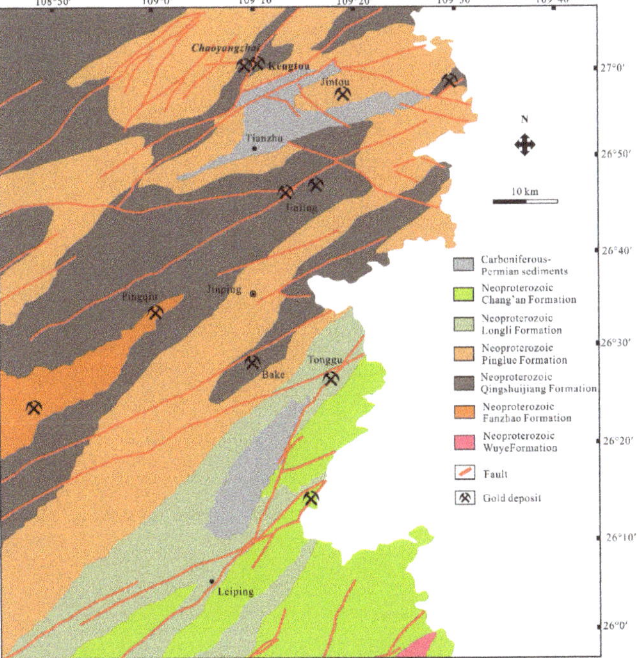


**Figure 2.** Geological map of the Southeast Guizhou, showing the location of Precambrian rocks, gold deposits (modified from [8,26]).

Multistage tectonic events since the Proterozoic has influenced the Southeast Guizhou [25]. During the Xuefeng movement, a series of NE-trending structures were formed, corresponding to the Jinning Orogeny [27–30]. Subsequently, several EW-trending basement rifts were formed during the Caledonian movement [27–29,31]. Subsequently, a series of NNE-trending faults were formed, which were caused by the subduction of the Pacific plate during the Yanshanian and Himalayan movements [27–29,31]. In addition, these NNE-trending faults overprint the EW- and NE-trending structures, and the current tectonic framework was established in this region [4,25]. The magmatic activity is not intense in this region, and only a few Neoproterozoic granitic plutons were intruded (Figure 1b).

## 2.2. Ore Deposit Geology

The Chaoyangzhai gold deposit, which was located at Tianzhu County, Southeast Guizhou, is one of the newly found medium to large scale gold deposits in this region, with an estimated gold reserve of 18,442 kg. The ore-bearing strata are the Neoproterozoic epimetamorphic rocks of Qingshuijiang Formation, Xiajiang Group, which are mainly composed of sandy- and tuffaceous-slates (Figures 3 and 4a,b). The sandy slate contains porphyroclastic quartz, plagioclase, and sericite (Figure 4c). The tuffaceous-slate is mainly composed of quartz as the phenocryst and particulate phenocryst as matrix (Figure 4d). The NE-trending faults are widespread in this deposit, and they are mostly trans-tensional normal faults and transpressional reverse thrust faults (Figure 3). These fractures are not only permeable structures but are also the ore-hosting structures.

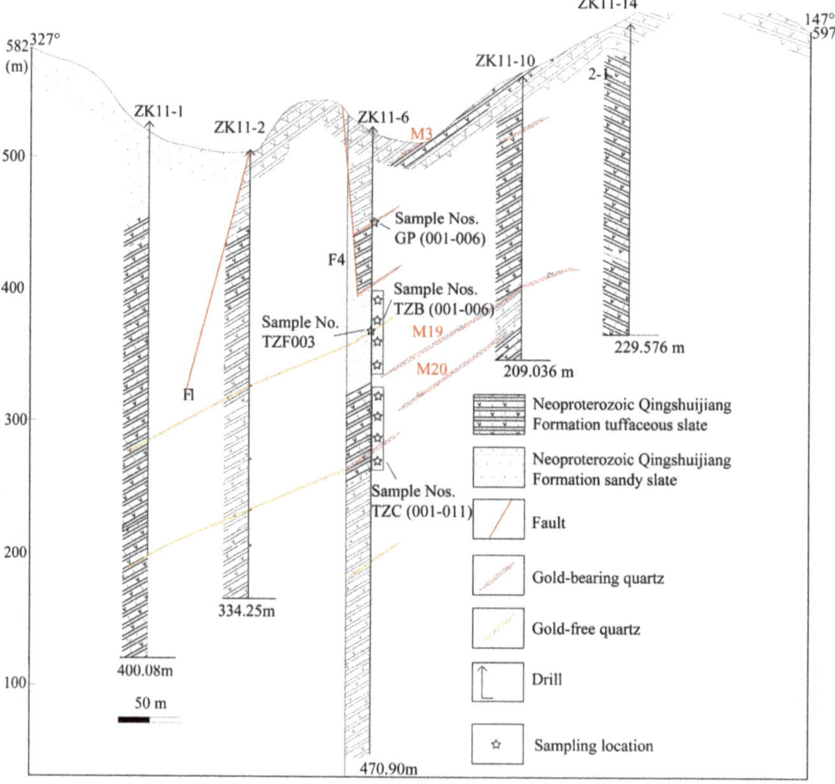


**Figure 3.** Geological Section of the Chaoyangzhai gold deposit, showing drill-core holes and depths.

A total of over fifty quartz veins are found in this deposit, including eight gold bearing veins. These ore veins are mainly NE-trending, with a length of 200–250 m, a thickness of 0.34–1.40 m (mean = 0.65 m), and an Au grade of 0.58–49.5 g/t (mean = 8.4 g/t). In addition, ore veins can occur in both of the sandy- and tuffaceous-slates (Figure 4a,b). The ore minerals are mainly composed of native gold, arsenopyrite, and pyrite (Figure 4e). Most of the native gold grains are 0.1 mm to 3.0 mm within the quartz veins, in granular and/or irregular form, and they co-exist with the sulfides (Figure 4e). Pyrite commonly occurs in the quartz veins, mostly co-existing with the arsenopyrite, with euhedral or subhedral (Figure 4f). Arsenopyrites are mainly disseminated and occurring in the quartz veins, with euhedral or subhedral, short-columnar and relatively coarse grains (0.5–4 mm) (Figure 4g,h). The gangue minerals include quartz, calcite, mica, and chlorite. In general, similar to those gold deposits in this region, hydrothermal alteration of the Chaoyangzhai gold deposit is subtle, and it mainly consists of silicification and carbonatization [7–9].

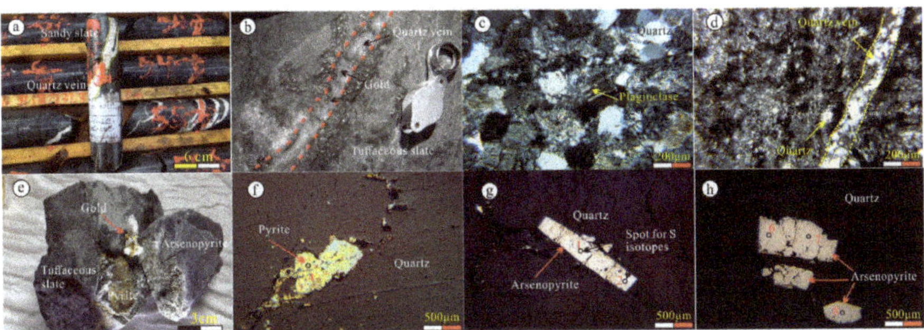

**Figure 4.** Morphology and mineral assemblages of the surrounding rocks, ore body and gold-bearing ore. (**a**) Quartz vein in the sandy-slate; (**b**) Gold-bearing Quartz vein in the tuffaceous slate; (**c**) Quartz and plagioclase in the sandy-slate; (**d**) Quartz veinlet in the tuffaceous slate; (**e**) Quartz, native gold, pyrite and arsenopyrite in the tuffaceous slate; (**f**) Pyrite in the quartz vein showing the location of the in situ sulfur isotopic analyses; (**g**) Clintheriform arsenopyrite in the quartz vein showing the spots of the in situ sulfur isotopic analyses; and, (**h**) Euhedral arsenopyrite in the quartz vein showing the spots of the in situ sulfur isotopic analyses.

## 3. Sampling and Analytical Methods

Samples of sandy slate (Sample Nos. TZB001–TZB006 and TZD002), tuffaceous slate (Sample Nos. TZC001–TZC011 and TZD001), gold-free quartz vein (Sample No. TZF003), and gold-bearing quartz veins (Sample Nos. GP1–GP10 and TZD003) were collected from the drill-cores. Figure 3 shows the sampling locations.

### 3.1. In Situ LA-ICP-MS Zircon U–Pb Dating

Three samples, which were selected for the zircon U–Pb dating, were tuffaceous slate (sample No. TZD001), sandy slate (sample No. TZD002), and gold-bearing quartz vein (sample No. TZD003). Firstly, the zircon grains were separated from these samples and then identified by hand picking under a binocular microscope, mounted in epoxy resin, and polished to expose the interiors. The transmitted and reflected light images of the zircon grains were photographed for documentation (not shown). Cathodoluminescence (CL) images of the zircons were taken while using a scanning micro-probe (JEOL JXA-8100, JEOL, Tokyo, Japan at CAS Key Laboratory of Crust–Mantle Materials and Environments in University of Science and Technology of China, Hefei, China). Zircon U–Pb dating was undertaken with an Agilent 7700 inductively coupled plasma-mass spectrometer (ICP-MS, Agilent, Santa Clara, CA, USA), which was combined with a Coherent 193 laser ablation (LA) system at Sample Solution Analytical Technology Co., Ltd, Wuhan, China. Two zircon standards, 91500 (1062 ± 4 Ma) [32] and GJ-1 (610.0 ± 1.7 Ma) [33], were used as the external standards for dating. Standard silicate glass

(NIST SRM610) was used for external standardization for trace element analysis, and 29Si was used for internal standardization (32.8% $SiO_2$ in zircon). The standard protocol correction method was used in analyzing the 91500 and GJ-1 standard zircons twice and once, respectively, after every five analyses. ICPMSDataCal software was used to process the raw ICP-MS data [34,35], and common Pb was corrected following [36]. Isoplot processed the concordia ages (Version 3.0; [37]).

*3.2. Whole Rock Major and Trace Elements Analyses*

ALS Geochemistry Laboratory in Guangzhou, China carried out whole rock major and trace elements analyses. Before the analyses, samples were crushed in a steel jaw crusher, and then powdered in an agate mill to grain size of 74 μm. The detailed methodologies for major element compositions are as follows: Loss of ignition (LOI) was determined after igniting sample powders at 1000 °C for 1 h. A calcined or ignited sample (0.9 g) was added to 9.0 g of Lithium Borate Flux ($Li_2B_4O_7$–$LiBO_2$), mixed well, and then fused in an auto fluxer at 1050 and 1100 °C. A flat molten glass disk was prepared from the resulting melt. A Panalytical Axios Max X-ray fluorescence (XRF, Panalytical, Almelo, The Netherlands) instrument analyzed this disk was then analyzed, with an analytical accuracy of ca. 1–5%. ICP-MS measured the trace element compositions (Perkin Elmer Elan 9000, Perkin, Waltham, MA, USA), with an analytical accuracy better than 5%.

*3.3. In Situ LA-MC-ICP-MS Sulfur Isotopic Analyses*

In situ sulfur isotopic analyses of pyrite and arsenopyrite were performed using Laser ablation system of a RESOlution M-50 laser ablation system (ASI, Australia), which was equipped with a 193 nm ArF CompexPro102 excimer laser (Coherent, Santa Clara, CA, USA) and Nu Plasma 1700 multi-collector inductively coupled plasma mass spectrometry (MC-ICP-MS, NP-1700, Nu Instruments, Wrexham, UK) in the State Key Laboratory of Continental Dynamics, Northwest University, Xi'an, China. The laser spot sizes of 25–37 μm were used at an energy density of 3.6 J/cm$^2$ and a repetition rate of 3 Hz. Each analysis included 30 s baseline and 60 s of ablation, with He gas (gas flows = 0.86 L/min) as the carrier gas during the analytical process. Sixteen Faraday cups and three ion counters were used to determine the sulfur isotopic compositions, with a H5 cup for $^{34}S$, an Ax cup for $^{33}S$, and a L4 cup for $^{32}S$. The in-house sulfur reference material (PY-4, $\delta^{34}S_{V\text{-}CDT}$ = 1.7 ± 0.3‰, [38]) was used for external standard bracketing. The detailed procedures for sulfide in-situ sulfur isotopic analyses were reported in [39,40].

**4. Results**

*4.1. LA-ICP-MS Zircon U–Pb Ages*

Table S1 presents the results of LA-ICP-MS zircon U–Pb ages.

Twenty-nine zircon grains from the tuffaceous slate were selected for the LA-ICP-MS U–Pb dating, and most of these selected zircon grains have internal oscillatory zoning, indicating a magmatic origin (Figure 5a,b, [41]). In addition, these zircon grains have Th and U contents of 160–3728 ppm and 126–3754 ppm, respectively, with Th/U ratios of 0.43–1.77. Most of the analyses are concordant, yielding the zircon U–Pb ages of 2731–246 Ma, with a major peak of 795 Ma (Figure 6a). Ten spots of zircon grains have youngest ages of 751–806 Ma, yielding a weighted mean age of 775 ± 13 Ma (MSWD = 0.64, Figure 6b).

Twenty-six zircon grains from the sandy slate sample, with internal oscillatory zoning, were selected for the zircon U–Pb dating (Figure 5c,d). These zircon grains have variable Th and U contents of 47.8–1219 ppm and 126–1176 ppm, respectively, with Th/U ratios of 0.43–3.12. Most of the analyses are concordant, yielding the zircon U–Pb ages of 2604–755 Ma, with a major peak of 815 Ma (Figure 6c). Six spots of zircon grains have the youngest ages of 796–756 Ma, yielding a weighted mean age of 777 ± 16 Ma (MSWD = 0.57, Figure 6d).

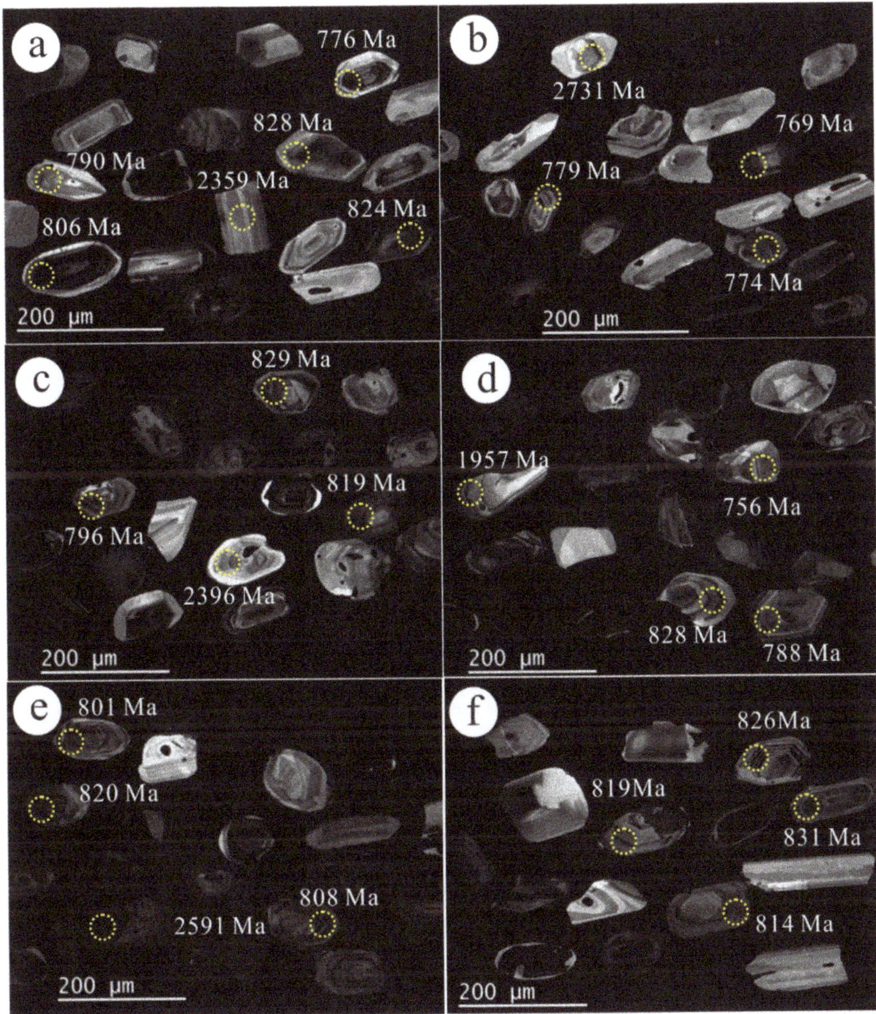



**Figure 5.** Cathodoluminescence (CL) images of zircon grains from the represented samples, showing the location of the spots for laser ablation-inductively coupled plasma-mass spectrometer (LA-ICP-MS) U–Pb dating. (**a**) and (**b**) are of tuffaceous slate; (**c**); and, (**d**) are of sandy slate; (**e**) and (**f**) are of gold-bearing quartz vein.

Twenty-nine zircon grains from the quartz sample, with internal oscillatory zoning, were selected for the zircon U–Pb dating (Figure 5e,f). These zircon grains have variable Th and U contents of 113–1722 ppm and 168–1442 ppm, respectively, with Th/U ratios of 0.51–2.35. Most of the analyses are concordant, yielding the zircon U–Pb ages of 2683–743 Ma, with a major peak of 810 Ma (Figure 6e). Five spots of zircon grains have youngest ages of 743–814 Ma, yielding a weighted mean age of 773 ± 18 Ma (MSWD = 0.97, Figure 6f).

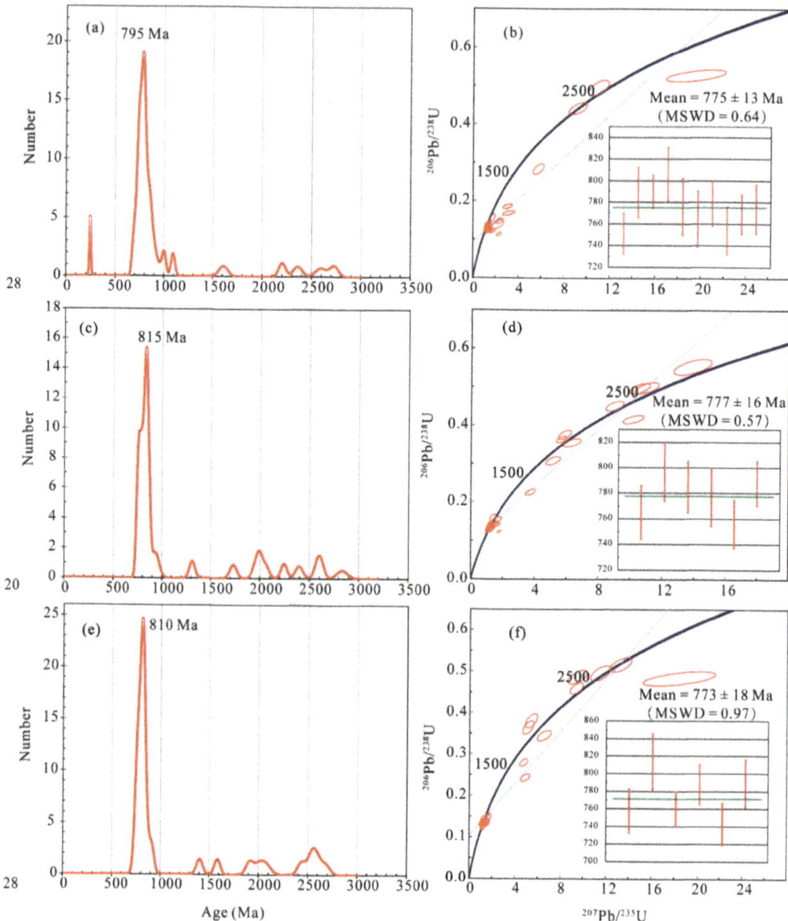

**Figure 6.** Historgams, Concordia and weighting mean ages of zircon grains from represented samples. (**a**) and (**b**) are of tuffaceous slate; (**c**) and (**d**) are of sandy slate; (**e**) and (**f**) are of gold-bearing quartz vein.

*4.2. Major and Trace Element Compositions*

The representative samples were analyzed for the major and trace element contents, and Tables S2 and S3 show the results. The sandy-slate samples have higher $TiO_2$, $Al_2O_3$, $K_2O$, and MgO contents, and lower $SiO_2$, and $Na_2O$ than those of the tuffaceous slates, with mean $SiO_2$ contents of 65.5 wt. % and 72.4 wt. %, $TiO_2$ contents of 0.63 wt. % and 0.34 wt. %, $Al_2O_3$ contents of 17.4 wt. % and 14.4 wt. %, MgO contents of 0.82 wt. % and 0.61 wt. %, $Na_2O$ contents of 2.65 wt. % and 3.0 wt. %, and $K_2O$ contents of 4.53 wt. % and 3.72 wt. %, for the sandy-slate and tuffaceous slate, respectively. The quartz sample contains a high $SiO_2$ content of 91.0 wt. %, with a little $Al_2O_3$ and $Fe_2O_3$.

These samples have variable trace element compositions, although they have similar primitive mantle normalized patterns with the depletion of Cs, Nb, Hf, Sr, and Y, and enrichment of Ba, Th, U, Rb, La, and Ce (Figure 7a). In addition, most of the sandy slate samples have a high content of Au (mostly ranging from 0.019 to 0.252 ppm), which are higher than those of the tuffaceous samples (mostly lower than 0.005 ppm). In terms of the rare earth elements (REEs) compositions of these samples, the slate samples have higher REE contents, with $\Sigma$REEs of 194–366 ppm (mean = 266 ppm), than those of the tuffaceous slate and quartz samples with $\Sigma$REEs of 113–318 ppm (mean = 194 ppm) and 39.3

ppm, respectively. In addition, the enrichment of light rare earth elements (LREEs) and depletion of heavy rare earth elements (HREEs) characterize these samples, and these samples show similar chondrite-normalized patterns (Figure 7b). Furthermore, the sandy slate, tuffaceous slate, and quartz samples have negative Eu anomalies, with δEu of 0.49–0.73 (mean = 0.58), 0.43–0.73 (mean = 0.50), and 0.70, respectively.

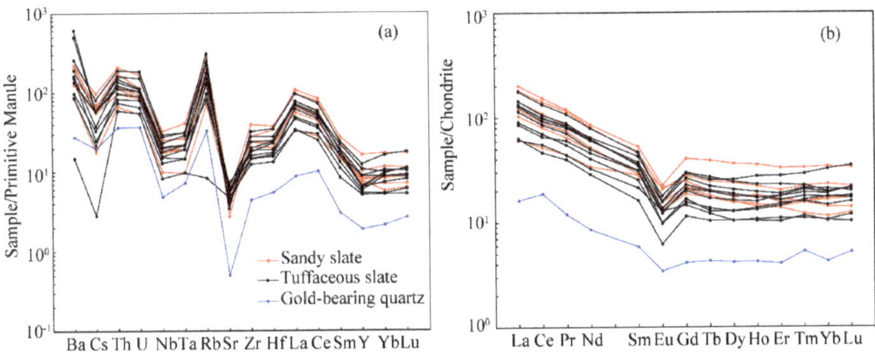


**Figure 7.** (**a**) Primitive mantle- and (**b**) chondrite-normalized patterns of samples from the Chaoyangzhai gold deposit. (a) and (b) are modified from [42,43], respectively.

Ten samples of gold-bearing quartz were assayed for the ore-forming metals analyses, and Table S3 displays the results. These gold-bearing quartz have variable Au (1.05–7.95 ppm), Cu (50–620 ppm), Pb (60–750 ppm), Zn (30–840 ppm), As (1500–7000 ppm), Sb (21.4–60.5 ppm), and Hg (0.02–0.07 ppm) contents (Figure 8).

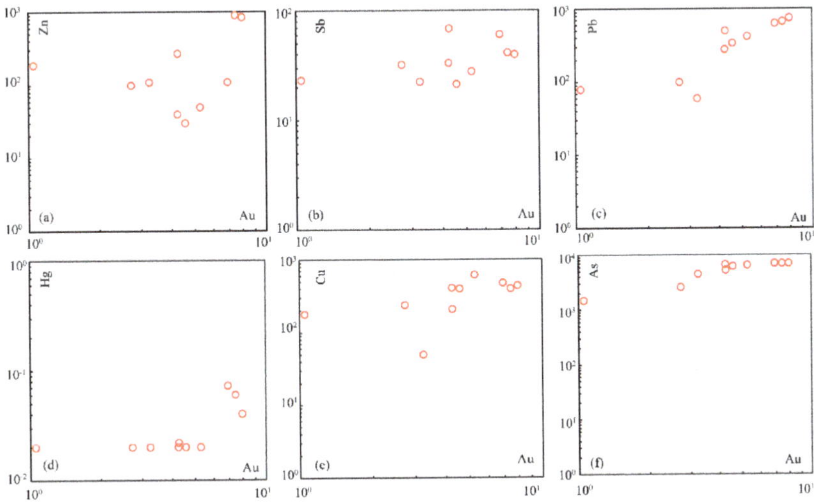


**Figure 8.** Geochemical plots of Au versus other ore-forming elements. All of the elements are given in ppm.

### 4.3. In Situ Sulfur Isotope Compositions

Table S4 presents the in situ LA-MC-ICP-MS analyses of sulfur isotopes. The measured $\delta^{34}S_{V\text{-CDT}}$ values of sulfides, which co-exit with the gold, range from +8.12‰ to +9.99‰, and from +9.78 to +10.78‰ for pyrite and arsenopyrite, respectively.

## 5. Discussion

### *5.1. Age and Source of Rocks from Qingshuijiang Formation*

The Southeast Guizhou province is one of the significant gold producers in China, and numerous small- to large-sized gold deposits were explored in this region [7–10,17,31]. These deposits share the common feature that the quartz vein type ore bodies occur in Neoproterozoic epimetamorphic rocks [8–10,23]. In addition, these rocks mainly belong to the Neoproterozoic Xiajiang Group, which consists of the Longli, Pinglue, Qingshuijiang, and Fanzhao Formations. Gold-bearing ore bodies could occur in almost all of the epimetamorphic rocks of these Formations, with the Bake and the Chaoyangzhai gold deposits in rocks of the Qingshuijiang Formation, the Pingqiu gold deposit in rocks of the Fanzhao Formation, the Tonggu gold deposit in rocks of the Longli Formation, and the Jintou gold deposit in rocks of the Pinglue Formation (Figure 2). Previous studies have revealed that the deposited ages of Longli, Pinglue, Qingshuijiang, and Fanzhao Formations are 725 Ma, 733 Ma, 773.8 Ma, and 774 Ma, respectively [44,45]. In this study, the youngest detrital zircon ages of the tuffaceous- and sandy-slates of the Qingshuijiang Formation are 775 ± 13 Ma and 777 ± 16 Ma, respectively, which are consistent with the previous studies, constraining the deposited ages of the Qingshuijiang Formation [44,45]. In addition, the peak ages of the tuffaceous- and sandy-slates of the Qingshuijiang Formation are 799 Ma and 815 Ma, respectively, which is consistent with the intense Neoproterozoic magmatism of South China, triggered by the break-up of the Rodinia supercontinent [45–48]

Previous studies have indicated that the turbidite-hosted gold deposits in Southeast Guizhou might have a genetic relationship with the surrounding Neoproterozoic rocks [7–10]. In addition, these rocks are enriched in gold, which the high gold content of 0.019–0.252 ppm in the sandy slate of the Qingshuijiang formation confirmed. Therefore, the determination of the source for these rocks is one of the crucial issues in the genesis of these gold deposits. The geochemical compositions of the clastic sedimentary rocks have been widely used in determining the source and tectonic setting of such rocks [49–52]. Based on the discrimination diagram that was proposed by [49], most of the samples in this study plot were in the field of felsic igneous provenance (Figure 9a). In addition, the ΣREE vs. La/Yb diagram, also confirm the conclusion, since almost all of the samples were plotted in the field of granite (Figure 9b). The occurrence of abundant magmatic zircon grains in these rocks also supported this conclusion. Therefore, we proposed that these epimetamorphic rocks of Qingshuijiang Formation might originate from a magmatic source.

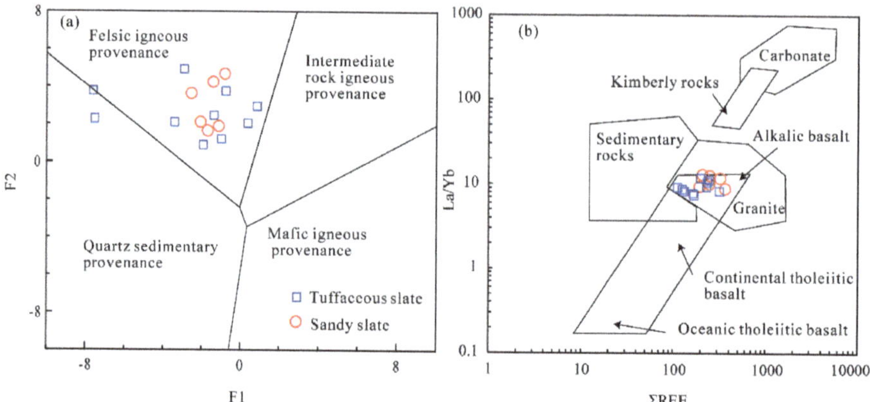


**Figure 9.** (**a**) F1 vs. F2 and (**b**) ΣREE vs. La/Yb plots for the rocks of Qingshuijiang Formation. Figure 9a,b are modified from [49] and [50], respectively. F1 = $-1.173\text{TiO}_2 + 0.607\text{Al}_2\text{O}_3 + 0.76\text{TFe}_2\text{O}_3 - 1.5\text{MgO} + 0.616\text{CaO} + 0.509\text{Na}_2\text{O} - 1.224\text{K}_2\text{O} - 9.09$; F2 = $0.445\text{TiO}_2 + 0.07\text{Al}_2\text{O}_3 - 0.25\text{TFe}_2\text{O}_3 - 1.142\text{MgO} + 0.438\text{CaO} + 1.475\text{Na}_2\text{O} + 1.426\text{K}_2\text{O} - 6.861$.

### 5.2. Source of the Sulfur and Gold

Sulfides show distinct sulfur isotopic compositions in different geological systems; therefore, sulfur isotopes can be used as a key tracer in reflecting the ore-forming material sources of metallic mineral deposits [39,40,53–56]. *In situ* sulfur isotope analyses have been widely used in economic geology, because they can provide added evidence regarding the source of deposits [38,39,54,55]. Based on the occurrence of arsenopyrite and pyrite and the absence of magnetite and/or sulfates in the quartz vein type ore bodies, we proposed that the ore-forming fluid were reduced and the sulfur isotopic compositions of the sulfides reflects that of the ore-forming fluid system [57,58]. As shown in Table S4, the $\delta^{34}S_{V\text{-CDT}}$ values of pyrite and arsenopyrite range from +8.12‰ to +9.99‰ and from +9.78 to +10.78‰, respectively, which is consistent with the $\delta^{34}S_{V\text{-CDT}}$ values of the epimetamorphic rocks of the Neoproterozoic Qingshuijiang Formation (+9.27‰ to +12.44‰), as reported by [8]. Together with the sulfur isotopic compositions of the sulfides from the nearby gold deposits in this region, sulfur isotopes of the Chaoyangzhai gold deposit fall outside of the range of granitic and basaltic rocks, but within the interval of the metamorphic rocks of the Qingshuijiang (Figure 10). In addition, the quartz vein has similar primitive mantle normalized trace element patterns and chondrite normalized rare earth element patterns to the ore-hosting tuffaceous- and sandy-slate, which also indicates that the metal source of this deposit is likely of the tuffaceous- and sandy-slate. Therefore, we proposed that the sulfur source might be from the ore-hosting metamorphic rocks of the Qingshuijiang Formation. Furthermore, the detrital zircon ages of the tuffaceous- and sandy-slate have peaks of 775 ± 13 Ma and 777 ± 16 Ma, respectively, which are consistent with the U–Pb age of zircon grains (773 ± 18 Ma) from the gold-bearing quartz vein. It is indicated that the fluids might likely bring in these zircon grains of the gold-bearing quartz vein when they circulated in the gold-rich rocks.

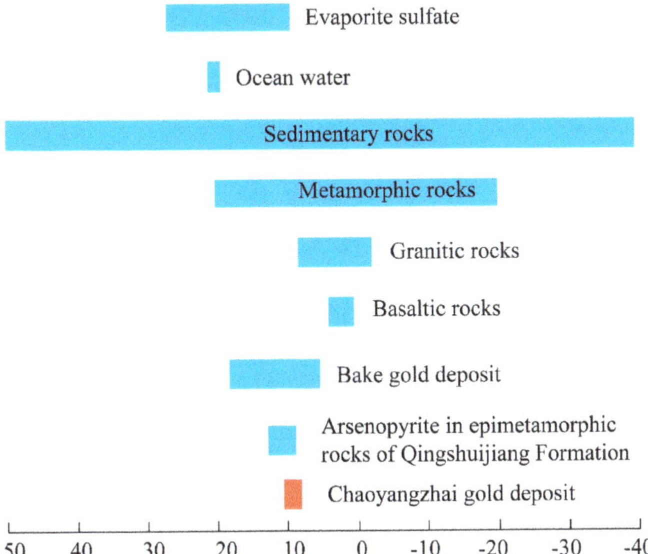


**Figure 10.** Sulfur isotopic compositions of the Chaoyangzhai gold deposit, showing some important sulfur reservoirs (modified from [59]). The sulfur isotopic data of arsenopyrite in epimetamorphic rocks of Qingshuijiang Formation and Bake deposit are from [8].

There are two types of metamorphic rocks of the Qingshuijiang Formation, the tuffaceous- and sandy-slate, however, which one is the source rocks of the Chaoyangzhai gold deposit? As shown in Table S2, the sandy-slate have a high content of Au (mostly ranging from 0.019 to 0.252 ppm), which is higher than those of the tuffaceous slate samples (mostly lower than 0.005 ppm). Thus, the most probable source of the gold for the Chaoyangzhai gold deposit is the sandy slates of the Qingshuijiang Formation. The gold deposits confirmed this conclusion, which were also hosted in the Qingshuijiang Formation, for example, Bake, Jinning, and Kengtou gold deposits in this region [8–10]. Consequently, we proposed that the metal source of Chaoyangzhai gold deposit was mostly derived from the sandy slates of the Neoproterozoic Qingshuijiang Formation.

*5.3. Genesis and Age of Turbidite-hosted Gold Deposits in Southeast Guizhou*

A notable feature of these turbidite-hosted gold deposits, including the Chaoyangzhai gold deposit, is the major paragenetic association of native gold, arsenopyrite, and pyrite, which indicates that Au should be deposited in an Au-saturated fluid [60]. In addition, the arsenopyrite and pyrite could be the Au-carriers in the ore-forming fluid, and it was confirmed by numerous cases worldwide [61–66]. [61] reported that the Au content of arsenopyrite could be up to 65 ppm in the Moshan gold deposit, Southeast Guizhou. The same features were also observed in sulfides of other gold deposits in Southeast Guizhou [8,67]. The obvious positive correlation of the Au and As in this study confirmed this conclusion (Figure 8f). Previous studies furthermore proposed that gold was likely transported as the $Au(HS)_2^-$ in the fluid of low temperatures, low salinities, and low oxygen fugacities [60,68,69]. The homogenization temperatures of the Chaoyangzhai gold deposit range from 145 to 319 °C, with a cluster of ca. 200 °C, whereas the calculated salinities range from 0.18 to 17.9 wt. % NaCl equiv. (average = 7.5 wt. % NaCl equiv) (unpublished data). These temperatures and salinities of the fluids are similar to those of the Pingqiu gold deposit, Southeast Guizhou [9]. Together with the absence of the sulfate minerals and magnetite, it was proposed that these ore-forming fluids were of low temperature, low salinity, and low oxygen fugacity. As far as the source of these ore-forming fluids, numerous studies have been carried out on this issue; however, different viewpoints were proposed [4,9,61,70–73]. These viewpoints can be classified into three groups: metamorphic fluid [9], magmatic-hydrothermal fluid [4,70], and mixed fluid [61]. The viewpoint of magmatic-hydrothermal model might be unlikely as an explanation for the fluid source of these gold deposits since no coeval granitic intrusions were found in the Southeast Guizhou. In addition, the reported H–O isotopic data of these gold deposit are mainly in the range of the metamorphic fluids, although groundwater might be involved in the formation of these gold deposits [9,74]. Therefore, the metamorphic devolatilization model appears to be the best explanation of the origin of ore-forming fluids, and these metamorphic fluids could extract the gold and other metals from epimetamorphic rocks. Subsequently, accompanied by the change of physico-chemical conditions, gold was deposited in the suitable fractures, leading to the large-scale gold mineralization in Southeast Guizhou.

In terms of the ore-forming ages of these deposits, the reported metallogenic ages of these deposits could be summarized in two groups: Ordovician-Silurian (450–410 Ma) and Triassic-Jurassic (240–140 Ma), which is consistent with Caledonian and Indosinian-Yanshanian deformation, respectively [10]. The debate regarding the timing of the ore-forming events of these deposits might be caused by the lack of suitable minerals for dating, since pyrite and arsenopyrite are the associated minerals in these gold deposits. Recently, several ages of these gold deposits were reported, with $^{40}Ar$-$^{39}Ar$ age of 425.7 ± 1.7 Ma for the Pingqiu gold deposit [9], Re-Os age of 400 ± 11 Ma for the Jinjing gold deposit [7], and Re-Os age of 412 ± 21 Ma for the Bake gold deposit [23]. In addition, these ages are consistent with the ages of other gold deposits in the West Jiangnan Orogen Belt, for example, the Woxi Au deposit (423.2 ± 1.2 Ma, [23]), Banxi Sb-Au deposit (422.2 ± 0.2 Ma, [75]), and Jinshan Au deposit (406 ± 25 Ma, [76]). These reported high precision ages of these gold deposits indicate that the Caledonian is a significant gold mineralization epoch in South China, recording the Caledonian orogeny and the formation of these orogenic gold deposits in Southeast Guizhou.

## 6. Conclusions

Based on the geological, geochemical, geochronological studies on the Chaoyangzhai gold deposit, we draw the following conclusions:

(1) LA-ICP-MS zircon U–Pb dating of the gold-bearing quartz, tuffaceous- and sandy-slates display similar characteristics of age distribution, indicating that the zircons in the gold-bearing quartz could originate from the surrounding tuffaceous- and sandy-slates.

(2) Rocks of the Qingshuijiang Formation might be sourced from a felsic igneous provenance.

(3) Similar geochemical patterns between the surrounding tuffaceous- and sandy-slate and gold-bearing quartz illustrate that the Chaoyangzhai gold deposit might be sourced from the surrounding tuffaceous- and sandy-slate, supported by the sulfur isotopes of the arsenopyrite and pyrite.

(4) The sandy-slates have higher Au contents than the tuffaceous slate, indicating that the gold might be sourced from the sandy-slate rather than tuffaceous slate.

**Supplementary Materials:** The following are available online at http://www.mdpi.com/2075-163X/9/4/235/s1, Table S1. LA-ICP-MS zircon U–Pb dating of the host rocks and gold-bearing quartz veins from the Chaoyangzhai gold deposit. Table S2. Major and trace element compositions of the rocks from the Chaoyangzhai gold deposit. Table S3. Ore-forming element contents of the ore (ppm). Table S4. Sulfur isotopic compositions of the sulfides from the Chaoyangzhai gold deposit.

**Author Contributions:** H.T. and J.C. wrote the paper; J.C. and X.Y. designed the experiments; H.T. and X.Y. took part in the field investigation.

**Funding:** This study is financially supported by grants from the National Key R&D Program of China (No. 2016YFC0600404).

**Conflicts of Interest:** The authors declare no conflict of interest.

## References

1. Horne, R.; Culshaw, N. Flexural-slip folding in the Meguma Group, Nova Scotia, Canada. *J. Struct. Geol.* **2001**, *23*, 1631–1652. [CrossRef]

2. Ramsay, W.; Bierlein, F.P.; Arne, D.C.; VandenBerg, A. Turbidite-hosted gold deposits of Central Victoria, Australia: their regional setting, mineralising styles, and some genetic constraints. *Ore Geol. Rev.* **1998**, *13*, 131–151. [CrossRef]

3. Windh, J. Saddle reef and related gold mineralization, Hill End gold field, Australia: Evolution of an auriferous vein system during progressive deformation. *Econ. Geol.* **1995**, *90*, 1764–1775. [CrossRef]

4. Lu, H.Z.; Wang, Z.G.; Chen, W.Y.; Wu, X.Y.; Zhu, X.Q.; Hu, R.Z. Turbidite hosted gold deposits in southeast Guizhou: Their structural control, mineralization characteristics, and some genetic constrains. *Miner. Depos.* **2006**, *4*, 369–387. (In Chinese)

5. Arne, D.C.; Bierlein, F.P.; Morgan, J.W.; Stein, H.J. Re–Os dating of sulfides associated with gold mineralization in Central Victoria, Australia. *Econ. Geol.* **2001**, *96*, 1455–1459. [CrossRef]

6. Molnar, F.; Middleton, A.; Stein, H.; O'Brien, H.; Lahaye, Y.; Huhma, H.; Pakkanen, L.; Johanson, B. Repeated syn- and post-orogenic gold mineralization events between 1.92 and 1.76 Ga along the Kiistala Shear Zone in the Central Lapland Greenstone Belt, northern Finland. *Ore Geol. Rev.* **2018**, *101*, 936–959. [CrossRef]

7. Wang, J.; Wen, H.; Li, C.; Zhang, J.; Ding, W. Age and metal source of orogenic gold deposits in Southeast Guizhou Province, China: Constraints from Re–Os and He–Ar isotopic evidence. *Geosci. Front.* **2019**, *10*, 581–593. [CrossRef]

8. Liu, A.; Jiang, M.; Ulrich, T.; Zhang, J.; Zhang, X. Ore genesis of the Bake gold deposit, southeastern Guizhou province, China: Constraints from mineralogy, in-situ trace element and sulfur isotope analysis of pyrite. *Ore Geol. Rev.* **2018**, *102*, 740–756. [CrossRef]

9. Liu, A.; Zhang, X.; Ulrich, T.; Zhang, J.; Jiang, M.; Liu, W. Geology, geochronology and fluid characteristics of the Pingqiu gold deposit, Southeastern Guizhou Province, China. *Ore Geol. Rev.* **2017**, *89*, 187–205. [CrossRef]

10. Xu, D.; Deng, T.; Chi, G.; Wang, Z.; Zou, F.; Zhang, J.; Zou, S. Gold mineralization in the Jiangnan Orogenic Belt of South China: Geological, geochemical and geochronological characteristics, ore deposit-type and geodynamic setting. *Ore Geol. Rev.* **2017**, *88*, 565–618. [CrossRef]

11. Goldfarb, R.J.; Groves, D.I. Orogenic gold: Common or evolving fluid and metal sources through time. *Lithos* **2015**, *233*, 2–26. [CrossRef]

12. Pitcairn, I.K.; Craw, D.; Teagle, D.A.H. Metabasalts as sources of metals in orogenic gold deposits. *Miner. Deposita* **2015**, *50*, 373–390. [CrossRef]

13. Phillips, G.N.; Powell, R. Origin of Witwatersrand gold: A metamorphic devolatilisation-hydrothermal replacement model. *Appl. Earth Sci.* **2011**, *120*, 112–129. [CrossRef]

14. Phillips, G.N.; Powell, R. Formation of gold deposits: a metamorphic devolatilization model. *J Metamorph. Geol.* **2010**, *28*, 689–718. [CrossRef]

15. Pang, J.T.; Dai, T.G. On the mineralization epoch of the Xuefeng gold metallogenic province. *Geol. Prospect.* **1998**, *34*, 37–41. (In Chinese)

16. Li, N.; Pirajno, F. Early Mesozoic Mo mineralization in the Qinling Orogen: An overview. *Ore Geol. Rev.* **2017**, *81*, 431–450. [CrossRef]

17. Tao, P.; Xiao, X.D.; Hu, C.L. The Au-bearing sedimentary sequences and theirimpact on the gold deposits in light metamorphic rock in the boundary of Hunan, Guizhou and Guangxi regions. *Geol. Sci. Technol. Inf.* **2009**, *28*, 100–104. (In Chinese)

18. Sun, J.; Shu, L.; Santosh, M.; Wang, L. Neoproterozoic tectonic evolution of the Jiuling terrane in the central Jiangnan orogenic belt (South China): Constraints from magmatic suites. *Precambrian Res.* **2017**, *302*, 279–297. [CrossRef]

19. Pan, X.; Hou, Z.; Zhao, M.; Chen, G.; Rao, J.; Li, Y.; Wei, J.; Ouyang, Y. Geochronology and geochemistry of the granites from the Zhuxi W–Cu ore deposit in South China: Implication for petrogenesis, geodynamical setting and mineralization. *Lithos* **2018**, *304*, 155–179. [CrossRef]

20. Wang, C.; Rao, J.; Chen, J.; Ouyang, Y.; Qi, S.; Li, Q. Prospectivity mapping for "Zhuxi-type" copper-tungsten polymetallic deposits in the Jingdezhen region of Jiangxi Province, South China. *Ore Geol. Rev.* **2017**, *89*, 1–14. [CrossRef]

21. Mao, J.; Xiong, B.; Liu, J.; Pirajno, F.; Cheng, Y.; Ye, H.; Song, S.; Dai, P. Molybdenite Re/Os dating, zircon U–Pb age and geochemistry of granitoids in the Yangchuling porphyry W–Mo deposit (Jiangnan tungsten ore belt), China: Implications for petrogenesis, mineralization and geodynamic setting. *Lithos* **2017**, *286*, 35–52. [CrossRef]

22. Huang, L.; Jiang, S. Highly fractionated S-type granites from the giant Dahutang tungsten deposit in Jiangnan Orogen, Southeast China: Geochronology, petrogenesis and their relationship with W-mineralization. *Lithos* **2014**, *202*, 207–226. [CrossRef]

23. Wang, J.; Wen, H.; Li, C.; Jiang, X.; Zhu, C.; Du, S.; Zhang, L. Determination of age and source constraints for the Bake quartz vein-type gold deposit in SE Guizhou using arsenopyrite Re–Os chronology and REE characteristics. *Geochem. J.* **2015**, *49*, 73–81. [CrossRef]

24. Deng, J.; Wang, Q. Gold mineralization in China: Metallogenic provinces, deposit types and tectonic framework. *Gondwana Res.* **2016**, *36*, 219–274. [CrossRef]

25. Lu, H.Z.; Wang, Z.G.; Wu, X.Y.; Chen, W.Y.; Zhu, X.Q.; Guo, D.J.; Hu, R.Z.; Keita, M. Turbidite-hosted gold Deposits in SE Guizhou Province, China: Their regional setting, structural control and gold mineralization. *Acta Geol. Sin.* **2005**, *79*, 98–105.

26. Tao, P.; Chen, Q.F.; Wang, L.; Hu, C.L. Location prediction of gold deposits for ore field grade in the Tianzhu-Jinping-Liping area, southeastern Guizhou, China: The method on potential assessment of mineral resources of low-exploration and study degree area. *J. Jilin Univ.* **2013**, *4*, 1235–1245. (In Chinese)

27. Goldfarb, R.J.; Taylor, R.D.; Collins, G.S.; Goryachev, N.A.; Orlandini, O.F. Phanerozoic continental growth and gold metallogeny of Asia. *Gondwana Res.* **2014**, *25*, 48–102. [CrossRef]

28. Zhou, T.H.; Goldfarb, R.J.; Phillips, G.N. Tectonics and distribution of gold deposits in China—an overview. *Miner. Deposita* **2002**, *37*, 249–282. [CrossRef]

29. Ren, J.S. The continental tectonics of China. *J. SE Asian Earth Sci.* **1996**, *13*, 197–204. [CrossRef]

30. Zeng, G.; Gong, Y.; Wang, Z.; Hu, X.; Xiong, S. Structures of the Zhazixi Sb–W deposit, South China: Implications for ore genesis and mineral exploration. *J. Geochem. Explor.* **2017**, *182*, 10–21. [CrossRef]

31. Liu, L.; Li, S.; Dai, L.; Suo, Y.; Liu, B.; Zhang, G.; Wang, Y.; Liu, E. Geometry and timing of Mesozoic deformation in the western part of the Xuefeng Tectonic Belt, South China: Implications for intra-continental deformation. *J. Asian Earth Sci.* **2012**, *49*, 330–338. [CrossRef]

32. Wiedenbeck, M.; Alle, P.; Corfu, F.; Griffin, W.L.; Meier, M.; Ober, F.; Von Quadt, A.; Roddick, J.C.; Speigel, W. Three natural zircon standards for U–Th–Pb, Lu–Hf, trace-element and REE analyses. *Geostand. Geoanal. Res.* **1995**, *19*, 1–23. [CrossRef]

33. Elhlou, S.; Belousova, E.; Griffin, W.L.; Pearson, N.J.; O Reilly, S.Y. Trace element and isotopic composition of GJ-red zircon standard by laser ablation. *Geochim. Cosmochim. Ac.* **2006**, *70*, A158. [CrossRef]

34. Liu, Y.; Gao, S.; Hu, Z.; Gao, C.; Zong, K.; Wang, D. Continental and Oceanic Crust Recycling-induced Melt-Peridotite Interactions in the Trans-North China Orogen: U–Pb Dating, Hf Isotopes and Trace Elements in Zircons from Mantle Xenoliths. *J. Petrol.* **2010**, *51*, 537–571. [CrossRef]

35. Liu, Y.; Hu, Z.; Gao, S.; Guenther, D.; Xu, J.; Gao, C.; Chen, H. In situ analysis of major and trace elements of anhydrous minerals by LA-ICP-MS without applying an internal standard. *Chem. Geol.* **2008**, *257*, 34–43. [CrossRef]

36. Andersen, T. Correction of common lead in U–Pb analyses that do not report $^{204}$Pb. *Chem. Geol.* **2002**, *192*, 59–79. [CrossRef]

37. Sláma, J.; Košler, J.; Condon, D.J.; Crowley, J.L.; Gerdes, A.; Hanchar, J.M.; Horstwood, M.S.A.; Morris, G.A.; Nasdala, L.; Norberg, N.; et al. Plešovice zircon—A new natural reference material for U–Pb and Hf isotopic microanalysis. *Chem. Geol.* **2008**, *249*, 1–35. [CrossRef]

38. Bao, Z.; Chen, L.; Zong, C.; Yuan, H.; Chen, K.; Dai, M. Development of pressed sulfide powder tablets for in situ sulfur and lead isotope measurement using LA-MC-ICP-MS. *Int. J. MASS Spectrom.* **2017**, *421*, 255–262. [CrossRef]

39. Yuan, H.; Liu, X.; Chen, L.; Bao, Z.; Chen, K.; Zong, C.; Li, X.; Qiu, J.W. Simultaneous measurement of sulfur and lead isotopes in sulfides using nanosecond laser ablation coupled with two multi-collector inductively coupled plasma mass spectrometers. *J. Asian Earth Sci.* **2018**, *154*, 386–396. [CrossRef]

40. Chen, L.; Chen, K.; Bao, Z.; Liang, P.; Sun, T.; Yuan, H. Preparation of standards for in situ sulfur isotope measurement in sulfides using femtosecond laser ablation MC-ICP-MS. *J. Anal. Atom Spectrom.* **2017**, *32*, 107–116. [CrossRef]

41. Hoskin, P.W.O.; Schaltegger, U. The composition of zircon and igneous and metamorphic petrogenesis. *Rev. Mineral. Geochem.* **2003**, *53*, 27–62. [CrossRef]

42. Sun, S.S.; McDonough, W.F. Chemical and isotopic systematics of oceanic basalts: Implications for mantle compositions and processes. In *Magmatism in the Ocean Basins*; Saunders, A.D., Norry, M.J., Eds.; Geological Society of London Special Paper: London, UK, 1989; Volume 32, pp. 313–345.

43. Taylor, S.R.; McLennan, S.M. *Continental Crust: Its Composition and Evolution. An Examination of the Geochemical Record Preserved in Sedimentary Rocks*; Blackwell Science Inc.: Boston, MA, USA, 1985; p. 312.

44. Wei, Y.N.; Jiang, X.S.; Cui, X.Z.; Zhuo, J.W.; Jiang, Z.F.; Cai, J.J. Deterital zircon U–Pb ages of the Neoproterzoic Qingshuijiang Formation in Southeastern Guizhou and their geological significance. *J. Miner. Petrol.* **2015**, *35*, 61–71.

45. Wang, X.; Li, X.; Li, Z.; Li, Q.; Tang, G.; Gao, Y.; Zhang, Q.; Liu, Y. Episodic Precambrian crust growth: Evidence from U–Pb ages and Hf–O isotopes of zircon in the Nanhua Basin, central South China. *Precambrian Res.* **2012**, *222*, 386–403. [CrossRef]

46. Wang, L.; Griffin, W.L.; Yu, J.; O Reilly, S.Y. Precambrian crustal evolution of the Yangtze Block tracked by detrital zircons from Neoproterozoic sedimentary rocks. *Precambrian Res.* **2010**, *177*, 131–144. [CrossRef]

47. Wang, J.; Shu, L.; Santosh, M. U–Pb and Lu–Hf isotopes of detrital zircon grains from Neoproterozoic sedimentary rocks in the central Jiangnan Orogen, South China: Implications for Precambrian crustal evolution. *Precambrian Res.* **2017**, *294*, 175–188. [CrossRef]

48. Ma, X.; Yang, K.; Li, X.; Dai, C.; Zhang, H.; Zhou, Q. Neoproterozoic Jiangnan Orogeny in southeast Guizhou, South China: Evidence from U–Pb ages for detrital zircons from the Sibao Group and Xiajiang Group. *Can. J. Earth Sci.* **2016**, *53*, 219–230. [CrossRef]

49. Roser, B.P.; Korsch, R.J. Provenance signatures of sandstone mudstone suites determined using discriminant function-analysis of major-element data. *Chem. Geol.* **1988**, *67*, 119–139. [CrossRef]

50. Allegre, C.J.; Minster, J.F. Quantitative models of trace-element behavior in magmatic processes. *Earth Planet. Sc. Lett.* **1978**, *38*, 1–25. [CrossRef]

51. Anani, C.Y.; Bonsu, S.; Kwayisi, D.; Asiedu, D.K. Geochemistry and provenance of Neoproterozoic metasedimentary rocks from the Togo structural unit, Southeastern Ghana. *J. Afr. Earth Sci.* **2019**, *153*, 208–218. [CrossRef]

52. Lu, Y.; Zhu, W.; Ge, R.; Zheng, B.; He, J.; Diao, Z. Neoproterozoic active continental margin in the northwestern Tarim Craton: Clues from Neoproterozoic (meta) sedimentary rocks in the Wushi area, northwest China. *Precambrian Res.* **2017**, *298*, 88–106. [CrossRef]

53. Bauer, M.E.; Burisch, M.; Ostendorf, J.; Krause, J.; Frenzel, M.; Seifert, T.; Gutzmer, J. Trace element geochemistry of sphalerite in contrasting hydrothermal fluid systems of the Freiberg district, Germany: Insights from LA-ICP-MS analysis, near-infrared light microthermometry of sphalerite-hosted fluid inclusions, and sulfur isotope geochemistry. *Miner. Deposita* **2019**, *54*, 237–262.

54. Hu, X.; Tang, L.; Zhang, S.; Santosh, M.; Spencer, C.J.; Zhao, Y.; Cao, H.; Pei, Q. In situ trace element and sulfur isotope of pyrite constrain ore genesis in the Shapoling molybdenum deposit, East Qinling Orogen, China. *Ore Geol. Rev.* **2019**, *105*, 123–136. [CrossRef]

55. Li, X.; Zhao, K.; Jiang, S.; Palmer, M.R. In-situ U–Pb geochronology and sulfur isotopes constrain the metallogenesis of the giant Neves Corvo deposit, Iberian Pyrite Belt. *Ore Geol. Rev.* **2019**, *105*, 223–235. [CrossRef]

56. Liang, J.; Li, J.; Sun, W.; Zhao, J.; Zhai, W.; Huang, Y.; Song, M.; Ni, S.; Xiang, Q.; Zhang, J.; et al. Source of ore-forming fluids of the Yangshan gold field, western Qinling orogen, China: Evidence from microthermometry, noble gas isotopes and in situ sulfur isotopes of Au-carrying pyrite. *Ore Geol. Rev.* **2019**, *105*, 404–422. [CrossRef]

57. Downes, P.M.; Seccombe, P.K.; Carr, G.R. Sulfur- and lead-isotope signatures of orogenic gold mineralisation associated with the Hill End Trough, Lachlan Orogen, New South Wales, Australia. *Miner. Petrol.* **2008**, *94*, 151–173. [CrossRef]

58. Jia, Y.F.; Li, X.; Kerrich, R. Stable isotope (O, H, S, C, and N) systematics of quartz vein systems in the turbidite-hosted Central and North Deborah gold deposits of the Bendigo cold field, Central Victoria, Australia: Constraints on the origin of ore-forming fluids. *Econ. Geol.* **2001**, *96*, 705–721. [CrossRef]

59. Hoefs, J. *Stable Isotope Geochemistry*; Spring: Berlin/Heidelberg, Germany, 2009; pp. 71–73.

60. Zhu, Y.; An, F.; Tan, J. Geochemistry of hydrothermal gold deposits: A review. *Geosci. Front.* **2011**, *2*, 367–374. [CrossRef]

61. Zhang, J.; Yu, D.L.; Zhang, X.Y.; Yang, Y.J. Geochemical characteristice and modes of occurrence of gold in the Moshan-Youmaao gold deposit of Tianzhu, Guizhou. *J. Guizhou Univ. Technol.* **1997**, *26*, 43–49. (in Chinese).

62. Zhang, H.; Zhu, Y. Mechanism of gold precipitation in the Gezigou gold deposit, Xinjiang, NW China: Evidence from fluid inclusions and thermodynamic modeling. *J. Geochem. Explor.* **2019**, *199*, 60–74. [CrossRef]

63. Song, G.X.; Cook, N.J.; Wang, L.; Qin, K.Z.; Ciobany, C.; Li, G.M. Gold behavior in intermediate sulfidation epithermal systems: A case study from the Zhengguang gold deposit, Heilongjiang Province, NE-China. *Ore Geol. Rev.* **2019**, *106*, 446–462. [CrossRef]

64. Kovalev, K.R.; Kalinin, Y.A.; Naumov, E.A.; Kolesnikova, M.K.; Korolyuk, V.N. Gold-bearing arsenopyrite in eastern Kazakhstan gold-sulfide deposits. *Russ. Geol. Geophys.* **2011**, *52*, 178–192. [CrossRef]

65. Cook, N.J.; Ciobanu, C.L.; Meria, D.; Silcock, D.; Wade, B. Arsenopyrite-Pyrite Association in an Orogenic Gold Ore: Tracing Mineralization History from Textures and Trace Elements. *Econ. Geol.* **2013**, *108*, 1273–1283. [CrossRef]

66. Salvi, S.; Velasquez, G.; Miller, J.M.; Beziat, D.; Siebenaller, L.; Bourassa, Y. The Pampe gold deposit (Ghana): Constraints on sulfide evolution during gold mineralization. *Ore Geol. Rev.* **2016**, *78*, 673–686. [CrossRef]

67. Zhao, J.; Liang, J.L.; Li, J.; Ni, X.S.; Xiang, Q.R. Genesis and metallogenic model of the Shuiyindong gold deposit, Guizhou Province: Evidences from high-resolution multi-element mapping and in situ sulfur isotopes of Au-carrying pyrites by NanoSIMS. *Earth Sci. Fron.* **2018**, *25*, 157–167. (in Chinese).

68. Fan, H.R.; Zhai, M.G.; Xie, Y.H.; Yang, J.H. Ore-forming fluids associated with granite-hosted gold mineralization at the Sanshandao deposit, Jiaodong gold province, China. *Miner. Deposita* **2003**, *38*, 739–750. [CrossRef]

69. Yoo, B.C.; Lee, H.K.; White, N.C. Mineralogical, fluid inclusion, and stable isotope constraints on mechanisms of ore deposition at the Samgwang mine (Republic of Korea)—A mesothermal, vein-hosted gold-silver deposit. *Miner. Deposita* **2010**, *45*, 161–187. [CrossRef]

70. Wu, W.M.; Liu, J.Z.; Wang, Z.P.; Yu, D.L. Indicative significance of H–O isotope and fluid inclusion in ore-forming fluid of Pingqiu gold mine, Guizhou province. *Gold Sci. Technol.* **2015**, *23*, 30–36.

71. Seward, T.M.; Williams-Jones, A.E.; Migdisov, A.A. The chemistry of metal transport and deposition by ore-forming hydrothermal fluids. In *Treatise on Geochemistry*, 2nd ed.; Turekian, K., Holland, H., Eds.; Elsevier: New York, NY, USA, 2014; Volume 13, pp. 29–57.

72. Williams-Jones, A.E.; Bowell, R.J.; Migdisov, A.A. Gold in solution. *Elements* **2009**, *5*, 281–287. [CrossRef]

73. Pal'yanova, G. Physicochemical modeling of the coupled behavior of gold and silver in hydrothermal processes: Gold fineness, Au/Ag ratios and their possible implications. *Chem. Geol.* **2008**, *255*, 399–413. [CrossRef]

74. Tian, H.D.; Huang, D.G.; Yu, Q.P. Analyses on gelogic characters and genesis of Kengtou gold deposit in Tianzhu, Southeast Guizhou. *Guizhou Geol.* **2011**, *28*, 265–271. (in Chinese).

75. Peng, J.T.; Hu, R.Z.; Zhao, J.H.; Fu, Y.Z.; Lin, Y.X. Scheelite Sm–Nd dating and quartz Ar–Ar daing of Au–Sb–W depos it in Woxi, west Hunan. *Chin. Sci. Bull.* **2003**, *48*, 1976–1981. (in Chinese). [CrossRef]

76. Wang, X.Z.; Liang, H.Y.; Shan, Q.; Cheng, J.P.; Xia, P. Metallogenic age of the Jinshan gold deposit and Caledonian gold mineralization in South China. *Geo. Rev.* **1999**, *45*, 19–25. (in Chinese).

MDPI

St. Alban-Anlage 66

4052 Basel

Switzerland

Tel. +41 61 683 77 34

Fax +41 61 302 89 18

www.mdpi.com

*Minerals* Editorial Office

E-mail: minerals@mdpi.com

www.mdpi.com/journal/minerals